INTRODUCTION TO OPERATIONAL MODAL ANALYSIS

INTRODUCTION TO OPERATIONAL MODAL ANALYSIS

Rune Brincker

Technical University of Denmark

Carlos E. Ventura

University of British Columbia, Canada

Library of Congress Cataloging-in-Publication Data

Brincker, Rune.
  Introduction to operational modal analysis / Rune Brincker, Technical University of Denmark, Denmark, Carlos Ventura, University of British Columbia, Canada.
       pages cm
  Includes bibliographical references and index.
  ISBN 978-1-119-96315-8 (cloth)
  1. Modal analysis. 2. Structural analysis (Engineering) I. Ventura, Carlos. II. Title.
  TA654.15.B75 2015
  624.1′71–dc23
                                                                    2015016355

A catalogue record for this book is available from the British Library.

Cover image: (Background) Darren Falkenberg/Getty (Graph) Courtesy of the author

Typeset in 9/11pt TimesLTStd by SPi Global, Chennai, India

1 2015

Contents

Preface

After many years of working on various aspects of operational modal analysis (OMA), conducting vibration tests, analyzing test data from a variety of structures, and giving courses to promote the use and understanding of OMA, we decided to write a book on this topic during a meeting at SVS in Aalborg in the summer 2003. Two years later, in the summer of 2005, we secluded ourselves for one month at the Ventura's family coffee farm in Guatemala where we prepared the first outline of the book and started on the writing process. At that time, we focused our efforts on some key areas, mainly testing, classical dynamics, and signal processing. However, due to a number of circumstances related to work commitments and personal and family affairs, our serious writing process was delayed until the summer of 2011 and continued on until the summer of 2014 when we finished this first edition of the book. It has been a long and demanding effort, but in the end we have prepared a book that reflects our personal views of OMA, both in terms of theory and practice.

It would be preposterous to say that we are the only specialists in the field qualified to write one of the first books on OMA, but both of us have extensive experience with this technology and we recognized that as a team, we were well qualified to write the first book that dealt in a formal manner with the theory behind OMA. We noticed that in the early 1990s people started to pay more attention to OMA and realized the advantages of OMA techniques to determine modal properties of structures. We also noticed that the theory behind OMA was not well understood and that many people were hesitant to use these techniques because of a lack of a clear understanding of why these work so well. So, in 2003 we decided to make an effort to "demystify OMA," and since then the need for such a book has just grown. When we look at our final product, it seems like we made a good decision to work on this book, and we are confident that it will help people working in the field of OMA.

This book is written to be used as a textbook by students, mainly graduate students and PhD students working in research areas where OMA is applied, but it can also be used by scientists and professionals as a reference book for the most important techniques presently being used to analyze vibration data obtained by using OMA testing techniques.

Some people might argue that the classical experimental modal analysis (EMA) and OMA are the same thing, and, therefore, there is no need for a special theory for OMA. But we have compelling reasons to disagree based on our understanding of the fundamental theory of OMA. Our opinion is that actually EMA and OMA are quite different; they have a different history, they use a different technology, they have a different theoretical background, and finally, their applications are different. OMA is indeed a special field that needs to be introduced properly, its mathematical basis and background need to be adequately explained, and good testing practices need to be introduced in order to obtain good data and meaningful results. This is why this book is needed.

The theory in this book is presented heuristically rather than rigorously, thus many mathematical details are omitted for the sake of brevity and conciseness. The aim is not to cover the whole subject in great detail but rather to present a consistent overview of the theories needed to understand the topic and to point out the what these theories have in common and how these theories can be implemented in practice.

This book is rich in mathematical equations that are needed for formulating the theory of OMA, but extensive derivations of equations and formulas are avoided. Our aim was to present each equation or formula in its simplest and clearest formulation. This is also a book rich in simple and clear explanations that will help the reader understand the background for the formulas and how to use them in an effective way in order to perform OMA.

During the writing process, we have been privileged to receive excellent advice from colleagues from around the world who also work on OMA techniques. Without all this advice, we would not have been able to complete the book in a manner that makes us proud of our efforts. We would like to thank all these colleagues for spending their time giving us feedback. We would like to thank Dr. Spilios Fassois and Dr. Nuno Maia for giving us feedback on Chapters 2 and 3, Dr. Anders Brandt and Dr. Henrik Karstoft for their feedback on Chapters 4 and 8, Dr. Manuel Lopez-Aenlle for his feedback on Chapters 5 and 12, Dr. Frede Christensen for his feedback on Chapter 5, Dr. George James and Dr. Lingmi Zhang for their feedback on Chapter 6, Dr. Lingmi Zhang for also giving us feedback on Chapters 9 and 10, Dr. James Brownjohn for his feedback on Chapter 7, and Dr. Bart Peeters for his feedback on Chapter 10. The many useful comments from our PhD students and coworkers, Martin Juul, Anela Bajric, Jannick B. Hansen, Peter Olsen, Anders Skafte, and Mads K. Hovgaard are very much appreciated. We would also like to thank Dr. Palle Andersen of SVS for providing insightful comments on the use of time-domain and frequency-domain techniques as implemented on the ARTeMIS program. The case studies presented in Chapter 11 are based on papers published by the authors and other colleagues. Special thanks are extended to Dr. Alvaro Cunha, Dr. Elsa Caetano, and Dr. Sven-Eric Rosenov for sharing with us the data sets for two of the case studies described in this chapter.

We like to thank our colleagues and friends who have encouraged us to write this book. And last, but not least, we would like to thank our wives, Henriette and Lucrecia for their unconditional support all these years and for having infinite patience with us while we struggled with the preparation of the various drafts of each chapter, and for their willingness to allow us to spend time working on this book rather than attending to family affairs – without their support and unconditional love, this book would have never been a reality.

June 2015
Rune Brincker
Carlos E. Ventura

1

Introduction

"Torture numbers and they'll confess to anything"

– Gregg Easterbrook

The engineering field that studies the modal properties of systems under ambient vibrations or normal operating conditions is called Operational Modal Analysis (OMA) and provides useful methods for modal analysis of many areas of structural engineering. Identification of modal properties of a structural system is the process of correlating the dynamic characteristics of a mathematical model with the physical properties of the system derived from experimental measurements.

It is fair to say that processing of data in OMA is challenging; one can even say that this is close to torturing the data, and it is also fair to say that fiddling around long enough with the data might lead to some strange or erroneous results that might look like reasonable results. One of the aims of this book is to help people who use OMA techniques avoid ending up in this situation, and instead obtain results that are valid and reasonable.

In OMA, measurement data obtained from the operational responses are used to estimate the parameters of models that describe the system behavior. To fully understand this process, one should have knowledge of classical structural mechanics, matrix analysis, random vibration concepts, application-specific simplifying assumptions, and practical aspects related to vibration measurement, data acquisition, and signal processing.

OMA testing techniques have now become quite attractive, due to their relatively low cost and speed of implementation and the recent improvements in recording equipment and computational methods. Table 1.1 provides a quick summary of the typical applications of OMA and how these compare with classical modal testing, also denoted experimental modal analysis (EMA), which is based on controlled input that is measured and used in the identification process.

The fundamental idea of OMA testing techniques is that the structure to be tested is being excited by some type of excitation that has approximately white noise characteristics, that is, it has energy distributed over a wide frequency range that covers the frequency range of the modal characteristics of the structure. However, it does not matter much if the actual loads do not have exact white noise characteristics, since what is really important is that all the modes of interest are adequately excited so that their contributions can be captured by the measurements.

Referring to Figure 1.1, the concept of nonwhite, but broadband loading can be explained as follows. The loading is colored, thus does not necessarily have an ideal flat spectrum, but the colored loads can be considered as the output from an imaginary (loading) filter that is loaded by white noise.

Introduction to Operational Modal Analysis, First Edition. Rune Brincker and Carlos Ventura.
© 2015 John Wiley & Sons, Ltd. Published 2015 by John Wiley & Sons, Ltd.
Companion Website: www.wiley.com/go/brincker

Table 1.1   General characteristics of structural response

	Mechanical engineering	Civil engineering

EMA	*Artificial excitation*
	Impact hammer
	Shakers (hydraulic,
	  electromechanical,
	  etc.)
	Controlled blasts
	Well-defined measured
	  input
		Artificial excitation
		Shakers, mainly hydraulic
		Drop weights
		Pull back tests
		Eccentric shakers and exciters
		Well defined, measured, or
		  unmeasured inputs
		Controlled blasts

OMA	*Artificial excitation*
	Scratching device
	Air flow
	Acoustic emissions
	Unknown signal,
	  random in time and
	  space
		Natural excitation
		Wind
		Waves
		Traffic
		Unknown signal, random in
		  time and space, with some
		  spatial correlation

Source: Adapted from American National Standard: "Vibration of Buildings – guidelines for the measurement of vibrations and their effects on buildings," ANSI S2.47-1990 (ASA 95-1990).

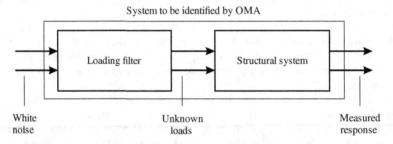

Figure 1.1   Illustration of the concept of OMA. The nonwhite noise loads are modeled as the output from a filter loaded by a white noise load

It can be proved that the concept of including an additional filter describing the coloring of the loads does not change the physical modes of the system, see Ibrahim et al. [1] and Sections 7.2.7 and 8.3.7. The coloring filter concept shows that in general what we are estimating in OMA is the modal model for "the whole system" including both the structural system and the loading filter.

When interpreting the modal results, this has to be kept in mind, because, some modes might be present due to the loading conditions and some might come from the structural system. We should also note that in practice we often estimate a much larger number of modes than the expected physical number of modes of the considered system.

This means that we need to find ways to justify which modes belong to the structural system, which modes might describe the coloring of the loading, and finally which modes are just noise modes that might not have any physical meaning. These kinds of considerations are important in OMA, and will be further illustrated later in this book.

We can conclude these first remarks by saying that OMA is the process of characterizing the dynamic properties of an elastic structure by identifying its natural modes of vibration from the operating responses. Each mode is associated with a specific natural frequency and damping factor, and these two parameters can be identified from vibration data from practically any point on the structure. In addition, each mode has a characteristic "mode shape," which defines the spatial distribution of movement over the entire structure.

1.1   Why Conduct Vibration Test of Structures?

Vibration measurements are made for a variety of reasons. They could be used to determine the natural frequencies of a structure, to verify analytical models of the structure, to determine its dynamic response under various environmental conditions, or to monitor the condition of a structure under various loading conditions. As structural analysis techniques continually evolve and become increasingly sophisticated, awareness grows of potential shortcomings in their representation of the structural behavior. This is prevalent in the field of structural dynamics. The justification and technology exists for vibration testing and analysis of structures.

Large civil engineering structures are usually too complex for accurate dynamic analysis by hand. It is typical to use matrix algebra based solution methods, using the finite element method of structural modeling and analysis, on digital computers. All linear models have dynamic properties, which can be compared with testing and analysis techniques such as OMA.

1.2   Techniques Available for Vibration Testing of Structures

Let us discuss in some detail the two main types of modal testing: the EMA that uses controlled input forces and the OMA that uses the operational forces.

Both forced vibration and in-operation methods have been used in the past and are capable of determining the dynamic characteristics of structures. Forced vibration methods can be significantly more complex than in-operation vibration tests, and are generally more expensive than in-operation vibration tests, especially for large and massive structures. The main advantage of forced vibration over in-operation vibration is that in the former the level of excitation and induced vibration can be carefully controlled, while for the latter one has to rely on the forces of nature and uncontrolled artificial forces (i.e., vehicle traffic in bridges) to excite the structure, sometimes at very low levels of vibration. The sensitivity of sensors used for in-operation vibration measurements is generally much higher than those required for forced vibration tests.

By definition, any source of controlled excitation being applied to any structure in order to induce vibrations constitutes a forced vibration test. In-operation tests that rely on ambient excitation are used to test structures such as bridges, nuclear power plants, offshore platforms, and buildings. While ambient tests do not require traffic shutdowns or interruptions of normal operations, the amount of data collected is significant and it can be a complex task to analyze this data thoroughly.

The techniques for data analysis are different. The theory for forced vibration tests of large structures is well developed and is almost a natural extension of the techniques used in forced vibration tests of mechanical systems. In contrast, the theory for ambient vibration tests still requires further development.

1.3   Forced Vibration Testing Methods

Forced vibration tests or EMA methods are generally used to determine the dynamic characteristics of small and medium size structures. In rare occasions, these methods are used on very large structures because of the complexity associated with providing significant levels of excitation to a large, massive structure. In these tests, controlled forces are applied to a structure to induce vibrations. By measuring the structure's response to these known forces, one can determine the structure's dynamic properties. The measured excitation and acceleration response time histories are used to compute frequency response functions (FRFs) between a measured point and the point of input. These FRFs can be used to determine the natural frequencies, mode shapes, and damping values of the structure using well-established methods of analysis. One can apply controlled excitation forces to a structure using several different methods. Forced vibrations encompass any motion in the structure induced artificially above the ambient level. Methods of inducing motion in structures include:

1. Mechanical shakers
   (a) Electro-magnetic
   (b) Eccentric mass
   (c) Hydraulic, including large shaking tables in laboratories
2. Transient loads
   (a) Pull-back and release, initial displacement
   (b) Impact, initial velocity
3. Man-excited motions
4. Induced ground motion
   (a) Underground explosions
   (b) Blasts with conventional explosives above the ground

The three most popular methods for testing structures are shaker, impact, and pull back or quick-release tests. A brief description of these methods follows:

1. *Shaker tests:* Shakers are used to apply forces to structures in a controlled manner to excite them dynamically. A shaker must produce sufficiently large forces, to effectively excite a large structure in a frequency range of interest. For very large structures, such as long-span bridges or tall buildings, the frequencies of interest are commonly less than 1 Hz. While it is possible to produce considerable forces with relatively small shakers at high frequencies, such as those used to test mechanical systems, it is difficult to produce forces large enough to excite a large structure at low frequencies. Although it is possible to construct massive, low frequency shakers, these are expensive to build, transport, and mount. In such cases, alternative methods to excite the structure are desirable.
2. *Impact tests:* Impact testing is another method of forced vibration testing. Mechanical engineers commonly use impact tests to identify the dynamic characteristics of machine components and small assemblies. The test article is generally instrumented with accelerometers, and struck with a hammer instrumented with a force transducer. While impact testing is commonly used to evaluate small structures, a number of problems may occur when this method is used to test larger structures. To excite lower modes of a large structure sufficiently, the mass of the impact hammer needs to be quite large. Not only is it difficult to build and use large impact hammers with force transducers, but the impact produced by a large hammer could also cause considerable local damage to the test structure.
3. *Pull back tests:* Pull back or quick-release testing has been used in some occasions for testing of large structures. This method generally involves inducing a prescribed temporary deformation to a structure and quickly releasing it, causing the structure to vibrate freely. Hydraulic rams, cables, bulldozers, tugboats, or chain blocks have been used to apply loads that produce a static displacement of the structure. The goal of this technique is to quickly release the load and record the free vibrations of the structure as it tends to return to its position of static equilibrium. The results from quick release tests

can be used to determine natural frequencies, mode shapes, and damping values for the structure's principal modes.

1.4 Vibration Testing of Civil Engineering Structures

What makes testing of large civil engineering structures different than testing mechanical systems? As we have just discussed, the obvious answer to this question is that the forces are larger and the frequencies are lower in large structures. But there is more than that. First, in general, analytical models of existing large structures are based on geometric properties taken from design or construction drawings and material properties obtained from small specimens obtained from the structure. A series of assumptions are also made to account for the surrounding medium and its interaction with the structure (such as soil-structure interaction in the case of buildings and bridges, and soil–water–structure interaction in the case of dams, wharves, and bridges) and the composite behavior of structural elements. This, in general, is not the case for mechanical systems. And second, in the field of mechanical engineering, there are a number of integrated systems that can handle very efficiently the experimental testing, system identification, and model refinement. These integrated systems are very sophisticated as they combine the results of several decades of research in the field. Due to their relatively small size, most mechanical specimens can be tested in laboratories under controlled conditions. There is no such advantage for the verification of dynamic models of large civil engineering structures.

During normal operating conditions, a building is subjected to ambient vibrations generated by wind, occupants, ventilation equipment, and so on. As we have argued earlier, a key assumption of the analysis of these ambient vibrations is that the inputs causing motion have "nearly" white noise characteristics in the frequency range of interest. This assumption implies that the input loads are not driving the system at any particular frequency and therefore any identified frequency associated with significant strong response reflects structural modal response. However, in reality, some of the ambient disturbances, such as, for instance, an adjacent machine operating at a particular frequency may drive the structure at that frequency. In this case, the deformed shapes of the structure at such driving frequencies are called operational modes or operational deflection shapes. This means that a crucial requirement of methods to analyze ambient vibration data is the ability to distinguish the natural structural modes from any imposed operational modes.

The integrated systems, developed for mechanical engineering applications are not practical and economical to test large civil engineering structures. Bridges form vital links in transportation networks and therefore a traffic shutdown required to conduct a forced vibration test would be costly. Controlled forced vibration tests of buildings may disturb the occupants and may have to be conducted after working hours, thus increasing the cost of the testing. Therefore, routine dynamic tests of bridges and buildings must be based on ambient methods, which do not interfere with the normal operation of the structure.

1.5 Parameter Estimation Techniques

The methods that have been developed for analyzing data from forced and in-operation vibration tests range from linear deterministic models to nonlinear stochastic models. The applications range from improving mathematical models of systems to damage detection, to identifying the input of a system for controlling its response. Parameter estimation methods using dynamic signals can be classified as

(a) time-domain methods
(b) frequency-domain methods
(c) joint frequency–time domain methods

The theory behind the first two methods is described in more detail in this book.

1.6   Brief History of OMA

Although very significant advances in OMA testing techniques have occurred since the early 1990s, there is a wealth of information about different uses of OMA since the 1930s. Even ancient history shows evidence of the use of the OMA concepts to better understand why and how structures vibrate.

Pythagoras is usually assumed to be the first Greek philosopher to study the origin of musical sound. He is supposed to have discovered that of two stretched strings fastened at the ends the higher note is emitted by the shorter one. He also noted that if one has twice the length the other, the shorter will emit a note an octave above the other. Galileo is considered the founder of modern Physics and in his book "Discourses Concerning Two New Sciences" in 1638: At the very end of the "First Day," Galileo has a very remarkable discussion of the vibration of bodies. He describes the phenomenon of sympathetic vibrations or resonance by which vibrations of one body can produce similar vibrations in another distant body. He also did an interesting comparison between the vibrations of strings and pendulums in order to understand the reason why sounds of certain frequencies appear to the ear to combine pleasantly whereas others are discordant.

Daniel Bernoulli's publication of the Berlin Academy in 1755 showed that it is possible for a string to vibrate in such a way that a multitude of simple harmonic oscillations are present at the same time and that each contributes independently to the resultant vibration, the displacement at any point of the string at any instant being the algebraic sum of the displacements for each simple harmonic at each node. This is what is called the Principle of "Coexistence," which is what we know today as the Superposition Principle. Today, we also refer to this as the method of Modal Superposition. Joseph Fourier's publication "Analytical Theory of Heat" in 1822 presents the development of his well-known theorem on this type of expansion. Isaac Newton in the second book of his "Principia" in 1687 made the first serious attempt to present a theory of wave propagation. John Strutt, 3rd Baron Rayleigh (1842–1919) through his investigations of sound and vibration provided the basis for modern structural dynamics and how mass, stiffness and damping are interrelated and determine the dynamic characteristics of a structural system.

The first studies on shocks and vibrations affecting civil engineering structures in the twentieth century were carried out at the beginning of the 1930s to improve the behavior of buildings during earthquakes. M.A. Biot introduced the concept of the shock spectrum to characterize the response of buildings to earthquakes and to compare their severity. G. Housner, refined the concept by defining it as the shock response spectrum (SRS) to clearly identify that it characterizes the response of a linear one-degree-of-freedom system subjected to a prescribed ground shaking. After the 1933 Long Beach earthquake in California, in 1935, D.S. Carder conducted tests of ambient vibrations in more than 200 buildings and applied rudimentary OMA techniques to determine the natural modes of vibrations of these buildings. The results of this investigation were used in the design codes to estimate natural frequencies of new buildings. The seminal work of M. Trifunac in 1972 showed that the analysis of ambient and forced vibrations led to the same results for practical engineering purposes.

The development of OMA techniques since the mid-1990s can be followed by reading the proceedings of the annual International Modal Analysis Conference (www.sem.org) and, most recently, those from the International Operational Modal Analysis Conference (www.iomac.dk).

1.7   Modal Parameter Estimation Techniques

In contrast to EMA, OMA testing does not require any controlled excitation. Instead, the response of the structure to "ambient" excitation sources such as wind, traffic on or beneath the structure, and microtremors is recorded. Many existing textbooks provide an extensive overview of input–output modal parameter estimation methods. See for instance, Heylen et al. [2] and Ewins [3]. In the operational case, ignoring the need to measure the input is justified by the assumption that the input does not contain any specific information, or expressed in other words, the input is approximately white noise. As with EMA, the measured time signals can be processed in the time domain or in the frequency domain. Since the forcing function is unknown, frequency response functions between force and response signals cannot be calculated. Instead, the analysis relies on correlation functions and spectral density functions estimated from the operational responses.

Further, since OMA sensors and cables can be expensive, a limited number of sensors are used and some of these sensors are roved over the structure to obtain several data sets. In order to be able to assemble mode shapes using the parts of the mode shape estimated by the different data sets, some of the response signals are declared as reference signals. The reference sensors are kept in the same place when recording all data sets while the remaining sensors are moved progressively over the structure.

A fast method to estimate modal parameters from OMA tests is the rather simple "peak-picking" frequency-domain technique. This technique has been used extensively over the years for all kinds of applications. The basic idea of the peak-picking technique is that when a structure is subjected to ambient excitations, it will have strong responses near its natural frequencies. These frequencies can be identified from the peaks in the power spectral densities (PSD) computed for the time histories recorded at the measurement points. This concept is illustrated in the example presented in Figure 1.2. The significant peaks of the PSDs for the OMA measurements conducted on a five-span bridge can be associated with the natural frequencies of vertical vibrations of this bridge.

The method has been widely used for many years. One practical implementation of this method was developed by Felber [4] and will be used here as an example of the application of early OMA techniques. In this implementation, the natural frequencies are determined as the peaks of the averaged normalized

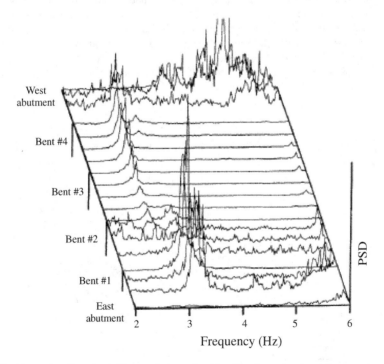

Figure 1.2 Example of peak-picking technique for identification of natural frequencies of vertical vibration of a five-span bridge (Source: courtesy of Felber [4])

power spectral densities (ANPSDs). It is then assumed that the coherence function computed for two simultaneously recorded response signals has values close to one at the natural frequencies; see Bendat and Piersol [5]. This also helps to decide which frequencies can be considered as natural frequencies for the structure. The components of the mode shapes are determined by the values of the transfer functions obtained at the natural frequencies. It should be noted that in the context of ambient testing, transfer function does not mean the ratio of response over force, but rather the ratio of the response measured at a location of the structure with respect to the reference sensor. Every transfer function yields a mode shape component at a natural frequency. In this method it is assumed, however, that the dynamic response at resonance is only determined, or controlled by one mode. The validity of this assumption increases when modes are well separated and when the modal damping is low (see more about this in Section 10.2).

Felber implemented a novel procedure based on this idea to expedite the modal identification of ambient vibration data, and this seminal work was the motivation for the development of interactive techniques for fast and efficient implementation of the peak-picking technique. Three programs were used for the process, one to generate the ANPSDs for the identification of natural frequencies, another to compute the transfer functions between sensors (program ULTRA) and the third to visualize and animate the mode shapes (program VISUAL). All three programs were used for preliminary analysis in the field and as well as for further analysis in the office. Figure 1.3 shows a flowchart of the implementation of the methodology developed by Felber.

Powerful time-domain techniques for operational identification have been developed over the years and have gained tremendous popularity in the last three decades. The Ibrahim Time Domain (ITD) was the first, Ibrahim and Milkulcik [6], and shortly after came the Polyreference, Vold et al. [7] and the eigen realization algorithm (ERA), Juan and Pappa [8]. The two last techniques were developed for multiple inputs, and the ITD can also be formulated for multiple inputs (see Section 9.4). What they have in common is that they assume that a free response function can be obtained from each record, and this is the basis for the implementation of the algorithms to deal with the responses to either single or multiple inputs using the correlation functions of the measured operational responses.

During the 1990s two important techniques became available: the stochastic subspace identification (SSI) technique by Van Overschee and De Moor [9] and the frequency-domain decomposition (FDD)

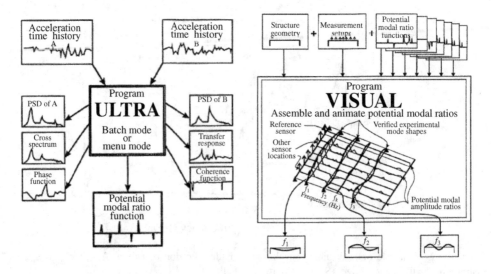

Figure 1.3   Example of peak-picking technique (Source: as developed by Felber [4])

technique by Brincker et al. [10]. And after the millennium, the PolyMax[1] technique by Guillaume et al. [11] was introduced to the engineering community.

It is interesting to note that most of the classical input–output methods can be readily modified to be used for the operational case. For instance, frequency response function (FRF) driven methods can be converted to spectrum-driven methods and impulse response function (IRF) driven methods can be used together with correlation functions instead.

The accelerations of a structure associated with ambient excitations are typically very small, generally in the order of tens to hundreds of milli-g's as shown in Table 1.2, and can vary considerably during acquisition, for instance, depending on whether a truck, a car, or no traffic is passing a bridge at a certain speed. This causes challenges to the sensors, the acquisition system and the identification algorithms that must be able to extract weakly excited modes from noisy data. The developments in recent years both on the acquisition side and on the identification side greatly enhanced the use of ambient vibration testing to estimate the modal parameters of a large civil engineering structure.

Next to hardware and software, the judgment and experience of the modal analyst plays a role in the success of the modal parameter estimation. Research into the automation of this process, so that any user interaction could be excluded, is certainly useful.

Table 1.2 General characteristics of structural response

Vibration excitation	Frequency range (Hz)	Displacement range (µm)	Velocity range (mm/s)	Acceleration range (µg)	Suggested measuring quantity
Traffic: road, rail, ground-borne	1–80	1–200	0.2–50	2–100	Velocity
Blasting vibration ground-borne	1–300	100–2500	0.2–500	2–500	Velocity
Pile driving ground-borne	1–100	10–50	0.2–50	2–200	Velocity
Machinery outside ground-borne	1–300	10–1000	0.2–50	2–100	Velocity/ Acceleration
Acoustic: traffic, machinery outside	10–250	1–1100	0.2–30	2–100	Velocity/ Acceleration
Air over pressure	1–40				Velocity
Machinery inside	1–1000	1–100	0.2–30	2–100	Velocity/ Acceleration
Human activities					
a. impact	0.1–100	100–500	0.2–20	2–500	Velocity
b. direct	0.1–12	100–5000	0.2–5.0	2–20	Acceleration
Earthquakes	0.1–30	$10–10^5$	0.2–400	2–2000	Velocity/ Acceleration
Wind	0.1–10	$10–10^5$			Acceleration
Acoustic inside	5–500				

Source: Adapted from American National Standard: "Vibration of Buildings – guidelines for the measurement of vibrations and their effects on buildings," ANSI S2.47-1990 (ASA 95-1990).

---

[1] PolyMax is a trademark of LMS International.

The operational modal technology allows the user to perform a modal analysis in an easier way and in many cases more effectively than traditional modal analysis methods. It can be applied for modal testing and analysis on a wide range of structures and not only for problems generally investigated using traditional modal analysis but also for those requiring load estimation, vibration level estimation, and fatigue analysis.

1.8   Perceived Limitations of OMA

It is commonly recognized that there are two primary apparent drawbacks in the modal identification on OMA data: (i) mass scaling of the mode shapes and (ii) lack of excitation of some modes. Because it is not possible to measure the input force when using ambient excitation, the identification process does not provide mass normalized mode shapes. This is a problem for damage detection techniques that require mass-normalized modes, such as flexibility-based damage methods. However, this limitation can be overcome as described in Section 12.6. The second limitation arises because the spectra and spatial locations of the input forces cannot be dictated, so some vibration modes may not be well excited and therefore may not be identifiable from the data.

When analyzing the results of a modal test, it is important to understand the statistical uncertainty on the results arising from random errors such as electrical noise, slight variations in testing conditions, environmental effects (such as temperature and wind), and so on. The ambient forces are often difficult to forecast, and this makes it also difficult to forecast the operational responses. Some guidelines exist to help the analyst deal with this limitation, such as the information presented in Table 1.2 and the approaches mentioned in Section 7.2.6.

1.9   Operating Deflection Shapes

What is an operating deflection shape (ODS)? Traditionally, an ODS has been defined as the deflection of a structure at a particular frequency of excitation. However, an ODS can be defined more generally as any motion of two or more points on a structure produced by an excitation. A shape is the motion of one point relative to all others. Motion is a vector quantity, which means that it has location and direction associated with it. This is also called a degree of freedom, or DOF.

An operating deflection shape contains both forced and resonant vibration components for two or more DOFs of a structure. In contrast, a mode shape characterizes only the resonant vibration at two or more DOFs.

In a time-domain ODS, all of the responses have to be measured simultaneously, or they have to be measured in a manner to ensure their correct magnitudes and phases relative to one another. ODSs can be obtained either from a set of time-domain responses, or from a set of frequency-domain functions that are computed from time-domain responses. Natural modes of vibration are different from ODSs in the following ways:

1. Each mode is defined for a specific natural frequency. An ODS can be defined at any frequency.
2. Modes are generally defined for linear, stationary structures. ODSs can be defined for nonlinear and nonstationary structures, although it is possible to define instantaneous mode shapes for nonlinear structures.
3. A mode shape can be associated with resonant vibrations of the structure. An ODS can be associated with both resonant and nonresonant vibrations.
4. Modes do not depend on forces or loads, but their superposition determines the structural response to the applied forces or loads. They are inherent properties of the structure (mass, stiffness and damping, and boundary conditions). ODSs depend on forces or loads, and will change if the loads change.

5. Modes only change if the material properties, geometrical properties, or boundary conditions change. ODSs will change if either the modes or the loads change.
6. Mode shapes do not have unique values or units – they can be dimensionless. ODSs do have unique values and units.

All experimental modal parameters are obtained by post-processing ODS measurements. At or near a resonance peak, the ODS is dominated by a mode. Therefore, at such frequencies, the ODS is approximately equal to the mode shape.

1.10   Practical Considerations of OMA

All of the well-known techniques of today can handle multiple input data, and the importance of this aspect is illustrated by a frequency-domain decomposition showing the interaction of the different singular values of the spectral density matrix, as well as, by a stochastic subspace identification technique where a stabilization diagram is being used. These two techniques represent two very different classes of identification, but they clearly illustrate what is believed to be a common tendency for all techniques: they work much better with multiple-input data.

In OMA, multiple-input data are – as we shall demonstrate throughout this book – naturally available by processing the operating response data in order to obtain correlation functions matrices and/or spectral density matrices and using the information in these matrices in a proper way.

In operational modal testing, this is normally done by just measuring the responses under the natural (ambient or operational) conditions. This means, for instance, that if a bridge is going to be tested, the bridge traffic and normal operation need not be interrupted during the test. On the contrary, the traffic will be used as the excitation source, and the natural response of the bridge to that loading – and to other natural loads acting on the structure at the same time – will be used to perform an operational modal identification.

Similarly, if an engine is going to be subjected to OMA testing, it is more desirable to perform such test with the engine running under normal operating conditions. The engine responses will be measured, and the operational identification will be performed under this load condition.

In cases where the number of sensors available is less than the desired number of measured DOFs, it will be necessary to use some of the sensors as references (they remain in the same points), and the remaining sensors will be roved over the structure. In many cases, the number of data sets obtained this way can be rather high. It is not unusual to have 20–30 data sets from a single test.

The number of sensors, their orientation, and the selection of reference sensors must be made during the test planning stage so that all modes of interest are clearly identifiable in all data sets. Special care has to be taken in cases where closely spaced modes are likely to exist. In such cases, the user must make sure that the two closely spaced modes are not only clearly visible in all data sets but also clearly distinguishable in all data sets.

As we shall see later, all current techniques for operational identification can be formulated for multiple-inputs. To clearly identify closely spaced modes, the loading must be multiple-input. The question is, however: In OMA where we do not control or even know the forces acting on the structure, how can the user be sure that the loading is multiple-input?

In order to answer this question, one has to make sure that at least one of two different types of loads produces a clear multiple-input:

• A loading that is moving over a large part of the structure.
• A distributed loading with a correlation length significantly smaller than the structure.

The first type of loading may result, for instance, when a car is crossing a bridge. The car passes over the bridge and thus loads the bridge at infinitely many points. Not only does this kind of loading provide multiple loading, it also helps us ensure that all modes that are sensitive to vertical loading will be excited. This is the ideal kind of loading. These issues are discussed more in detail in Chapter 7.

The second kind of loading results, for instance, when wind is acting along the height of a building or waves are loading an offshore structure. Such loading is random in time and space, but as there is a correlation time at a fixed point, there is also a correlation length at a fixed time. To make sure that the wind load on a building is multiple-input, the correlation length of the wind loading must be significantly smaller than the width and height of the building. The same can be said about traffic on a road nearby a building. If the road is close to the building, then the traffic is actually loading the structure in many points; however, if the road is at a fair distance from the building, then a car passing by would produce a single wave that will be propagated toward the structure. In this case, the building could be considered to be excited by a single-input load.

This explains why it is not desirable that the source of excitation for a structure being tested originates from a single source. The same applies to scaled models of structures being tested in the laboratory. If we scale a building down to 1/30 or 1/50 of its original size but keep the distance to the traffic source the same as for the prototype, or keep the correlation length for the simulated wind loading, then we may get a loading that resembles a single input, and in this case the identification becomes a difficult task.

For such cases, it would then be desirable to provide some kind of artificial loading that resembles a multiple input. This requirement is much easier to satisfy when compared to the work required to setup a forced vibration test, in which shakers have to be installed, forces have to be controlled and measured, and so on. For the OMA test, we just need to make sure that the loading is reasonable random in time and space. Let some cats chase some rats on the structure, let it rain on it, let somebody walk on it, or somebody drive a cart on it. All of these are examples of suitable random loads necessary for good testing practices in OMA.

One could say that in operational testing there are in fact only two main rules to be followed:

- Make sure that you have multiple input loads.
- Make sure that you record good quality operating response data.

The first rule was explained earlier, but a good question to answer may be: how can the user check that? The answer is that one of the easiest ways to check this is to perform a singular value decomposition of the spectral matrix of the measured data and plot the singular values as a function of frequency. If a family of singular values is observed so that several of them participate in defining a resonant peak of the structure, this serves as a strong indication of multiple-input loading. If the excitation was from a single input case, then only the first singular value curve will clearly express the operational response, and all the other singular values will be clearly distinct only relating to the noise in the data.

The second rule is of general applicability for all experimental cases – get good data, because if you do not, then you are going to have a hard time performing a modal identification. In OMA, this is especially important, and since responses are often weak, measures must be taken to secure good signal-to-noise ratios in the signal. Not only do you need to ensure that your data has a good signal-to-noise ratio, you also have to check and remove outliers and dropouts, and have good signal processing tools to handle spikes and all that terrible things that can happen when taking measurements. Furthermore, it is not enough to have good-quality data. You also need to collect detailed information about the test being performed. As it appears, many things might go wrong in the testing part, and this is why we have dedicated a rather extensive chapter (Chapter 7) to discuss all the technical issues involved with OMA testing.

For many people, OMA identification is related to time-domain identification. There seem to be an expectation that if one is performing an operational modal identification, then one is only working in the time domain. The truth is that there is no "natural domain," both time domain and frequency domain work well with good data, and perform poorly if one has bad data.

The range of structures where modal analysis can be applied extends dramatically when we move to the operational way of doing things. You can even say the bigger the better, because – as we have learned in the first section – the bigger the structure, the more independent inputs exist, and as we have argued, the better is the identification of the dynamic properties of the structure.

The application of operational modal is only limited by the possibility to actually measure the response. For very large structures that might still be a significant problem simply because with the measurement technology of today, it is costly to manage all those miles of cables. However, since we are moving in the direction of wireless digital sensors, cabling problems are disappearing and that means that in the future we will be able to apply operational modal on virtually any kinds of large structures. Testing very small structures still remain a challenge, but with further development of the laser technology, it is expected that soon affordable systems for response measurements of tiny components will be available.

OMA has more applications than traditional modal. Let us try to explain the range of possibilities. Some of them are questionable since they have not been tried much, but anyway the potential is there, which cannot be said about traditional modal testing.

First of all, OMA testing can be used for all the applications, which is home ground for EMA. Now when we can also get the scaling factors on the mode shapes, there are no limitations. Accuracy is also not an issue, OMA can do as well. Maybe traditional modal is more competitive in some cases because you get the scaling factor "for free", controlling the input is easy if the structure is medium size and so on. However, there is a range of applications where traditional modal can hardly compete with OMA. Here are some examples.

One of the well-known applications is continuous monitoring of structures. In this field, traditional modal is out of question because we cannot go the structure and install an exciter, drop-load or conduct a hammer test every time we are going to take a measurement to see if the structure is still ok. The obvious solution is to install a measurement system there, take measurements whenever it is needed, and then – do your OMA. As measurements systems become cheaper and more accurate, to acquire data from long distance (for instance over the Internet) is becoming easier to perform, and therefore more appealing to the owners and operators of the facilities because of the potential savings related to maintenance and safety.

This application is more or less well known. But what is not so well known is that OMA can also be used for estimation of loads, vibration levels, and fatigue. These three very important applications are directly related to the fact that in OMA, we store the measured raw time series. We keep the full information about the measured responses. As a result, by measuring the response in only a relatively small number of points, then after an OMA of the recorded data – as it is shown in Sections 12.5.1 and 12.5.2 – the corresponding response in any point of the structure can be estimated. This is of vital importance, for instance, for vibration level effects estimation. The vibration levels in an office building do not have to be measured in all the offices in the building to know them. You can tell the vibration level in hundreds of offices just by taking measurements in a few points.

Similarly one can solve one of the greatest challenges in fatigue analysis: to find the real stress history at any point of the structure. By conducting an analysis like this, the engineer can perform an improved fatigue analysis by-passing all the uncertainty from loading modeling and from relating the loading with the stress history. Since a large uncertainty is removed, many structures that are already "dead" due to standardized fatigue analysis might be able to have their operational lives substantially extended. By making this kind of analysis under different loading conditions, and by building a statistical knowledge data base on the frequency of these loading conditions, a much better prognosis for the future damage accumulations might be made.

1.11 About the Book Structure

Although OMA is not necessarily a new field, many technologies related to OMA have developed fast in the last few years, and different opinions and ideas are still being debated by the scientific community. A key issue to consider is if a completely different framework should be used to formulate the theories

Table 1.3 Organization of this book

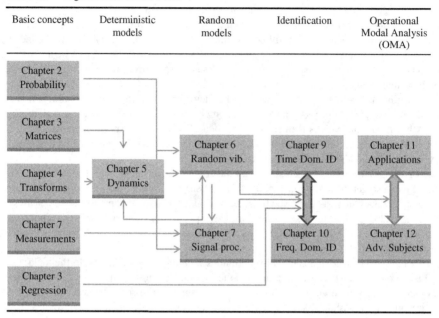

Basic concepts	Deterministic models	Random models	Identification	Operational Modal Analysis (OMA)
Chapter 2 Probability				
Chapter 3 Matrices				
Chapter 4 Transforms	Chapter 5 Dynamics	Chapter 6 Random vib.	Chapter 9 Time Dom. ID	Chapter 11 Applications
Chapter 7 Measurements		Chapter 7 Signal proc.	Chapter 10 Freq. Dom. ID	Chapter 12 Adv. Subjects
Chapter 3 Regression				

behind OMA, like using a stochastic framework instead of using a deterministic framework, as is done for traditional modal analysis.

This book is written based on the assumption that we use a stochastic framework when dealing with data and signal processing. But we recognize that a deterministic framework can be used once the data has been processed, and the information contained in the correlation functions and/or spectral density functions has a physical meaning.

This book is organized in four main sections over 12 chapters as shown in Table 1.3. The first section is about basic concepts that are fundamental for a clear understanding of OMA. The five chapters that are part of this section are about probability, matrices, transforms, measurements, and regression. These topics are covered in detail in many other books. We could have omitted this material in the book, but we felt that it was necessary to provide a quick review of these concepts in order to help the reader gain a better understanding of what are the most important concepts that are needed to better appreciate the advantages and limitations of OMA.

The second section helps the reader gain a better understanding of the vibration of structures based on deterministic models. Again this material could have been omitted, but we are presenting this material with the same reasons as before – to build a solid basis to understand why OMA works so well.

In the third section, the stochastic framework for the formulation of a random model of the structure is presented. This includes models for random vibrations such as the theoretical forms of the correlation and spectral matrices. It also presents the different ways of processing operational response data so that correlation and spectral matrices can be estimated. This material brings us close to the core of OMA. Once this is understood, the only thing that remains is to estimate a model that fits the data. To do this, we typically need to perform a regression either in the time domain or in the frequency domain and the final basic concept comes into play.

The last four chapters in the book should – as it is indicated in Table 1.3 – always be seen as interacting together with each other. For instance, one should always consult the advanced subjects list to see if some special problems are present, and OMA should always be seen as related to both time-domain and frequency-domain techniques. One should never limit the analysis to one of them, but seek enough information from the two domains so that the OMA results can be validated.

The response data used in OMA are generally provided in discrete form as a consequence of the analog-to-digital conversion that happens when data is acquired. Because the data are discrete, it is common practice to present the basics of modal identification using formulas in discrete time.

In this book, however, we have chosen a different approach, that is, to present all the fundamental concepts using continuous time formulations. However, we recognize that in certain cases, it would be more effective to present fundamental concepts using a discrete time formulation. For instance, let us consider the question of calculating the average value of a signal $y(t)$ over a certain time interval of duration T. When dealing with the average value of a signal, in this book, we will prefer to use the continuous time definition given by

$$\mu = \frac{1}{T} \int_0^T y(t)dt \tag{1.1}$$

The corresponding formula in discrete time is obtained by taking the signal $y(t)$ and sampling it at a time step Δt. The resulting discrete signal (time series) is $y(n) = y(n\Delta t)$ and the average of this time series can be obtained by replacing the integral with a sum in the Eq. (1.1). This leads to the discrete time equation

$$\mu = \frac{1}{N\Delta t} \sum_{n=1}^{N} y(n)\Delta t = \frac{1}{N} \sum_{n=1}^{N} y(n) \tag{1.2}$$

In this formula, we have exchanged the time interval T with the number of data points N in the interval. Some people might think that the resulting formula provides the same information as the continuous time expression given by Eq. (1.1). However, in Eq. (1.2), we have actually lost the information that the average is calculated over a certain time interval T unless we recover this information by remembering that $T = N\Delta t$ taking into account that a sampling interval Δt must be defined.

In this book, we have decided to use continuous time formulations unless the circumstances force us to use a discrete formulation. When the reader does not need to worry about the sampling time step or remember the implications of relations like $T = N\Delta t$ we do not want him to do so. We believe that this keeps things simpler and closer to the basic ideas.

We invite the reader to follow us in the coming chapters of this book. And we also invite the reader to download the MATLAB toolbox that we have developed as part of this book to help the reader understand better the concepts discussed herein by performing numerical analysis of OMA test data. The toolbox can be obtained from www.wiley.com/go/brincker.

References

[1] Ibrahim, S.R., Asmussen, J.C. and Brincker, R.: *Modal parameter identification from responses of general unknown random inputs.* In IMAC XIV – 14th International Modal Analysis Conference, 1996.

[2] Heylen, W., Lammens, S. and Sas, P.: *Modal analysis theory and testing.* Department of Mechanical Engineering, Katholieke Universiteir Leuven, 2007.

[3] Ewins, D.J.: *Modal testing, theory practice and application.* 2nd edition. Research Studies Press Ltd., 2000.

[4] Felber, A.J.: *Development of a hybrid bridge evaluation system.* PhD. Thesis, University of British Columbia, Vancouver, Canada, 1993.

[5] Bendat, J.S. and Piersol, A.G.: *Engineering applications of correlation and spectral analysis.* 2nd edition. John Wiley & Sons, Inc, 1993.

[6] Ibrahim, S.R. and Milkulcik, E.C.: *A method for direct identification of vibration parameters from the free response.* Shock Vib. Bull., V. 47, p. 183–196, 1977.

[7] Vold, H., Kundrat, J., Rocklin, G.T. and Russell, R.: *A multi-input modal estimation algorithm for mini-computers.* SAE Paper Number 820194, 1982.

[8] Juang, J.N. and Pappa, R.S.: *An eigen system realization algorithm for modal parameter identification and modal reduction*. J. Guidance, V. 8, No. 5, p. 620–627, 1985.

[9] Overschee, P. and De Moor, B.: *Subspace identification for linear systems, theory, implementation, application*. Kluwer Academic Publishers, 1996.

[10] Brincker R., Zhang L., and Andersen P.: *Modal identification from ambient responses using frequency domain decomposition*. In Proceedings of IMAC 18, the International Modal Analysis Conference, p. 625–630, San Antonio, TX, USA, February 2000.

[11] Guillaume P., Verboven P., Vanlanduit S, Van der Auweraer H. and Peeters B.: *A poly-reference implementation of the least-squares complex frequency domain-estimator*. In Proceedings of the 21st International Modal Analysis Conference, Kissimmee (Florida), February 2003.

2

Random Variables and Signals

"I will never believe that God plays dice with the universe"

– Albert Einstein

From the colophon of this chapter, one could infer that modeling what is happening as a random event might not be the best way of doing it, it is rather like doing the second best – the best is to model what is really going on.

If we measure an event with great detail and gather enough information of what is really happening, then the theories of random signals that we use in this book would be unnecessary as we simply would not need them, because knowing the reality is better than modeling something close to the reality.

We have to accept that it is nearly impossible to know in detail, for instance, the forces acting on a bridge when it is loaded by external forces emanating from traffic, from the turbulent flow of the wind, from micro seismic activity in the ground and, maybe even from waves due to water flowing underneath the bridge. And we will have to accept that using a random modeling approach is a practical way to deal with the lack of information of how the structure is being excited and its experienced response.

In this chapter, we will build on common knowledge of random variables and try to generalize this to random signals. For the specific case that we are dealing with, we do not know the inherent probability characteristics (the density function), but only know what we have measured. We will argue that since nearly all random responses have Gaussian characteristics, and that normally we do not care about the mean values of the measured signals, all important information is contained in the correlation functions.

In this chapter, we will follow the theories discussed in well-known textbooks on this subject matter such as those by Newland [1] and Thompson [2]. It should be noted also that the material presented here has been inspired by more formal mathematical texts such as that by Papoulis and Pillai [3].

So, let us deal with the basic concepts first.

2.1 Probability

2.1.1 Density Function and Expectation

In this book, we are dealing with signals that correspond to observations of variables as a function of time. For instance, we can consider the signal $x(t)$, representing data about the quantity X as a function of time, see right plot in Figure 2.1. We can then establish a distinction – as it is normally done in probability theory and statistics – between the variable X and the actual observed signal $x(t)$, which is also denoted as a realization of X.

Introduction to Operational Modal Analysis, First Edition. Rune Brincker and Carlos Ventura.
© 2015 John Wiley & Sons, Ltd. Published 2015 by John Wiley & Sons, Ltd.
Companion Website: www.wiley.com/go/brincker

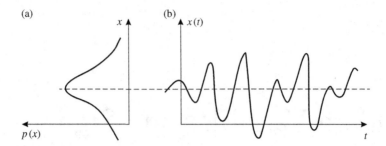

Figure 2.1   (a) The PDF for the corresponding random variable X. (b) The considered signal $x(t)$

Recognizing that we may not know the exact behavior of the quantity X and that the variation of this variable is uncertain, we can then say that the variable is random (or stochastic), and we describe the variable by its probability density function (PDF) $p(x)$. Different realizations x of the stochastic variable X have different probabilities associated with it; see Figure 2.1a.

More precisely, the PDF $p(x)$ is related to the probability of the event that the random variable X is in the small interval $[x; x + dx]$. This probability is defined by

$$\Pr[X \in [x; x + dx]] = p(x)\,dx \tag{2.1}$$

and is represented graphically on Figure 2.2a. Since the probability $\Pr[X \in [-\infty; \infty]] = 1$ we have that the PDF must fulfill the condition

$$\int_{-\infty}^{\infty} p(x)\,dx = 1 \tag{2.2}$$

Once we know the PDF, we can calculate the expectation of the random variable and all possible functions of it. In particular, we have the mean value $\mu = \mu_x$, which is the first moment of $p(x)$ and is also denoted as the expectation of X

$$\mathrm{E}[X] = \int_{-\infty}^{\infty} x p(x)\,dx = \mu \tag{2.3}$$

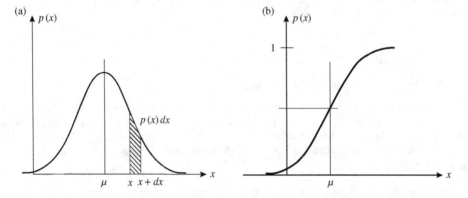

Figure 2.2   (a) Density function for the random variable X. (b) Distribution function for the random variable X

which is a measure of the location of the majority of the probability density. In Figure 2.2a, we show a typical PDF where the mean value is located close to the point with maximum density. The variance $\sigma^2 = \sigma_x^2$ of the variable X is given by the second moment of the PDF

$$E\left[(X - \mu_x)^2\right] = \int_{-\infty}^{\infty} (x - \mu_x)^2 p(x)\, dx = \sigma^2 \tag{2.4}$$

This equation defines the well-known, standard deviation σ, which represents a measure of the width of the PDF. In general, for any function $f(\cdot)$, we can define the expectation of the function of the random variable as

$$E\left[f(X)\right] = \int_{-\infty}^{\infty} f(x) p(x)\, dx \tag{2.5}$$

At this point, it is practical to introduce the concept of stationarity. We will say that a signal is stationary if the PDF of the underlying stochastic variable X is not a function of time. That is, it does not change with time. This also means that all possible expectations of X are not a function of time. Real data are always more or less nonstationary – but for practical purposes we will, as a main rule, always assume stationarity of the data as an approximation.

The integral of the density function defines the distribution function

$$P(x) = \int_{-\infty}^{x} p(u)\, du \tag{2.6}$$

that is equal to the probability $\Pr\left[X \leq x\right]$. This is illustrated graphically by the right plot in Figure 2.2.

Example 2.1 Estimating the density and distribution function

In this example, the distribution function and the PDF for the first channel of data of the first data set of the Heritage Court Building data is estimated. See Chapter 11 for details about this test case.

It is simpler to estimate the distribution function than the density function. First step is to read the considered data, normalize the data by subtracting the mean and dividing by the standard deviation. Then we simply sort these data in ascending order and define the normalized and sorted data as the variable $x(n)$ to be plotted on the horizontal axis while on the vertical axis we plot the data $y(n) = n/N$ where $n \in [1; N]$ and N is the number of data points. This produces the top plot of Figure 2.3.

The density function shown at the bottom of Figure 2.3 is obtained by numerical differentiation of the distribution function. As it appears, the distribution function seems very smooth, whereas the density function shows more scatter. This is a normal situation in density function estimations.

2.1.2 Estimation by Time Averaging

In most practical cases, we do not know the density function $p(x)$, we only know the signal $x(t)$ that for each time instant is a realization of the random variable X. This means that we cannot directly use the aforementioned formulas to calculate the characteristics of the signal because they are all based on the assumption that the density function is already known. We need a way to solve this problem. The solution is to go back to the definition given by Eq. (2.1) and assume that the signal is known in the time

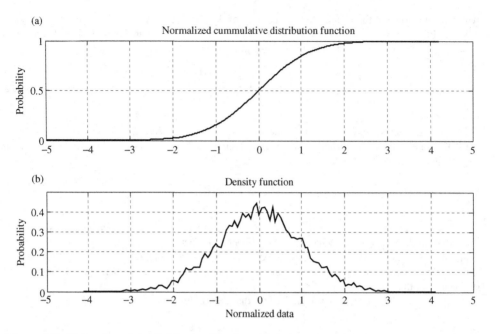

Figure 2.3 Estimation of the distribution function and the PDF for the first data set of the Heritage Court Building data, see Appendix B for details on this case

interval $[0; T]$ and that the time length T of the considered realization $x(t)$ is long enough so that the density function can be estimated using the following equation

$$\Pr\left[x(t) \in \left[x_n; x_n + \Delta x_n\right]\right] = p(x_n)\,\Delta x_n = \frac{\sum_k \Delta t_{n,k}}{T} \qquad (2.7)$$

where $\sum_k \Delta t_{n,k}$ denotes the summation of all the small time segments $\Delta t_{n,k}$ where $x(t) \in \left[x_n; x_n + \Delta x_n\right]$. This is illustrated in Figure 2.4. Using the discrete form of Eq. (2.3) and combining it with Eq. (2.7) the mean value is given by

$$E[x(t)] = \sum_{n=-\infty}^{\infty} x_n p(x_n)\,\Delta x_n = \sum_{n=-\infty}^{\infty} x_n \frac{\sum_k \Delta t_{n,k}}{T} \qquad (2.8)$$

The set of all the time segments $\Delta t_{n,k}$ is a double indexed quantity but constitute the total time interval $[0; T]$ in which the signal is known. We can arrange the time segments $\Delta t_{n,k}$ to form a new set of time intervals δt_m that is arranged in chronological order so that the time t_k to the end of the time interval δt_m is given by $t_k = \sum_{m=1}^{k} \delta t_m$. In this way, we can carry out the double summation in Eq. (2.8) as a single summation over the new set of time segments. By using Eq. (2.8) on the new set of time segments and letting the time segments approach zero, we get the following integral expression that is valid over the time interval $[0; T]$,

$$E[x(t)] = \lim_{\delta t \to 0} \sum_m x(t_m) \frac{\delta t_m}{T} = \frac{1}{T} \int_0^T x(t)\,dt \qquad (2.9)$$

Equation (2.9) illustrates the general and simple principle of time averaging, which is assumed to be identical to Eq. (2.3) for $T \to \infty$. When this is the case and the time averaging gives the same result as

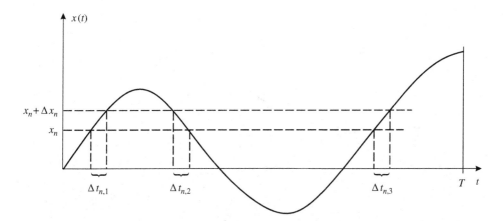

Figure 2.4 Each interval $[x_n; x_n + \Delta x_n]$ on the axis for the random variable $x(t)$ defines a number of associated time intervals $\Delta t_{n,k}$

when using the density function averaging, it is normal to say that the stochastic system that has generated the signal is ergodic. We will assume that all the systems that we will treat in this book are ergodic with respect to all averaging processes. Similarly for the variance we have

$$E\left[\left(x(t) - \mu_x\right)^2\right] = \frac{1}{T} \int_0^T \left(x(t) - \mu_x\right)^2 dt \tag{2.10}$$

and for any function $f(x)$ we have that

$$E\left[f(x(t))\right] = \frac{1}{T} \int_0^T f(x(t)) \, dt \tag{2.11}$$

2.1.3 Joint Distributions

By generalizing Eq. (2.1), the joint density function for two random variables X, Y is defined as

$$\Pr\left[X \in [x; x + dx] \text{ and } Y \in [y; y + dy]\right] = p_{xy}(x, y) \, dx dy \tag{2.12}$$

As illustrated in Figure 2.5, the value of the corresponding 2-dimensonal density function represents the probability that the random variables are in a small area $dxdy$ in the x, y-plane, which describe the location $X \in [x; x + dx]$ and $Y \in [y; y + dy]$. The marginal density of the random variable X is found by integrating over the other variable

$$p_x(x) = \int_{-\infty}^{\infty} p_{xy}(x, y) \, dy \tag{2.13}$$

and similarly

$$p_y(y) = \int_{-\infty}^{\infty} p_{xy}(x, y) \, dx \tag{2.14}$$

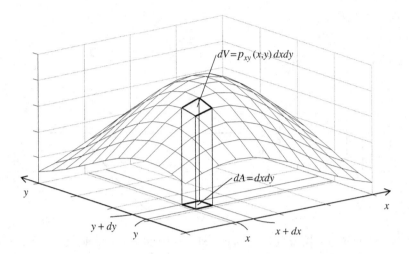

Figure 2.5 Joint density function $p_{xy}(x, y)$ for the random variables X and Y. The probability that both variables are inside the infinitesimal area $dA = dx\,dy$ is equal to the infinitesimal volume $dV = p_{xy}(x, y)\,dx\,dy$

The conditional density probability $p_{x|y}(x)$ is defined in terms of the classical probabilities

$$p_{x|y}(x)\,dx \times p_y(y)\,dy = p_{xy}(x, y)\,dxdy \tag{2.15}$$

This expression can be simplified by discarding the infinitesimal intervals. This allows us to isolate the conditional density and express it as follows:

$$p_{x|y}(x) = \frac{p(x, y)}{p_y(y)} \tag{2.16}$$

We can see that if $p(x, y) = p_x(x) p_y(y)$ then

$$p_{x|y}(x) = p_x(x) \tag{2.17}$$

and we can then say that X, Y are independent variables. In general, all random variables are related through one encompassing joint density function (one in which everything depends on everything) so that the independence between the two variables is a special case – or more precisely – a simplification that is often used in probabilistic theories. It is a simplification that we shall not use in this book because in OMA the dependence between variables contains an essential part of the physical properties of the problem and cannot be avoided.

By using time averaging, we do not have to bother about the fact that the variables might be dependent. Since the formulas are exactly the same as before (we still just use Eqs. (2.9–2.11)), the dependency is automatically taken into account. This is worth noticing, because by using time averaging we avoid the need to justify that variables are independent, we avoid the large errors that arise from such assumptions, because all dependence in between all variables is taken properly into account.

The covariance that gives us a simple measure of the dependence between two variables is the expectation of the multiplication of the two variables

$$\mathrm{cov}[X, Y] = \mathrm{E}[XY] = \int_{-\infty}^{\infty}\int_{-\infty}^{\infty} xy p_{xy}(x, y)\,dxdy \tag{2.18}$$

However, it is more convenient for the theme of this book to calculate the covariance using time averaging as follows:

$$\text{cov}\left[x(t),y(t)\right] = E\left[x(t)\,y(t)\right] = \frac{1}{T}\int_0^T x(t)\,y(t)\,dt \tag{2.19}$$

It should be noted that correlation is a similar quantity but with the mean values removed from the signal

$$\text{cor}\left[x(t),y(t)\right] = E\left[(x(t)-\mu_x)(y(t)-\mu_y)\right] = \frac{1}{T}\int_0^T \left(x(t)-\mu_x\right)\left(y(t)-\mu_y\right)dt \tag{2.20}$$

We should note that since we seldom use the mean value of recorded data and we normally remove the mean from the data at the very beginning of any analysis, we shall not distinguish between covariance and correlation, and in the following we will only talk about "correlation."

In this book, we will mainly use time averaging on the related time series. So from this point on, we will not distinguish between the realizations $x(t),y(t)$ and their corresponding random variables X,Y.

2.2   Correlation

2.2.1   Concept of Correlation

The concept of correlation was introduced in the preceding section. Correlation is a key issue in OMA because the first thing we do in OMA is to estimate correlation functions – as these contain all information of interest to us in case the signals are Gaussian. Before we go into these matters, it is useful to consider a simple example to illustrate the importance and simplicity of the concept.

In Figure 2.6, we show two cases of observations x_n, y_n of the random variables X, Y. As said before, the notation that we have introduced does not make a clear distinction between the data and the random variables, so we will just use x, y for both.

For one of the cases – the case to Figure 2.6a – there is obviously a weak or no dependence between the two variables. For the other case to Figure 2.6b, there is a clear dependence, because we see that large values of x introduce an increased probability for higher values of y. Since for the case to the right there is a clear dependence, it is natural to describe this tendency by introducing the simple linear model

$$y = ax \tag{2.21}$$

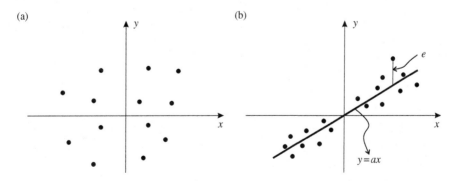

Figure 2.6   Two cases of distributed variables where the left case (a) has little or no correlation and the case to the right (b) has a clear correlation, thus in the case to the right it is natural to introduce the linear model $y = ax$ and then try to minimize the error e

where we have assumed that the mean value is zero for the variables x, y (this is why the line is going through the origin of the x, y coordinate system). In order to find a reasonable value for the slope a, we minimize the expectation of the square error $e^2 = (y - ax)^2$ that is

$$E\left[e^2\right] = E\left[(y - ax)^2\right] = E\left[y^2\right] + a^2 E\left[y^2\right] - 2aE\left[xy\right] \tag{2.22}$$

Assuming the expected square error to be smooth around the minimum, we know that for the value of a that minimizes the expected variance, the derivative of this quantity with respect to the parameter a is zero. We can now find the optimal value of the slope from

$$\frac{\partial}{\partial a} E\left[e^2\right] = 0 + 2aE\left[x^2\right] - 2E\left[xy\right] = 0 \tag{2.23}$$

and since $E\left[x^2\right] = \sigma_x^2$ we obtain

$$a = \frac{E\left[xy\right]}{\sigma_x^2} \tag{2.24}$$

This leads to the normalized linear model

$$\frac{y}{\sigma_y} = \frac{E\left[xy\right]}{\sigma_x \sigma_y} \frac{x}{\sigma_x} \tag{2.25}$$

where the slope in the normalized equation is known as the correlation coefficient:

$$\rho_{xy} = \frac{E\left[xy\right]}{\sigma_x \sigma_y} \tag{2.26}$$

As it appears, it is the correlation that gives us the essential information about the slope of the straight line. This idea will be further developed in all the subsequent analyses and discussions in this book.

2.2.2 Autocorrelation

Considering a single time varying signal $x(t)$ as shown in Figure 2.7, it is natural to wonder how adjacent points are correlated. It must be expected that if the points are close, then the correlation is high, and as the separation between points increases then the correlation is lower – and finally if the points are far

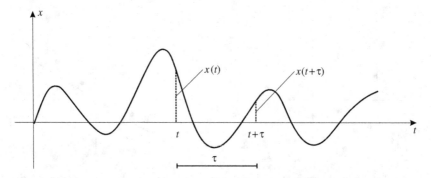

Figure 2.7 Autocorrelation is simply introduced as the correlation between the considered variable $x(t)$ at time t and time $t + \tau$

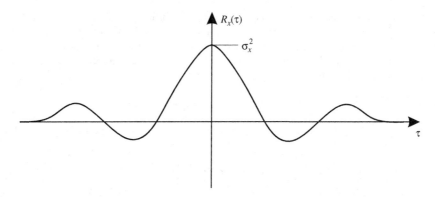

Figure 2.8 The autocorrelation function is symmetric, the initial value is equal to the variance and for large values of the time lag the function approaches zero

apart – then the correlation is for practical purposes zero. To investigate this issue, we consider two points $x(t)$ and $x(t+\tau)$ with a time separation τ in between them (see Figure 2.7). By taking the variable x as $x(t)$ and the variable y as $x(t+\tau)$ Eq. (2.26) can be used to define the autocorrelation function defined as

$$R_x(\tau) = \mathrm{E}[x(t)x(t+\tau)] \tag{2.27}$$

We see that for $\tau = 0$ the correlation function is equal to the variance σ_x^2. Using Eq. (2.26) one obtains the normalized correlation function

$$\rho_x(\tau) = \mathrm{E}[x(t)x(t+\tau)]/\sigma_x^2 \tag{2.28}$$

that might be useful for plotting purposes, but in OMA it is normal practice to stay with the version given by Eq. (2.27). For $\tau \to \infty$, the two variables $x(t)$ and $x(t+\tau)$ in Eq. (2.27) becomes independent, and therefore $R_x(\tau)$ approaches $\mathrm{E}[x(t)]\,\mathrm{E}[x(t+\tau)]$ that is equal to μ_x^2 assuming the signal to be stationary. We see that if the process is zero mean – as we will normally assume – then the autocorrelation approaches zero for large values of the time lag τ.

In general, if the random signal $x(t)$ is stationary, that is, if the density function does not depends on time, then all expectations are also independent of time as we have argued before. Thus, the correlation function given by Eq. (2.27) is independent of the time t, and we can introduce any time shift without changing the result. Specifically we can take the time shift $-\tau$, which leads to the symmetry relation

$$R_x(\tau) = \mathrm{E}[x(t-\tau)x(t)] = R_x(-\tau) \tag{2.29}$$

Figure 2.8 illustrates a typical time variation of the autocorrelation function. The autocorrelation is in practice obtained by using time averaging as explained in the preceding sections

$$R_x(\tau) = \frac{1}{T}\int_0^T x(t)x(t+\tau)\,dt \tag{2.30}$$

2.2.3 Cross Correlation

Cross correlation is simply defined by generalizing Eq. (2.27) to two random signals

$$\begin{aligned} R_{xy}(\tau) &= \mathrm{E}\left[x(t)y(t+\tau)\right] = \mathrm{E}\left[x(t-\tau)y(t)\right] \\ R_{yx}(\tau) &= \mathrm{E}\left[y(t)x(t+\tau)\right] = \mathrm{E}\left[y(t-\tau)x(t)\right] \end{aligned} \tag{2.31}$$

It should be noted that the definition given here is consistent with that given commonly in application oriented books such as Newland [1] and Thompson [2]. In contrast, the more theoretical oriented texts such as Papoulis and Pillai [3] often like to define the cross-correlation function as $R_{xy}(\tau) = \mathrm{E}\left[x(t+\tau)y(t)\right]$.[1] As before, we assume that the signals are stationary and by taking the time shift $-\tau$ we obtain the symmetry relation

$$R_{xy}(\tau) = \mathrm{E}\left[x(t-\tau)y(t)\right] = R_{yx}(-\tau) \tag{2.32}$$

Again, the cross-correlation functions are, in practice, calculated using time averaging as follows:

$$R_{xy}(\tau) = \frac{1}{T}\int_0^T x(t)y(t+\tau)\,dt$$

$$\tag{2.33}$$

$$R_{yx}(\tau) = \frac{1}{T}\int_0^T y(t)x(t+\tau)\,dt$$

Working with measured data, we will normally have several measurements, say N measurements, arranged in a response vector $\mathbf{y}(t) = \left\{y_1(t), y_2(t), \ldots, y_N(t)\right\}^T$. In this case, we can use a matrix formulation covering both the autocorrelation and cross correlation to form the correlation function (CF) matrix

$$\mathbf{R}(\tau) = \mathrm{E}\left[\mathbf{y}(t)\mathbf{y}^T(t+\tau)\right] \tag{2.34}$$

The diagonal elements are autocorrelation functions, and off-diagonal elements are cross-correlation functions. The symmetry relation (2.32) now reads

$$\mathbf{R}^T(\tau) = \mathrm{E}\left[\mathbf{y}(t+\tau)\mathbf{y}^T(t)\right] = \mathrm{E}\left[\mathbf{y}(t)\mathbf{y}^T(t-\tau)\right] = \mathbf{R}(-\tau) \tag{2.35}$$

Again, in practice, the CF matrix is calculated by time averaging as

$$\mathbf{R}(\tau) = \frac{1}{T}\int_0^T \mathbf{y}(t)\mathbf{y}^T(t+\tau)\,dt \tag{2.36}$$

The correlation matrix $\mathbf{C}$ for the time series $\mathbf{y}(t)$ vector is given by

$$\mathbf{C} = \mathrm{E}\left[\mathbf{y}(t)\mathbf{y}^T(t)\right] = \mathbf{R}(0) \tag{2.37}$$

In practical applications, we often deal with a large number of signals, but the number of random variables controlling the signals is often smaller than the number of signals.[2] This can be illustrated by thinking of a number of signals as created by a linear combination of independent random signals. If we consider a set of zero mean independent stochastic signals $\mathbf{e}(t)$ with unit variance, then the correlation matrix of these signals is the unity matrix $\mathrm{E}\left[\mathbf{e}(t)\mathbf{e}^T(t)\right] = \mathbf{I}$. We can create a linear combination of these independent signals by multiplying the signal vector $\mathbf{e}(t)$ by a matrix $\mathbf{V}$

$$\mathbf{y}(t) = \mathbf{V}\mathbf{e}(t) \tag{2.38}$$

---

[1] This has some impact on how we can interpret correlation functions and their applications for identification, see Section 6.2.5.
[2] Due to strong correlation between the signals introduced for instance by sensors placed close to each other and thus measuring basically the same signals. This is further discussed in Section 7.2.1 of Chapter 7.

and the correlation matrix of the resultant signal is

$$\mathbf{C} = \mathrm{E}\left[\mathbf{y}\left(t\right)\mathbf{y}^{T}\left(t\right)\right] = \mathbf{V}\mathrm{E}\left[\mathbf{e}\left(t\right)\mathbf{e}^{T}\left(t\right)\right]\;\mathbf{V}^{T} = \mathbf{V}\mathbf{V}^{T} \tag{2.39}$$

This also means that for any signal, the number of independent components can be evaluated by performing a singular value decomposition (SVD)[3] of the correlation matrix

$$\mathbf{C} = \mathbf{U}\mathbf{S}\mathbf{U}^{T} \tag{2.40}$$

and then use the results to estimate the number of independent signals by counting the nonzero singular values in the diagonal matrix $\mathbf{S}$ and then taking the corresponding singular vectors in the matrix $\mathbf{U}$ defining the sub matrix $\mathbf{U}_{s}$. The transformation $\mathbf{V}$ can then be estimated as

$$\hat{\mathbf{V}} = \mathbf{U}\sqrt{\mathbf{S}} \tag{2.41}$$

It should be noted that from the aforementioned considerations, it follows that the covariance matrices are always positive definite symmetric matrices,[4] thus a possible covariance matrix can be found as any symmetric matrix where all eigenvalues are forced to be nonnegative.

2.2.4   Properties of Correlation Functions

First we will consider how the derivative of a correlation function relates to the derivative of the underlying signal. Given Eq. (2.32), we have for the derivative of the correlation function

$$\frac{d}{d\tau}R_{xy}\left(\tau\right) = \mathrm{E}\left[x\left(t\right)\dot{y}\left(t+\tau\right)\right] \tag{2.42}$$

We can shift the signals in time and differentiate again

$$\frac{d^{2}}{d\tau^{2}}R_{xy}\left(\tau\right) = \mathrm{E}\left[\dot{x}\left(t-\tau\right)\dot{y}\left(t\right)\right] = -R_{\dot{x}\dot{y}}\left(\tau\right) \tag{2.43}$$

and we then have the simple relations

$$\ddot{R}_{xy}\left(\tau\right) = -R_{\dot{x}\dot{y}}\left(\tau\right)$$
$$\ddot{R}_{x}\left(\tau\right) = -R_{\dot{x}}\left(\tau\right) \tag{2.44}$$

For the autocorrelation case, Eq. (2.42) leads to

$$\dot{R}_{yy}\left(\tau\right) = \mathrm{E}\left[y\left(t\right)\dot{y}\left(t+\tau\right)\right] \tag{2.45}$$

We see that the correlation between the signal $y\left(t\right)$ and its derivative $\dot{y}\left(t\right)$ is equal to $\dot{R}_{yy}\left(0\right)$, and we know that the autocorrelation function is symmetric and therefore its derivative must be zero at time lag zero, that is. $\dot{R}_{yy}\left(0\right) = 0$. We can therefore conclude that $\mathrm{E}\left[y\left(t\right)\dot{y}\left(t\right)\right] = 0$, thus, a signal $y\left(t\right)$ is always uncorrelated with its own derivative $\dot{y}\left(t\right)$.

Another important property is related to the way we calculate the correlation function given by Eq. (2.33). Exchanging the variable of integration $-\theta = t + \tau$ we get

$$R_{xy}\left(\tau\right) = \frac{1}{T}\int_{-T-\tau}^{-\tau} x\left(-\theta - \tau\right)y\left(-\theta\right)d\theta \tag{2.46}$$

---

[3] See Section 3.2.4.
[4] Strictly speaking, nonnegative definite since some eigenvalues might be zero.

Now taking into account that the processes are stationary and assuming that the data is also available out-side of the data segment $t \in [0; T]$ we can shift the integration limit upward $T + \tau$ and shift the argument in the time series by τ and we have

$$R_{xy}(\tau) = \frac{1}{T} \int_0^T x(-\theta) y(\tau - \theta) d\theta \qquad (2.47)$$

which is a convolution if we either assume that the data is periodic with period T or assume that $T \to \infty$. The general result is that the correlation function $R_{xy}(\tau)$ can be calculated as a convolution between the signal $x(-t)$ and the signal $y(t)$

$$R_{xy}(\tau) = x(-t) * y(t) \qquad (2.48)$$

which is of significant practical importance because convolutions can be easily computed in the frequency domain.[5]

2.3   The Gaussian Distribution

2.3.1   Density Function

Even though we do not directly use the density functions for OMA, it is useful to consider the Gaussian density function. The reason is that we shall normally assume that our signals have zero mean and are Gaussian or close to Gaussian. In such cases, all properties are contained in the correlation functions. The univariate Gaussian density function is given by

$$p(x) = \frac{1}{\sqrt{2\pi\sigma^2}} e^{-\frac{(x-\mu)^2}{2\sigma^2}} \qquad (2.49)$$

For data with zero mean it reduces to

$$p(x) = \frac{1}{\sqrt{2\pi\sigma^2}} e^{-x^2/2\sigma^2} \qquad (2.50)$$

The corresponding density function for multivariate zero mean data is given by

$$p(x) = \frac{1}{(2\pi)^{M/2} |\mathbf{C}|} e^{-\mathbf{x}^T \mathbf{C}^{-1} \mathbf{x}/2} \qquad (2.51)$$

where x is the multivariate random variable $\mathbf{x}^T = \{x_1, x_2, \dots, x_M\}$ and $|\mathbf{C}|$ is the determinant of the correlation matrix. The Gaussian univariate distribution is totally described by its mean and standard deviation, or in case of zero mean multivariate data, by its correlation matrix.

2.3.2   The Central Limit Theorem

The central limit theorem explains why the Gaussian distribution plays a central role from a practical point of view.

The central limit theorem states that a sum of many independent and identically distributed random variables will tend to be Gaussian distributed as the number of variables approaches infinity.

---

[5] For the theoretical background, see the Chapter 4 on transforms and for the practical application of the result see Chapter 8 on signal processing.

More precisely, let us consider the set of random variables $\{x_1, x_2, \ldots, x_M\}$ of size M – that is, a set of independent and identically distributed random variables with mean values μ and variances σ^2. A linear combination of these random variables is given by

$$y = \sum_{m=1}^{M} a_m x_m \tag{2.52}$$

The central limit theorem states that for large values of M the distribution of y is approximately Gaussian with mean $\mu_y = \mu \sum a_m$ and variance $\sigma_y^2 = \sigma^2 \sum a_m^2$ and is approaching the Gaussian distribution for $M \rightarrow \infty$. The true strength of the theorem is that y approaches the Gaussian distribution regardless of the type of the distributions of individual random variables in the linear combination (2.52).

For instance, in structural dynamics, we have that the response $y(t)$ is the convolution between the force $x(\tau)$ and the impulse response function $h(t)$[6] and this relationship can be expressed as

$$y(t) = \int_{-\infty}^{\infty} h(t - \tau) x(\tau) \, d\tau \tag{2.53}$$

Writing this formula for a signal sampled with a time step Δt, the aforementioned equation becomes a summation of the form

$$y(n) = \sum_{k=-\infty}^{\infty} h(n - k) x(k) \, \Delta t \tag{2.54}$$

If we take only the part where the coefficients $h(n-k)$ are nonvanishing, we can rewrite the aforementioned expression as

$$y(n) = \sum_{k=n-N_m}^{n} h(n - k) x(k) \, \Delta t \tag{2.55}$$

where the N_m is the number of sample points that should be included in order to reasonably to reflect the memory of the system. This shows us that for a white noise input[7] a dynamic response is always a summation of the form given by Eq. (2.52) where the individual load contributions $x(k)$ might be non-Gaussian, but where the resulting response will be approximately Gaussian as predicted by the central limit theorem.

This is the reason why nearly all random responses are Gaussian, regardless of the kind of loading being considered. This also helps explaining the other name for the Gaussian distribution: the "normal" distribution. The normal case is that random responses are Gaussian distributed, and this helps us understand what we need to do to extract all physical information from a random signal; we just need to calculate the correlation, because the correlation fully describes the Gaussian distribution in case of a zero mean signal.

The central limit theorem also explains why lightly damped structures such as large bridges or buildings tend to show responses more close to Gaussian than responses from more heavily damped systems such as cars. The more lightly damped the structure is, the longer is the memory time, and thus, the larger is the number of terms in Eq. (2.55).

Example 2.2   Gaussian response properties of a non-Gaussian input

The response of a 2-DOF system is simulated using the `fftsim.m` function in the OMA toolbox. The natural frequencies were chosen to be $f_1, f_2 = 2.1, 2.5$ Hz, the damping is $\varsigma_1, \varsigma_2 = 0.035, 0.020$, the mode

---

[6] See Chapter 5 on classical dynamics.

[7] This means that different loading components $x(k)$ are independent. The argument can be generalized to dependent variables, but to comprehend the principle, the present argument is sufficient.

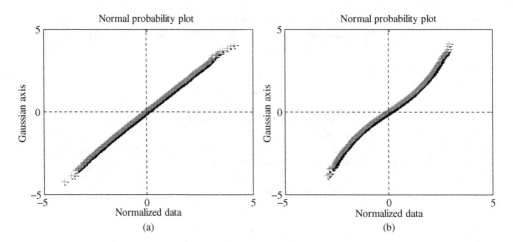

Figure 2.9 Gaussian probability plots of a system loaded by uniformly distributed white noise for the case of low damping (a) and high damping (b). As it appears when damping is low, the response is close to Gaussian even though the input is non-Gaussian as predicted by the central limit theorem

shapes to $\mathbf{b}_1^T = \{1\ 0\}$ and $\mathbf{b}_2^T = \{0\ 1\}$ and the sampling step was chosen to $\Delta t = 0.15$ s. The loading on the two DOF's were simulated as a white noise sequence with a rectangular distribution between -0.5 and 0.5.

The Gaussian probability plot of the response is shown in Figure 2.9a. As it appears the response is very close to Gaussian even though the input is strongly non-Gaussian. This is due to the central limit theorem and the fact that because of the relatively low damping there are a sufficient number of terms in Eq. (2.55) to ensure the Gaussian properties of the response.

The corresponding Gaussian probability plot of the response for the case of damping ten times higher is shown in Figure 2.9b. The higher damping reduces the number of terms in Eq. (2.55), and the result is that the limited number of terms is no longer enough to secure the Gaussian properties of the response.

2.3.3 Conditional Mean and Correlation

In this section, we will consider some closed-form solutions for the conditional expectation and variance of Gaussian distributed variables.

Let $\mathbf{x}$ and $\mathbf{y}$ be two Gaussian vectors of length M with the elements $\mathbf{x}^T = \{x_1, x_2, \cdots x_M\}$ and $\mathbf{y}^T = \{x_1, x_2, \cdots x_M\}$. The total set of variables are jointly distributed with the expectation and correlation matrix

$$\mathrm{E}\left[\begin{Bmatrix} \mathbf{x} \\ \mathbf{y} \end{Bmatrix}\right] = \begin{Bmatrix} \mathbf{\mu}_x \\ \mathbf{\mu}_y \end{Bmatrix}; \quad \mathrm{E}\left[\begin{Bmatrix} \mathbf{x} \\ \mathbf{y} \end{Bmatrix}\begin{Bmatrix} \mathbf{x}^T & \mathbf{y}^T \end{Bmatrix}\right] = \begin{bmatrix} \mathbf{C}_{xx} & \mathbf{C}_{xy} \\ \mathbf{C}_{yx} & \mathbf{C}_{yy} \end{bmatrix} \tag{2.56}$$

We now define a new vector $\mathbf{z}$ as the linear combination of $\mathbf{x}$ and $\mathbf{y}$ as follows

$$\mathbf{z} = \mathbf{x} - \mathbf{B}\mathbf{y} \tag{2.57}$$

where $\mathbf{B}$ is a $M \times M$ matrix. Since all variables are Gaussian, $\mathbf{z}$ will be independent of $\mathbf{y}$ if and only if the covariance between $\mathbf{z}$ and $\mathbf{y}$ vanish, that is using Eq. (2.57) we find the solution for the matrix $\mathbf{B}$

$$\mathrm{E}\left[\mathbf{z}\mathbf{y}^T\right] = 0 \quad \Rightarrow \quad \mathbf{C}_{xy} - \mathbf{B}\mathbf{C}_{yy} = 0 \quad \Rightarrow \quad \mathbf{B} = \mathbf{C}_{xy}\mathbf{C}_{yy}^{-1} \tag{2.58}$$

This allows us to calculate the conditional mean value using the fact that $\mathbf{z}$ is independent of $\mathbf{y}$ and is given by Eq. (2.57)

$$
\begin{aligned}
\boldsymbol{\mu}_{x|y} &= \mathrm{E}\left[\mathbf{x}\,|\mathbf{y}\right] \\
&= \mathrm{E}\left[\mathbf{z} + \mathbf{B}\mathbf{y}\,|\mathbf{y} = \mathbf{y}_0\right] \\
&= \mathrm{E}\left[\mathbf{z}\right] + \mathbf{B}\mathbf{y}_0 \\
&= \boldsymbol{\mu}_x - \mathbf{B}\boldsymbol{\mu}_y + \mathbf{B}\mathbf{y}_0 \\
&= \boldsymbol{\mu}_x + \mathbf{B}\left(\mathbf{y}_0 - \boldsymbol{\mu}_y\right)
\end{aligned}
\tag{2.59}
$$

Similarly, the conditional correlation matrix can be obtained as

$$
\begin{aligned}
\mathbf{C}_{xx|y} &= \mathrm{E}\left[\mathbf{x}\mathbf{x}^T\,|\mathbf{y}\right] \\
&= \mathrm{E}\left[(\mathbf{z} + \mathbf{B}\mathbf{y})(\mathbf{z} + \mathbf{B}\mathbf{y})^T\,|\mathbf{y} = \mathbf{y}_0\right] \\
&= \mathrm{E}\left[\mathbf{z}\mathbf{z}^T\,|\mathbf{y} = \mathbf{y}_0\right] \\
&= \mathrm{E}\left[\mathbf{z}\mathbf{z}^T\right]
\end{aligned}
\tag{2.60}
$$

The expectation $\mathrm{E}\left[\mathbf{z}\mathbf{z}^T\right]$ can be expressed using Eq. (2.57) as

$$
\begin{aligned}
\mathrm{E}\left[\mathbf{z}\mathbf{z}^T\right] &= \mathrm{E}\left[\mathbf{x}\mathbf{x}^T - \mathbf{B}\mathbf{y}\mathbf{x}^T - \mathbf{x}\mathbf{y}^T\mathbf{B}^T + \mathbf{B}\mathbf{y}\mathbf{y}^T\mathbf{B}^T\right] \\
&= \mathbf{C}_{xx} - \mathbf{B}\mathbf{C}_{yx} - \mathbf{C}_{xy}^T\mathbf{B}^T + \mathbf{B}\mathbf{C}_{yy}^T\mathbf{B}^T
\end{aligned}
\tag{2.61}
$$

Since $\mathbf{C}_{yy}$ and $\mathbf{C}_{yy}^{-1}$ are symmetric and since we have the solution given by Eq. (2.58) for the matrix $\mathbf{B}$ we obtain

$$
\begin{aligned}
\mathrm{E}\left[\mathbf{z}\mathbf{z}^T\right] &= \mathrm{E}\left[\mathbf{x}\mathbf{x}^T - \mathbf{B}\mathbf{y}\mathbf{x}^T - \mathbf{x}\mathbf{y}^T\mathbf{B}^T + \mathbf{B}\mathbf{y}\mathbf{y}^T\mathbf{B}^T\right] \\
&= \mathbf{C}_{xx} - \mathbf{B}\mathbf{C}_{yx} - \mathbf{C}_{xy}\mathbf{C}_{yy}^{-1}\mathbf{C}_{xy}^T + \mathbf{C}_{xy}\mathbf{C}_{yy}^{-1}\mathbf{C}_{yy}\mathbf{C}_{yy}^{-1}\mathbf{C}_{xy}^T \\
&= \mathbf{C}_{xx} - \mathbf{B}\mathbf{C}_{yx}
\end{aligned}
\tag{2.62}
$$

The conditional correlation matrix can then be expressed as

$$
\begin{aligned}
\mathbf{C}_{xx|y} &= \mathrm{E}\left[\mathbf{z}\mathbf{z}^T\right] \\
&= \mathbf{C}_{xx} - \mathbf{B}\mathbf{C}_{yx}
\end{aligned}
\tag{2.63}
$$

The derivations for the conditional mean in this section are needed in order to formulate the basic equations for the random decrement technique discussed in Chapter 8 (see Section 8.4.4). Furthermore, the conditional variance can be used to formulate closed-form solutions for the variance of the random decrement functions. The derivations in this section are also useful for the derivation of the stochastic subspace identification technique discussed in Chapter 9 (see Section 9.5).

References

[1] Newland, D.E.: *An introduction to random vibrations, spectral & wavelet analysis*. 3rd edition. Longman, 1997.
[2] Thompson, W.T.: *Theory of vibration with applications*. 2nd edition. George Allen & Unwin Ltd., 1981.
[3] Papoulis, A. and Pillai, S.U.: *Probability, random variables and stochastic processes*. 4th Edition. McGraw-Hill, 2002.

3

Matrices and Regression

"Science is facts; just as houses are made of stone, so is science made of facts; but a pile of stones is not a house, and a collection of facts is not necessarily science."

– Jules Henri Poincaré

In the field of dynamics of structures, most of the analysis methods are formulated in terms of vectors and matrices. For instance, system properties such as mass and stiffness are described by mass and stiffness matrices; the natural frequencies and the mode shapes of vibration are obtained through the solution of an eigenvalue problem; and the mode shapes are actually vectors. Because matrices and vectors are not only used for dynamic analysis of discrete systems, but also for system identification of the dynamical properties of systems, which in some cases makes use of regression analysis, it is important to have a clear understanding of how matrices and vectors should be properly utilized. The first part of this chapter is devoted to the fundamental concepts of vector and matrices as a tool for dynamic analysis of structures.

The intent of this chapter is not to provide a detailed treatment of the theory of matrices, vectors and regression, but rather to introduce the concepts related to vectors, matrices, and regression in a coherent manner suitable for their practical use in subsequent chapters.

As stated by the message in the colophon, facts are not necessarily science, and this can be used to argue that measurements are not necessarily information – or at least not the kind of information that we want or need to have. In order to extract the information that we need from the measurements obtained during a test, we need to use these measurements in a manner that we can perform an estimation of that needed information. This is often done by regression analysis, and, in turn, regression is done using vectors and matrices. Therefore this subject is also important in OMA and the last part of the chapter is devoted to regression as a tool for analysis of experimental data.

The material in this book is based on classical texts such as the Numerical recipes by Press et al. [1], Pandit and Wu [2] on time series modeling, Golub and Van Loan [3] on matrix computations, Lawson and Hanson [4] on solving least square problems, Lancaster and Salkauskas [5] on curvefitting, and finally a more new text by Simon [6] on optimal state estimation.

3.1 Vector and Matrix Notation

In modal analysis, the traditional way of writing matrices and vectors is by using brackets where a vector is written as $\{X\}$ or $\{x\}$ and a matrix is written as $[A]$ or $[a]$. The authors prefer to use a different notation that makes equations easier to read; therefore, in this book matrices are represented by upper case bold variables while vectors are represented by lower case bold variables.

Introduction to Operational Modal Analysis, First Edition. Rune Brincker and Carlos Ventura.
© 2015 John Wiley & Sons, Ltd. Published 2015 by John Wiley & Sons, Ltd.
Companion Website: www.wiley.com/go/brincker

Thus, $\mathbf{A}$ is a matrix and $\mathbf{a}$ is a vector, and – unless otherwise indicated – $\mathbf{a}$ is a column vector. For a matrix $\mathbf{A}$, the bracket notation is mainly used to indicate the element a_{rc} of row r and column c as follows:

$$\mathbf{A} = \left[a_{rc}\right] \tag{3.1}$$

The row index r is defined over the N_r number of rows, and similarly the column index c is defined over the N_c number of columns. If $\mathbf{A}$ is a diagonal matrix, then the elements of the diagonal will be represented as

$$\mathbf{A} = \left[a_r\right] \tag{3.2}$$

In this case, the matrix is assumed to be square and the elements are given by

$$a_{rc} = \begin{cases} a_r; & r = c \\ 0; & r \neq c \end{cases} \tag{3.3}$$

Similarly, the element of row r of a column vector is given as

$$\mathbf{a} = \left\{a_r\right\} \tag{3.4}$$

The elements of the transpose $\mathbf{A}^T$ of $\mathbf{A}$ are

$$\mathbf{A}^T = \left[a_{cr}\right] \tag{3.5}$$

and for vectors, the transpose means the corresponding row vector. The elements of the Hermitian matrix $\mathbf{A}^H$ of $\mathbf{A}$ correspond to the complex conjugate of the transpose of $\mathbf{A}$

$$\mathbf{A}^H = \left[a_{cr}^*\right] \tag{3.6}$$

Let $\mathbf{A}$, $\mathbf{B}$ and $\mathbf{C}$, be three matrices. If matrix $\mathbf{B}$ has the same number of columns as the number of rows of matrix $\mathbf{C}$ then the matrix product $\mathbf{A} = \mathbf{B}\mathbf{C}$ means that

$$a_{rc} = \sum_n b_{rn} c_{nc} \tag{3.7}$$

Matrix multiplication and transpose follow the well-known rules

$$(\mathbf{AB})^T = \mathbf{B}^T \mathbf{A}^T \tag{3.8}$$

The inverse $\mathbf{A}^{-1}$ of a matrix $\mathbf{A}$ is the matrix that multiplied with $\mathbf{A}$ produces the identity matrix $\mathbf{I}$

$$\mathbf{A}^{-1}\mathbf{A} = \mathbf{A}\mathbf{A}^{-1} = \mathbf{I} \tag{3.9}$$

The inverse is only defined for square matrices and the identity matrix is defined as the unitary matrix that leaves any square matrix $\mathbf{A}$ unchanged, that is

$$\mathbf{I}\mathbf{A} = \mathbf{A} \tag{3.10}$$

The inverse of the product of two matrices is given by (if the inverses exist) the well-known rule

$$(\mathbf{AB})^{-1} = \mathbf{B}^{-1}\mathbf{A}^{-1} \tag{3.11}$$

If the elements of two matrices are functions of time then the convolution of these two matrices is expressed as $\mathbf{A}(t) = \mathbf{B}(t) * \mathbf{C}(t)$, which means that

$$a_{rc}(t) = \sum_n b_{rn}(t) * c_{nc}(t) = \sum_n \int_{-\infty}^{\infty} b_{rn}(t-\tau) c_{nc}(\tau) d\tau \tag{3.12}$$

Since convolution is a linear operation, matrix convolution follows the same rules as for matrix operations. For instance, the transpose of the convolution of two matrices is given by

$$(\mathbf{A}(t) * \mathbf{B}(t))^T = \mathbf{B}^T(t) * \mathbf{A}^T(t) \tag{3.13}$$

In this book, we are dealing with complex vectors and matrices and we will make the general assumption that vectors and matrices are complex valued. Examples of complex vectors are, for instance, mode shapes of a system with general damping[1] and examples of complex matrices are the transfer function matrices and frequency response function (FRF) matrices.

3.2 Vector and Matrix Algebra

This section presents a brief overview of the most important properties of vectors and matrices that are relevant to the material treated in this book.

3.2.1 Vectors and Inner Products

Two vectors $\mathbf{a}$ and $\mathbf{b}$ can be added by adding their row elements

$$\mathbf{a} + \mathbf{b} = \left\{a_r + b_r\right\} \tag{3.14}$$

Multiplication of a vector by a scalar quantity means that each element of the vector is multiplied by the scalar quantity

$$c\mathbf{a} = \left\{ca_r\right\} \tag{3.15}$$

A vector $\mathbf{b}$ can be formed as a linear combination of a number of vectors, like

$$\mathbf{b} = c_1\mathbf{a}_1 + c_2\mathbf{a}_2 + \cdots c_N\mathbf{a}_N = \sum_{n=1}^{N} c_n\mathbf{a}_n \tag{3.16}$$

and we then say that $\mathbf{b}$ is a vector that is linear dependent upon the set of vectors that has created $\mathbf{b}$. If none of vectors in the set $\mathbf{a}_1, \mathbf{a}_2, \cdots \mathbf{a}_N$ is linear dependent upon the other ones, then we say that the set is a linear independent set of vectors.

All possible linear combinations of a linearly independent set of vectors define a vector space and the number of vectors in the linearly independent set defines the dimension of the vector space. If we consider a subset of the linearly independent set of vectors defining a vector space, we say that the subset defines a subspace of the first vector space.

The inner product between two vectors $\mathbf{a}$ and $\mathbf{b}$ – also called the dot product $\mathbf{a} \cdot \mathbf{b}$ – is given by

$$\mathbf{a} \cdot \mathbf{b} = \sum_n a_n b_n \tag{3.17}$$

and we see from this expression that the inner product is always a scalar. We will write the inner product as

$$\mathbf{a} \cdot \mathbf{b} = \mathbf{a}^T\mathbf{b} \tag{3.18}$$

If the vectors are complex valued, then we will sometimes use the complex conjugate of the first vector to express the inner product as

$$\mathbf{a} \cdot \mathbf{b} = \mathbf{a}^H\mathbf{b} \tag{3.19}$$

---

[1] See Chapter 5 on classical dynamics.

In this case, the inner product is in general a complex scalar. The inner product of a vector with itself defines its length $|\mathbf{a}|$

$$|\mathbf{a}|^2 = \mathbf{a}^H \mathbf{a} \tag{3.20}$$

It is easy to prove that $\mathbf{a}^H \mathbf{a}$ is always real and positive, and so is the length as defined earlier. The generalized interior angle θ between two vectors is defined by

$$|\mathbf{a}^H \mathbf{b}| = |\mathbf{a}| \, |\mathbf{b}| \cos(\theta) \tag{3.21}$$

3.2.2 Matrices and Outer Products

One of the most useful properties of matrices is the ability to transform vectors. Let a matrix $\mathbf{T}$ be a transformation from the vector $\mathbf{a}$ to the vector $\mathbf{b}$, thus

$$\mathbf{b} = \mathbf{T a} \tag{3.22}$$

Often we will describe the matrix $\mathbf{T}$ as a collection of column vectors

$$\mathbf{T} = \left[\mathbf{u}_1, \mathbf{u}_2, \ldots \right] \tag{3.23}$$

and the transformation (3.22) can also be written

$$\mathbf{b} = \mathbf{T a} = \sum_{n=1}^{N} a_n \mathbf{u}_n \tag{3.24}$$

From Eq. (3.24), it can be seen that a matrix defines the new vector $\mathbf{b}$ as a linear combination of the column vectors of the transformation matrix. Therefore, the number of components of the column vectors must be equal to the number of components of the vector $\mathbf{b}$, and if the column vectors constitute a linear independent set, the number of column vectors N cannot be larger than the number of components of the vector $\mathbf{b}$. On the other hand, if the number of column vectors is smaller than the number of components of the vector $\mathbf{b}$, then the vector space spanned by the column vectors is a subspace of the largest possible vector space to be spanned by the column vectors. In order to construct a transformation matrix that just has the proper size to describe all possible transformations of the vector $\mathbf{a}$ to the vector $\mathbf{b}$, the transformation matrix must be square.

Alternatively, we can describe the matrix as a collection of row vectors

$$\mathbf{T} = \left[\mathbf{v}_1, \mathbf{v}_2, \ldots \right]^T = \begin{bmatrix} \mathbf{v}_1^T \\ \mathbf{v}_2^T \\ \vdots \end{bmatrix} \tag{3.25}$$

and the transformation (3.22) can also be written as

$$\mathbf{b} = \mathbf{T a} = \begin{Bmatrix} \mathbf{v}_1^T \mathbf{a} \\ \mathbf{v}_2^T \mathbf{a} \\ \vdots \end{Bmatrix} \tag{3.26}$$

A diagonal matrix has the properties that it scales the components of a vector. If $\mathbf{D}$ is a diagonal matrix with diagonal elements d_n, then the corresponding transformation to Eq. (3.22) is

$$\mathbf{b} = \mathbf{D a} = \left[d_n \right] \mathbf{a} = \left\{ d_n a_n \right\} \tag{3.27}$$

Similarly, a diagonal matrix multiplied from the right to any matrix $\mathbf{T}$ scales its column vectors

$$\mathbf{TD} = \begin{bmatrix} \mathbf{u}_1, \mathbf{u}_2, \ldots \end{bmatrix} \begin{bmatrix} d_n \end{bmatrix} = \begin{bmatrix} \mathbf{u}_1 d_1, \mathbf{u}_2 d_2, \ldots \end{bmatrix} \tag{3.28}$$

and multiplied from the left scales its row vectors

$$\mathbf{DT} = \begin{bmatrix} d_n \end{bmatrix} \begin{bmatrix} \mathbf{v}_1^T \\ \mathbf{v}_2^T \\ \vdots \end{bmatrix} = \begin{bmatrix} \mathbf{v}_1^T d_1 \\ \mathbf{v}_2^T d_2 \\ \vdots \end{bmatrix} \tag{3.29}$$

An important subject in matrix theory is the matrix rank. The rank of a matrix can be defined in different ways. In this book, we will define the matrix rank in terms of outer products. The outer product $\mathbf{a} \otimes \mathbf{b}$ between the two vectors $\mathbf{a}$ and $\mathbf{b}$ is defined as the matrix

$$\mathbf{T} = \mathbf{a} \otimes \mathbf{b} = \mathbf{a}\mathbf{b}^T \tag{3.30}$$

In this notation, the elements of $\mathbf{T}$ are given by

$$\begin{bmatrix} t_{rc} \end{bmatrix} = \begin{bmatrix} a_r b_c \end{bmatrix} \tag{3.31}$$

If a matrix is given by a single outer product as in Eq. (3.30), then it is said to be rank one, since it is defined by a single row and a single column vector. This concept can be easily generalized, because any matrix can be created as a linear combination of outer products.[2] Generalizing to complex vectors where the outer product also normally includes the complex conjugation transpose of the last vector leads to

$$\mathbf{T} = \sum_{r=1}^{R} \alpha_r \mathbf{u}_r \mathbf{v}_r^H \tag{3.32}$$

For nonzero coefficients α_r and assuming the vector sets $\mathbf{u}_r$, $\mathbf{v}_r$ to be linearly independent, this defines a matrix of rank R. A matrix is said to have full rank if R is equal to the number of columns or rows, whichever is the smallest (and represented as min (N_r, N_c)). If the rank is smaller than the min (N_r, N_c) then the matrix is said to be rank deficient. It is easy to see that the rank is equal to the dimension of the column vectors – or the row vectors which is the same.

If in the inner product given by Eq. (3.17) the right-hand vector is given as the transformation of a third vector, then the inner product is generalized to

$$\mathbf{a}^H \mathbf{b} = \mathbf{a}^H \mathbf{T}\mathbf{c} \tag{3.33}$$

which is known as the Hermitian form of the inner product. In this book, we will also denote this as the inner product of $\mathbf{a}$ and $\mathbf{c}$ over the matrix $\mathbf{T}$. We see immediately that if the matrix is Hermitian ($\mathbf{T}^H = \mathbf{T}$), then the two vectors can trade places just like the classical definition in Eq. (3.18) without changing the result

$$\mathbf{a}^H \mathbf{T}\mathbf{c} = \left(\mathbf{a}^H \mathbf{T}\mathbf{c} \right)^H = \mathbf{c}^H \mathbf{T}\mathbf{a} \tag{3.34}$$

Again if we assume the matrix to be Hermitian, it follows from the definition that the inner product of any vector with itself is always a real number.

A matrix is said to be positive definite if for any vector $\mathbf{a}$ the inner product with itself is always positive as it is for the simple inner product (3.17)

$$\mathbf{a}^H \mathbf{T}\mathbf{a} > 0 \tag{3.35}$$

---

[2] See Section 3.2.4 on singular value decomposition.

3.2.3   Eigenvalue Decomposition

The eigenvalue problem arises from the case where the linear transformation of a vector $\mathbf{v}$ defined by the square matrix $\mathbf{T}$ creates a vector that is proportional to $\mathbf{v}$, thus

$$\mathbf{Tv} = e\mathbf{v} \tag{3.36}$$

The vector is then said to be an eigenvector for the matrix $\mathbf{T}$ with eigenvalue e. If the matrix $\mathbf{T}$ is symmetric, then it defines a set of eigenvectors with simple orthogonality properties.

Let us consider two different eigenvectors $\mathbf{v}_n$ and $\mathbf{v}_m, m \neq n$ then

$$\begin{aligned} \mathbf{Tv}_n &= e_n \mathbf{v}_n \\ \mathbf{Tv}_m &= e_m \mathbf{v}_m \end{aligned} \tag{3.37}$$

Pre-multiplying both sides of the top equation by $\mathbf{v}_m^T$ and both sides of the bottom equation by $\mathbf{v}_n^T$ and dividing the resulting equations by their respective eigenvalues, we see that the resulting right-hand sides are equal because $\mathbf{v}_m^T \mathbf{v}_n = \mathbf{v}_n^T \mathbf{v}_m$. Subtracting the two resulting equations leads to

$$\frac{1}{e_n} \mathbf{v}_m^T \mathbf{Tv}_n - \frac{1}{e_m} \mathbf{v}_n^T \mathbf{Tv}_m = 0 \tag{3.38}$$

Now if the matrix $\mathbf{T}$ is symmetric, then $\mathbf{v}_m^T \mathbf{Tv}_n = \mathbf{v}_n^T \mathbf{Tv}_m$ and

$$\left(\frac{1}{e_n} - \frac{1}{e_m} \right) \mathbf{v}_m^T \mathbf{Tv}_n = 0 \tag{3.39}$$

If the eigenvalues are nonzero and different, then the eigenvectors are orthogonal with respect to the matrix $\mathbf{T}$

$$\mathbf{v}_m^T \mathbf{Tv}_n = 0 \tag{3.40}$$

We can conclude that the eigenvectors of a symmetric matrix with distinct and nonzero eigenvalues define a set of orthogonal vectors, and therefore constitute a natural basis for the vector space spanned by the eigenvectors. We see from Eq. (3.36) that the eigenvectors are known except for a scaling constant. This means that we can scale the eigenvectors in any arbitrary manner. Scaling them so that $\mathbf{b}_n^H \mathbf{b}_n = 1$, then we see from Eq. (3.36) that

$$\mathbf{v}_n^T \mathbf{Tv}_n = e_n \tag{3.41}$$

Equations (3.40) and (3.41) can be expressed by the following single equation

$$\left[\mathbf{v}_n \right]^T \mathbf{T} \left[\mathbf{v}_n \right] = \left[e_n \right] \tag{3.42}$$

where we have put the eigenvectors side by side as columns in the matrix $\left[\mathbf{v}_n \right]$ and we have arranged the eigenvalues in the diagonal matrix $\left[e_n \right]$. If the matrix $\mathbf{T}$ is complex and Hermitian, similar arguments can be followed to obtain the generalized expression

$$\left[\mathbf{v}_n \right]^H \mathbf{T} \left[\mathbf{v}_n \right] = \left[e_n \right] \tag{3.43}$$

For the complex and Hermitian case, we can write any vector $\mathbf{a}$ in the vector space spanned by the eigenvectors as a linear combination of the eigenvectors

$$\mathbf{a} = \alpha_1 \mathbf{v}_1 + \alpha_2 \mathbf{v}_2 + \cdots = \left[\mathbf{v}_n \right] \boldsymbol{\alpha} \tag{3.44}$$

and the inner product $\mathbf{a}^H \mathbf{T} \mathbf{a}$ is given by

$$\mathbf{a}^H \mathbf{T} \mathbf{a} = \alpha^H \left[\mathbf{v}_n\right]^H \mathbf{T} \left[\mathbf{v}_n\right] \alpha = \alpha^H \left[e_n\right] \alpha$$
$$= \sum_m e_n \alpha_n^* \alpha_n = \sum_m e_n |\alpha_n|^2 \qquad (3.45)$$

From this expression, we see that a Hermitian matrix always has real eigenvalues and that the matrix is positive definite if and only if all eigenvalues are positive.

In general, the eigenvalue equation (3.36) can also be written as

$$\mathbf{T} \left[\mathbf{v}_n\right] = \left[\mathbf{v}_n\right] \left[e_n\right] \qquad (3.46)$$

In this expression, we have used the scaling properties of the diagonal matrix introduced in Eq. (3.28). Multiplying by the inverse of the eigenvector matrix from the right, we have the so-called eigenvalue decomposition (EVD) of the matrix $\mathbf{T}$

$$\mathbf{T} = \left[\mathbf{v}_m\right] \left[e_m\right] \left[\mathbf{v}_m\right]^{-1} \qquad (3.47)$$

This also means that

$$\left[\mathbf{v}_m\right]^{-1} \mathbf{T} \left[\mathbf{v}_m\right] = \left[e_m\right] \qquad (3.48)$$

Comparing this expression with Eq. (3.42), we see that for a Hermitian matrix – if we scale the eigenvectors to unit length – the eigenvector matrix is unitary

$$\left[\mathbf{v}_m\right]^{-1} = \left[\mathbf{v}_m\right]^H \qquad (3.49)$$

and we can also write the EVD as

$$\mathbf{T} = \left[\mathbf{v}_m\right] \left[e_m\right] \left[\mathbf{v}_m\right]^H \qquad (3.50)$$

This means that the $N \times N$ Hermitian matrix $\mathbf{T}$ is expressed as a series of outer products (see Eq. (3.32))

$$\mathbf{T} = \sum_{n=1}^{N} e_n \mathbf{v}_n \mathbf{v}_n^H \qquad (3.51)$$

Therefore, a Hermitian matrix has full rank if and only if all eigenvalues are nonzero.

For square matrices, it is reasonable to talk about the function of a matrix. This is because a function of a matrix can be defined from its power series. It makes sense as well to talk about a matrix $\mathbf{T}$ raised to the arbitrary power p. From the recursive application of Eq. (3.47), we obtain

$$\mathbf{T}^p = \left[\mathbf{v}_m\right] \left[e_m\right]^p \left[\mathbf{v}_m\right]^{-1} = \left[\mathbf{v}_m\right] \left[e_m^p\right] \left[\mathbf{v}_m\right]^{-1} \qquad (3.52)$$

If an arbitrary function is defined by its power series as

$$f(x) = \sum_n c_n x^n \qquad (3.53)$$

we can have the direct representation of the function of the matrix as

$$f(\mathbf{T}) = \left[\mathbf{v}_m\right] \left[f(e_m)\right] \left[\mathbf{v}_m\right]^{-1} \qquad (3.54)$$

More specifically, we have the general expression for the inverse of a square matrix

$$\mathbf{T}^{-1} = \left[\mathbf{v}_m\right] \left[1/e_m\right] \left[\mathbf{v}_m\right]^{-1} \qquad (3.55)$$

This equation, and especially its counterpart where the last matrix is replaced by the transpose (this is possible according to Eq. (3.49)), is closely related to the application of the singular value decomposition (SVD) for inversion of matrices that are not square.

3.2.4   Singular Value Decomposition

The SVD can be seen as a direct generalization of the EVD of the Hermitian matrix given by Eq. (3.50). The SVD of a general matrix $\mathbf{A}$ is expressed as

$$\mathbf{A} = \mathbf{USV}^H \tag{3.56}$$

where $\mathbf{U}$ and $\mathbf{V}$ are unitary matrices holding the columns vectors called the singular vectors

$$\begin{aligned} \mathbf{U} &= \begin{bmatrix} \mathbf{u}_n \end{bmatrix} \\ \mathbf{V} &= \begin{bmatrix} \mathbf{v}_n \end{bmatrix} \end{aligned} \tag{3.57}$$

and $\mathbf{S}$ is a diagonal matrix holding the real and nonnegative singular values of $\mathbf{A}$ in descending order starting with the highest value at the upper left corner

$$\mathbf{S} = \begin{bmatrix} s_n \end{bmatrix} \tag{3.58}$$

We shall only think about the SVD in its simplest form where $\mathbf{S}$ is square and the number of vectors in $\mathbf{U}$ and $\mathbf{V}$ are the same; the number of rows in $\mathbf{A}$ is given by the number of components in the singular vectors $\mathbf{u}_n$, and the number of columns in $\mathbf{A}$ is given by the number of components in the singular vectors $\mathbf{v}_n$.

Carrying out the multiplications in Eq. (3.56), it is easy to see that the SVD leads directly to a series of outer products as given by Eq. (3.32)

$$\mathbf{A} = \sum_{r=1}^{R} s_r \mathbf{u}_r \mathbf{v}_r^H \tag{3.59}$$

The rank R is then directly given by the number of nonzero singular values.

If a $N \times N$ matrix has a rank that is smaller than the size, that is $R < N$, then we say that the matrix is rank deficient, and the matrix cannot be immediately inverted. In some cases, a square matrix might have full rank but some of the singular values might be so small that they can be considered "nearly equal the zero." In this case, the matrix might be difficult to invert and we say that the matrix have deficiency problems. A way to quantify the deficiency problem is to calculate the so-called condition number, which is the ratio between the largest and smallest value of the matrix given by s_1/s_N.

Assuming that we have a matrix where the number of rows is larger than the number of columns, $N_r > N_c$ – also known as a matrix related to an overdetermined problem – an $N_r \times N_r$ Hermitian matrix $\mathbf{T}$ can be obtained by

$$\mathbf{T} = \mathbf{A}^H \mathbf{A} \tag{3.60}$$

Using Eq. (3.51) and the fact that the singular matrices are unitary we have

$$\mathbf{T} = \left(\mathbf{USV}^H \right)^H \mathbf{USV}^H = \mathbf{VS}^2 \mathbf{V}^H \tag{3.61}$$

We can see that the matrix $\mathbf{T}$ has the eigenvectors $\mathbf{v}_n$ and the eigenvalues s_n^2.

The SVD and the EVD are two closely related mathematical ways of decomposing matrices – the EVD decomposes square matrices and SVD do a similar decomposition for rectangular matrices, and the two decompositions are related through Eq. (3.61).

3.2.5   Block Matrices

In dynamic analysis, it is common to work with matrices that are assembled from other matrices. For instance, given the four $N \times N$ matrices $\mathbf{A}_1, \mathbf{A}_2, \mathbf{A}_3, \mathbf{A}_4$ we can form the following $2N \times 2N$ matrix

$$\mathbf{A} = \begin{bmatrix} \mathbf{A}_1 & \mathbf{A}_2 \\ \mathbf{A}_3 & \mathbf{A}_4 \end{bmatrix} \tag{3.62}$$

and we say that the matrix $\mathbf{A}$ is partitioned into four block matrices $\mathbf{A}_1, \mathbf{A}_2, \mathbf{A}_3, \mathbf{A}_4$ with two block rows and two block columns. In general, we have

$$
\mathbf{A} = \begin{bmatrix} \mathbf{A}_{11} & \mathbf{A}_{12} & \cdots \\ \mathbf{A}_{21} & \ddots & \\ \vdots & & \mathbf{A}_{rc} \end{bmatrix} \tag{3.63}
$$

Here r denotes the block row entry and c the block column entry.

If a partitioned matrix is constant along its block anti-diagonals (or block skew-diagonals), that is $\mathbf{A}_{rc} = \mathbf{A}_{r+1,c-1} = \mathbf{A}_{r-1,c+1}$ then the matrix is called a block Hankel matrix.

If a partitioned matrix is constant along its block diagonals, then it is called a block Toeplitz matrix.

Example 3.1   Data matrix organized as a Hankel matrix

It is common to organize a data set $\mathbf{y}(n\Delta t) = \mathbf{y}(n)$ sampled with a time step Δt in a block Hankel matrix. For instance, if this is done with two block rows like in the Ibrahim time domain technique,[3] then one could write

$$
\mathbf{H} = \begin{bmatrix} \mathbf{y}(1) & \mathbf{y}(2) & \mathbf{y}(3) & \cdots \\ \mathbf{y}(2) & \mathbf{y}(3) & \mathbf{y}(4) & \cdots \end{bmatrix} \tag{3.64}
$$

Now multiplying the Hankel matrix by its own transpose and dividing by the number of block columns C, that is $\mathbf{H}\mathbf{H}^T/C$, we have that the resulting matrix is a block matrix where the four different block entries define the block matrices

$$
r, c = \begin{cases} 1, 1: & \sum_n \mathbf{y}(n)\,\mathbf{y}^T(n)\,/C \\[2mm] 1, 2: & \sum_n \mathbf{y}(n)\,\mathbf{y}^T(n+1)\,/C \\[2mm] 2, 1: & \sum_n \mathbf{y}(n+1)\,\mathbf{y}^T(n)\,/C \\[2mm] 2, 2: & \sum_n \mathbf{y}(n)\,\mathbf{y}^T(n)\,/C \end{cases} \tag{3.65}
$$

that are different estimates of the correlation function matrix[4] $\mathbf{R}_{yy}(k)$ where the time lag is given by $k\Delta t$

$$
\mathbf{H}\mathbf{H}^T/C = \begin{bmatrix} \mathbf{R}_{yy}(0) & \mathbf{R}_{yy}(1) \\ \mathbf{R}_{yy}(-1) & \mathbf{R}_{yy}(0) \end{bmatrix} \tag{3.66}
$$

We see that the resulting matrix containing the block correlation matrices is in fact a block Toeplitz matrix. This example is an illustration of how common ways of manipulating measured data leads to matrices of the kind mentioned earlier.

3.2.6   Scalar Matrix Measures

One of the most important scalar measures of a square matrix is the determinant. For a formal definition of the concept, we will refer the reader to the available literature on matrix analysis. Most of the software for matrix analysis includes algorithms that easily calculate this quantity.

---

[3] See Section 9.4.

[4] See Chapter 8 on signal processing for the reason why this is the case.

The determinant of the matrix is denoted by $\det(\mathbf{A})$ or $|\mathbf{A}|$. It can be shown that for the two square matrices $\mathbf{A}$ and $\mathbf{B}$ we have

$$\det(\mathbf{AB}) = \det(\mathbf{A})\det(\mathbf{B}) \tag{3.67}$$

And for the $N \times N$ matrix it can be shown that the determinant can be calculated as the product of its eigenvalues

$$\det(\mathbf{A}) = \prod_{n=1}^{N} e_n \tag{3.68}$$

As it is well known from elementary linear algebra, if the square matrix has one or more eigenvalues that are zero, then the determinant is zero, and the matrix is singular and cannot be inverted.

Another scalar measure of a square matrix $\mathbf{A} = \left[a_{rc}\right]$ is the trace, which is simply defined as

$$\text{tr}(\mathbf{A}) = \sum_{n=1}^{N} a_{nn} \tag{3.69}$$

The trace has the simple property of being equal to the sum of the eigenvalues

$$\text{tr}(\mathbf{A}) = \sum_{n=1}^{N} e_n \tag{3.70}$$

For the case of two square matrices $\mathbf{A}$ and $\mathbf{B}$ we have

$$\text{tr}(\mathbf{AB}) = \text{tr}(\mathbf{BA}) \tag{3.71}$$

and the obvious relations $\text{tr}(\mathbf{A} + \mathbf{B}) = \text{tr}(\mathbf{A}) + \text{tr}(\mathbf{B})$, $\text{tr}\left(\mathbf{A}^T\right) = \text{tr}\left(\mathbf{A}^T\right)$ and $\text{tr}(\alpha\mathbf{A}) = \alpha\,\text{tr}(\mathbf{A})$.

Sometimes it is useful to have a scalar measure that can be interpreted as the "magnitude" of a vector or a matrix. Such a scalar measure is denoted as a norm if it fulfills some requirements like that the scalar measure is always nonnegative, that if we multiply by the vector or matrix by a constant, then the scalar measure is proportional to the constant, and that if the scalar measure is zero it means that we are dealing with a zero vector or zero matrix. A commonly used vector norm is the p-norm defined for the vector $\mathbf{a}$ as

$$\|\mathbf{a}\|_p = \left(\sum_r |a_r|^p \right)^{1/p} \tag{3.72}$$

The p-norm reduces to the sum of magnitudes for $p = 1$, $\|\mathbf{a}\|_1 = \sum_r |a_r|$ and to the Euclidean norm for $p = 2$, $\|\mathbf{a}\|_2 = \sqrt{\sum_r |a_r|^2}$ that is equal to the length of the vector, Eq. (3.20). For $p \to \infty$ the p-norm reduces to the maximum magnitude norm $\|\mathbf{a}\|_\infty = \max |a_r|$. A matrix norm can be based on the vector norms taking the sum of the norms of the row or column vectors. A commonly used norm is the Frobenius norm defined for the matrix $\mathbf{A} = \left[a_{rc}\right]$ as

$$\|\mathbf{A}\|_F = \sqrt{\sum_r \sum_c |a_{rc}|^2} \tag{3.73}$$

This norm can be shown to be equal to the square root of the sum of the singular values of the matrix. Another commonly used norm is the spectral norm that is equal to the largest singular value of the matrix.

3.2.7 Vector and Matrix Calculus

In this section, we extend the common calculus of scalar functions to the case of vectors and matrices. If a matrix has elements that are a function of time $\mathbf{A}(t) = [a_{rc}(t)]$ then derivation and integration are defined as

$$\frac{d}{dt}\mathbf{A}(t) = \dot{\mathbf{A}}(t) = [\dot{a}_{rc}(t)]$$

$$\int \mathbf{A}(t)\,dt = \left[\int a_{rc}(t)\,dt\right]$$

(3.74)

To apply the classical calculus to the vector and matrix algebra is not a straightforward process. However, we will need this for certain cases. For instance, it is not clear how we should define the time derivative of the inverse of a matrix like $(d/dt)\mathbf{A}^{-1}(t)$. As an example on how to solve such problem, we will use that $\mathbf{A}(t)\mathbf{A}^{-1}(t) = \mathbf{I}$ is a constant matrix, so its derivative is zero. By following the classical rules for differentiation, one obtains

$$\frac{d}{dt}\left(\mathbf{A}(t)\mathbf{A}^{-1}(t)\right) = \dot{\mathbf{A}}(t)\mathbf{A}^{-1}(t) + \mathbf{A}(t)\frac{d}{dt}\left(\mathbf{A}^{-1}(t)\right) = 0$$

(3.75)

From this expression, we can isolate the derivative we are looking for

$$\frac{d}{dt}\mathbf{A}^{-1}(t) = -\mathbf{A}^{-1}(t)\dot{\mathbf{A}}(t)\mathbf{A}^{-1}(t)$$

(3.76)

An important case to consider is when we need to take the derivative of a scalar function $f(\mathbf{x})$ with respect to its variables gathered in the column vector $\mathbf{x} = \{x_n\}$. In this case, we define the derivative of the scalar function with respect to the vector argument as the following column vector:[5]

$$\frac{\partial}{\partial \mathbf{x}}f(\mathbf{x}) = \left\{\frac{\partial f}{\partial x_n}\right\}$$

(3.77)

By following this idea, we can extend to the case where the scalar function $f(\mathbf{A})$ is a function of a matrix $\mathbf{A} = [a_{rc}]$ and we can define similarly the derivative with respect to the matrix as the new matrix

$$\frac{\partial}{\partial \mathbf{A}}f(\mathbf{A}) = \left[\frac{\partial f}{\partial a_{rc}}\right]$$

(3.78)

The chain rule also applies to matrix (and vector) derivation. For instance, the time derivate of a matrix squared is

$$\frac{d}{dt}\mathbf{A}^2(t) = \frac{d}{dt}(\mathbf{A}(t)\mathbf{A}(t)) = \dot{\mathbf{A}}(t)\mathbf{A}(t) + \mathbf{A}(t)\dot{\mathbf{A}}(t)$$

(3.79)

If the matrix and its derivative commute, that is $\dot{\mathbf{A}}(t)\mathbf{A}(t) = \mathbf{A}(t)\dot{\mathbf{A}}(t)$ then by recursive derivation we get the rule

$$\frac{d}{dt}\mathbf{A}^n(t) = n\mathbf{A}^{n-1}(t)\dot{\mathbf{A}}(t)$$

(3.80)

A special case to consider is when only the eigenvalues depend on time, see Eq. (3.54), and the end result is the same as given by Eq. (3.80)

$$\frac{d}{dt}\mathbf{A}^n(t) = \frac{d}{dt}[\mathbf{v}_m][e_m^n(t)][\mathbf{v}_m]^{-1} = n[\mathbf{v}_m][e_m^{n-1}(t)][\dot{e}_m(t)][\mathbf{v}_m]^{-1} = n\mathbf{A}^{n-1}(t)\dot{\mathbf{A}}(t)$$

(3.81)

---

[5] Some authors like to define the vector derivate as a row vector; however, it is does not matter for the final results if we use one or the other definition.

Considering the simple inner product like in Eq. (3.17) between the two vectors $\mathbf{x}$ and $\mathbf{y}$

$$\mathbf{x} \cdot \mathbf{y} = \mathbf{x}^T \mathbf{y} \tag{3.81}$$

We can now differentiate the resultant scalar with respect to the vector $\mathbf{x}$, and using the definition (3.75) we obtain

$$\frac{\partial}{\partial \mathbf{x}} \mathbf{x}^T \mathbf{y} = \frac{\partial}{\partial \mathbf{x}} \sum_n x_n y_n = \{y_n\} = \mathbf{y} \tag{3.83}$$

Similarly, we find that

$$\frac{\partial}{\partial \mathbf{y}} \mathbf{x}^T \mathbf{y} = \frac{\partial}{\partial \mathbf{y}} \sum_n x_n y_n = \{x_n\} = \mathbf{x} \tag{3.84}$$

If we do the same exercise on the generalized inner product of the two real vectors $\mathbf{x}$ and $\mathbf{y}$ over a matrix $\mathbf{A}$, as defined in Eq. (3.33), then we obtain

$$\frac{\partial}{\partial \mathbf{y}} \mathbf{x}^T \mathbf{A} \mathbf{y} = \frac{\partial}{\partial \mathbf{y}} \sum_r \sum_c x_r a_{rc} y_c = \left\{ \frac{\partial}{\partial y_k} \sum_r \sum_c x_r a_{rc} y_c \right\} = \sum_r x_r a_{rk}$$
$$= \mathbf{A}^T \mathbf{x} \tag{3.85}$$

and similarly

$$\frac{\partial}{\partial \mathbf{x}} \mathbf{x}^T \mathbf{A} \mathbf{y} = \frac{\partial}{\partial \mathbf{x}} \sum_r \sum_c x_r a_{rc} y_c = \left\{ \frac{\partial}{\partial x_k} \sum_r \sum_c x_r a_{rc} y_c \right\} = \sum_c a_{kc} y_c$$
$$= \mathbf{A} \mathbf{y} \tag{3.86}$$

If we take the derivative of the length of a vector squared like in Eq. (3.35), then we have

$$\frac{\partial}{\partial \mathbf{x}} \mathbf{x}^T \mathbf{A} \mathbf{x} = \frac{\partial}{\partial \mathbf{x}} \sum_r \sum_c x_r a_{rc} x_c = \left\{ \frac{\partial}{\partial x_k} \sum_r \sum_c x_r a_{rc} x_c \right\} = \sum_c a_{kc} x_c + \sum_r x_r a_{rk}$$
$$= \mathbf{A} \mathbf{x} + \mathbf{A}^T \mathbf{x} = \left(\mathbf{A} + \mathbf{A}^T \right) \mathbf{x} \tag{3.87}$$

Or if the matrix $\mathbf{A}$ is symmetric

$$\frac{\partial}{\partial \mathbf{x}} \mathbf{x}^t \mathbf{A} \mathbf{x} = 2 \mathbf{A} \mathbf{x} \tag{3.88}$$

The partial derivative of a vector with respect to another vector can be used to obtain meaningful results such as $\partial/\partial \mathbf{x} (\mathbf{A} \mathbf{x}) = \mathbf{A}$. The reader can find more information about this subject in several textbooks (see for instance Simon [6]).

These results for the calculus of vectors and matrices are important for a simple way of presenting the least squares problem, because the derivation of the least solution done in the following in Eqs. (3.93–3.95) is based directly hereon.

3.3 Least Squares Regression

3.3.1 Linear Least Squares

The linear least squares (LS) regression problem arises from the situation where we have a set of observations x_n, y_n of the variable x (called the basis variable) and the variable y (called the response variable) that we like to fit with a set of functions. We are assuming that a relation exists between the two variables and that this relation can be expressed by a set of functions $f_n(x)$ – often called the basis functions – such that

$$y(x) = a_1 f_1(x) + a_2 f_2(x) + \cdots \tag{3.89}$$

When using the actual data set x_n, y_n the problem we are facing is to find an estimate of the coefficients a_n. To find this estimate we want to use the least squares principle. We say that the problem is a "linear" least squares problem because the relation in Eq. (3.89) is linear with respect to the coefficients a_n. Gathering the basis functions in the vector $\mathbf{f}(x) = \{f_n(x)\}$ and the coefficients in the vector $\mathbf{a} = \{a_n\}$ we can also write Eq. (3.89) in matrix form

$$y(x) = \mathbf{a}^T \mathbf{f}(x) = \mathbf{f}^T(x) \mathbf{a} \tag{3.90}$$

or formulating Eq. (3.89) for the actual data set x_n, y_n we naturally define the response vector $\mathbf{y} = \{y(x_n)\}$ and the matrix often called the design matrix as

$$\mathbf{X} = [X_{rc}] = [f_c(x_r)]$$

$$= \begin{bmatrix} f_1(x_1) & f_2(x_1) & \cdots \\ f_1(x_2) & f_2(x_2) & \cdots \\ \vdots & \vdots & \ddots \end{bmatrix} \tag{3.91}$$

Finally, we can express Eq. (3.89) for the actual data set using the following matrix equation

$$\mathbf{y} = \mathbf{Xa} \tag{3.92}$$

For this equation to be meaningful, at least the number of data point in the data set x_n, y_n must be as large as the number of unknowns in the parameter vector $\mathbf{a}$. However, we will assume here that the number of data points is considerably larger than the number of unknowns (as is the case from data obtained from experiments), and so Eq. (3.92) is an overdetermined problem that cannot be immediately solved, but must be solved approximately. We can use LS to solve this problem.

Actually, since $\mathbf{y}$ are the observed responses, and $\mathbf{Xa}$ is a model describing an assumed relationship between x and y there will always be a difference between the left-hand side and the right-hand side of Eq. (3.92); thus a noise vector $\boldsymbol{\varepsilon}$ is added in order to represent this situation more adequately

$$\mathbf{y} = \mathbf{Xa} + \boldsymbol{\varepsilon} \tag{3.93}$$

The idea of the LS estimation is to minimize the sum of the squared error calculated as

$$\boldsymbol{\varepsilon}^T \boldsymbol{\varepsilon} = (\mathbf{y} - \mathbf{Xa})^T (\mathbf{y} - \mathbf{Xa}) \tag{3.94}$$

Using the definition of the partial derivative of a scalar with respect to a vector given by Eq. (3.77), we minimize the total squared error assuming that the minimum is flat and thus all derivatives of the error with respect to the unknown parameters should be zero at the optimal point. We then have the condition

$$\frac{\partial}{\partial \mathbf{a}} \boldsymbol{\varepsilon}^T \boldsymbol{\varepsilon} = 0 \tag{3.95}$$

and by using Eq. (3.94) we have

$$\frac{\partial}{\partial \mathbf{a}} \boldsymbol{\varepsilon}^T \boldsymbol{\varepsilon} = \frac{\partial}{\partial \mathbf{a}} \left(\mathbf{y}^T \mathbf{y} - \mathbf{y}^T \mathbf{Xa} - \mathbf{a}^T \mathbf{X}^T \mathbf{y} + \mathbf{a}^T \mathbf{X}^T \mathbf{Xa} \right) \tag{3.96}$$

The first term becomes zero because $\mathbf{y}^T \mathbf{y}$ is a constant, the second term is found from Eq. (3.85) to give $-\mathbf{X}^T \mathbf{y}$ and the same result is found from Eq. (3.86) for the third term. Finally, the last term can be found realizing that $\mathbf{X}^T\mathbf{X}$ is symmetric, which according to Eq. (3.88) gives the result $2\mathbf{X}^T\mathbf{X}\mathbf{a}$. Using this result in Eq. (3.96) leads to the equations called the normal equations:

$$-2\mathbf{X}^T\mathbf{y} + 2\mathbf{X}^T\mathbf{X}\mathbf{a} = 0$$

$$\Downarrow \qquad\qquad\qquad\qquad (3.97)$$

$$\mathbf{X}^T\mathbf{X}\mathbf{a} = \mathbf{X}^T\mathbf{y}$$

Since the matrix $\mathbf{X}^T\mathbf{X}$ is a quadratic and symmetric matrix that normally will have full rank and can be immediately inverted, the LS parameter estimate is found to be

$$\hat{\mathbf{a}} = \left(\mathbf{X}^T\mathbf{X}\right)^{-1}\mathbf{X}^T\mathbf{y} \qquad\qquad (3.98)$$

We see that the least square solution gives us the solution to Eq. (3.92). The solution is also found by premultiplying Eq. (3.92) by the design matrix transposed and then inverting the square matrix $\mathbf{X}^T\mathbf{X}$ obtained from the right-hand side of Eq. (3.92). The matrix $\left(\mathbf{X}^T\mathbf{X}\right)^{-1}\mathbf{X}^T$ is also known as the pseudo inverse $\mathbf{X}^+$ to the design matrix $\mathbf{X}$. We write this as

$$\mathbf{X}^+ = \left(\mathbf{X}^T\mathbf{X}\right)^{-1}\mathbf{X}^T \qquad\qquad (3.99)$$

At this point it is worth commenting on the inversion of the matrix $\mathbf{X}^T\mathbf{X}$ in Eqs. (3.98/3.99) and about Eq. (3.107). The matrix is independent of the data, it only depends on the choice of basis functions and basis variable values. If the basis functions are chosen so that they can be clearly distinguished when calculated at the basis variable values, then the rank of the matrix $\mathbf{X}$ will be equal to the number of basis functions. This means that the resulting matrix $\mathbf{X}^T\mathbf{X}$ normally is ending up without any deficiency problems; this can be checked by calculating the condition number of the matrix. The end result is that using a reasonable choice of basis functions and basis variable values will lead to a matrix $\mathbf{X}^T\mathbf{X}$ that can be inverted without any problems. Thus in a case where bad estimates of the parameters are obtained, this is most likely not due to problems inverting the matrix $\mathbf{X}^T\mathbf{X}$, but due to nonoptimal choice of basis functions and/or to excessive noise on the response variables – or both; see Example 3.3.

It can be shown that if the matrix $\mathbf{X}^T\mathbf{X}$ has full rank, then the pseudo inverse calculated using Eq. (3.99) is equal to inverting the matrix using the SVD

$$\mathbf{X} = \mathbf{U}\mathbf{S}\mathbf{V}^H \qquad\qquad (3.100)$$

In this case the pseudo inverse is obtained by inverting the singular values and using the fact that the matrices $\mathbf{U}$ and $\mathbf{V}$ are unitary[6]

$$\mathbf{X}^+ = \mathbf{V}\mathbf{S}^{-1}\mathbf{U}^H \qquad\qquad (3.101)$$

Even though the matrix is normally not expected to have deficiency problems as explained earlier, it might be the case sometimes if a large set of basis functions are being used. If the matrix $\mathbf{X}^T\mathbf{X}$ has deficiency problems (one or more of the singular values are zero or close to zero), then it might a better choice to use

$$\mathbf{X}^+ = \mathbf{U}\mathbf{S}^+\mathbf{V}^H \qquad\qquad (3.102)$$

where $\mathbf{S}^+$ means inverting only the nonzero singular values leaving the zeroes in place, or in case of near-to-zero singular values, forcing the corresponding singular values of $\mathbf{S}^+$ to zero. See more about this way of inverting close-to-singular matrices in Golub and Loan [3].

---

[6] Note that this is a generalization of Eq. (3.55) that is only valid for square matrices whereas Eq. (3.101) is valid for any matrix.

3.3.2 Bias, Weighting and Covariance

It is relatively easy to verify that the linear LS estimator given by Eq. (3.98) is unbiased if the expectation of the response is equal to the proposed model, that is

$$E[\mathbf{y}] = \boldsymbol{\mu}_y = \mathbf{Xa} \tag{3.103}$$

From Eq. (3.93), we see that this implies that the errors $\boldsymbol{\varepsilon}$ must be zero mean and it is then natural to say that the value of the parameter vector $\mathbf{a}$ that satisfies Eq. (3.103) is the true value. The LS estimate $\hat{\mathbf{a}}$ of the parameter vector is given by Eq. (3.98). If the expectation of the estimate $E[\hat{\mathbf{a}}]$ is equal to the true value $\mathbf{a}$, then we will say that the estimator is unbiased. From Eqs. (3.98–3.103), we see that this is the case because

$$\begin{aligned} E[\hat{\mathbf{a}}] &= (\mathbf{X}^T\mathbf{X})^{-1}\mathbf{X}^T E[\mathbf{y}] \\ &= (\mathbf{X}^T\mathbf{X})^{-1}\mathbf{X}^T\mathbf{Xa} \\ &= \mathbf{a} \end{aligned} \tag{3.104}$$

However, since Eq. (3.103) is not always satisfied, bias might be present. This is often caused by the fact that noise is not only present on the response variable y, but is also present on the basis variable x in some cases. This is illustrated in Example 3.2.

If the errors $\boldsymbol{\varepsilon} = \{\varepsilon_k\}$ are not equally distributed, but rather are having a varying standard deviation σ_k so that the weighted error vector $\{\varepsilon_k/\sigma_k\}$ has constant variance, then a weighting is often introduced by defining the diagonal matrix

$$\mathbf{W} = [1/\sigma_k^2] \tag{3.105}$$

The normal equations generalized from Eq. (3.97) now become

$$\mathbf{X}^T\mathbf{WXa} = \mathbf{X}^T\mathbf{Wy} \tag{3.106}$$

and the LS solution generalized from (3.96) becomes

$$\hat{\mathbf{a}} = (\mathbf{X}^T\mathbf{WX})^{-1}\mathbf{X}^T\mathbf{Wy} \tag{3.107}$$

The statistical errors that often prevent satisfying Eq. (3.103) are often concentrated at the ends of the basis vector. That is, for small and large values of x, they might sometimes reduce the bias by applying a weighting that reduces the influence of the information at the ends, see Example 3.2.

If we assume that the errors in the error vector are zero mean as expressed by Eq. (3.103), but in general correlated, and with nonconstant variance, then the covariance matrix of the error vector can be used to characterize the statistical properties of the errors

$$\mathbf{C}_\varepsilon = E[\boldsymbol{\varepsilon}\boldsymbol{\varepsilon}^T] \tag{3.108}$$

The same covariance matrix also describes the response vector because the model term $\mathbf{Xa}$ in Eq. (3.93) is not stochastic and we have

$$E\left[(\mathbf{y} - \boldsymbol{\mu}_y)(\mathbf{y} - \boldsymbol{\mu}_y)^T\right] = \mathbf{C}_\varepsilon \tag{3.109}$$

Using that for any constant matrix $\mathbf{A}$ the transformation $\mathbf{Ay}$ leads to the covariance

$$E\left[(\mathbf{A}(\mathbf{y} - \boldsymbol{\mu}_y))(\mathbf{A}(\mathbf{y} - \boldsymbol{\mu}_y))^T\right] = \mathbf{A}E\left[(\mathbf{y} - \boldsymbol{\mu}_y)(\mathbf{y} - \boldsymbol{\mu}_y)^T\right]\mathbf{A}^T = \mathbf{A}\mathbf{C}_\varepsilon\mathbf{A}^T \tag{3.110}$$

considering that $\mathbf{X}^T\mathbf{W}\mathbf{X}$ and its inverse are symmetric matrices and making the choice for the weighting matrix

$$\mathbf{W} = \mathbf{C}_\varepsilon^{-1} \tag{3.111}$$

which is the generalized version of the weighting matrix given by Eq. (3.105) – the covariance matrix for the estimated parameter vector from Eq. (3.107) is given as

$$
\begin{aligned}
\mathbf{C}_a &= \mathrm{E}\left[\left(\hat{\mathbf{a}} - \boldsymbol{\mu}_a\right)\left(\hat{\mathbf{a}} - \boldsymbol{\mu}_a\right)^T\right] \\
&= \left(\left(\mathbf{X}^T\mathbf{W}\mathbf{X}\right)^{-1}\mathbf{X}^T\mathbf{W}\right)\mathbf{C}_\varepsilon\left(\mathbf{X}^T\mathbf{W}\mathbf{X}\right)^{-1}\mathbf{X}^T\mathbf{W}\right)^T \\
&= \left(\mathbf{X}^T\mathbf{W}\mathbf{X}\right)^{-1}\mathbf{X}^T\mathbf{W}\mathbf{X}\left(\mathbf{X}^T\mathbf{W}\mathbf{X}\right)^{-1} \\
&= \left(\mathbf{X}^T\mathbf{C}_\varepsilon^{-1}\mathbf{X}\right)^{-1}
\end{aligned}
\tag{3.112}
$$

In the case of uniformly distributed and uncorrelated errors with the common variance σ_ε^2, the covariance matrix of the error reduces to

$$\mathbf{C}_\varepsilon = \sigma_\varepsilon^2\mathbf{I} \tag{3.113}$$

leading to the initial weighting matrix given by Eq. (3.105). The covariance matrix for the estimated parameter vector then reduces to

$$\mathbf{C}_a = \sigma_\varepsilon^2\left(\mathbf{X}^T\mathbf{X}\right)^{-1} \tag{3.114}$$

Once we have estimated the parameter covariance $\mathbf{C}_a$, we can also obtain the uncertainty on the fitting function given by Eq. (3.90), where the response and the parameter vector are now assumed to be stochastic

$$\hat{y}(x) = \mathbf{f}^T(x)\hat{\mathbf{a}} \tag{3.115}$$

Assuming that the parameter vector estimate is unbiased (expectation of the estimated parameter vector is equal to the true value), then taking the expectation of Eq. (3.115) brings us back to the deterministic version of Eq. (3.115) given by Eq. (3.90)

$$\mathrm{E}\left[\hat{y}(x)\right] = \mu_y(x) = \mathbf{f}^T(x)\mathbf{a} \tag{3.116}$$

Using Eqs. (3.113/3.114) the variance on the fitting function is then obtained as follows:

$$
\begin{aligned}
\sigma_y^2(x) &= \mathrm{E}\left[\left(\hat{y}(x) - \mu_y(x)\right)^2\right] \\
&= \mathrm{E}\left[\left(\mathbf{f}^T\hat{\mathbf{a}} - \mathbf{f}^T\mathbf{a}\right)\left(\mathbf{f}^T\hat{\mathbf{a}} - \mathbf{f}^T\mathbf{a}\right)^T\right] \\
&= \mathbf{f}^T(x)\mathrm{E}\left[\left(\hat{\mathbf{a}} - \mathbf{a}\right)\left(\hat{\mathbf{a}} - \mathbf{a}\right)^T\right]\mathbf{f}(x) \\
&= \mathbf{f}^T(x)\mathbf{C}_a\mathbf{f}(x)
\end{aligned}
\tag{3.117}
$$

Thus, the variance on the fitting function is simply the inner product of the basis function vector over the parameter covariance matrix.

Example 3.2 Simple linear regression

In order to illustrate the principle of LS estimation, we will consider the simplest possible case; the case of linear regression where a straight line is fitted to a set of data. Also we will use this simple case to illustrate the problem of bias introduced when uncertainty is present on the basis variable.

We consider here the case where the response data is linearly dependent on the base variables, that is

$$y = a_1 x + a_2 \qquad (3.118)$$

Based on a number of points $x_1, x_2, \ldots$ we can then simulate the corresponding responses by

$$y_k = a_1 x_k + a_2 + \varepsilon \qquad (3.119)$$

where ε are normal distributed zero mean variables with a certain variance, in this case defined to have a standard deviation equal to 0.15. The parameters in Eq. (3.118) were chosen to be $a_1 = a_2 = 1$. The mean value of the base variables is centered at $x = 0$. Around the point of gravity of the data typical values of the response is around 1, thus we can say that the errors are about 15% of the response.

Figure 3.1a shows the simulated data as described earlier together with the straight line estimate found by using linear regression based on Eq. (3.98). In this case, we built the response vector $\mathbf{y} = \{y_k\}$ and the design matrix

$$\mathbf{X} = \begin{bmatrix} x_1 & 1 \\ x_2 & 1 \\ \vdots & \vdots \end{bmatrix} \qquad (3.120)$$

So that our model takes the form

$$\mathbf{y} = \mathbf{Xa} = \begin{bmatrix} x_1 & 1 \\ x_2 & 1 \\ \vdots & \vdots \end{bmatrix} \begin{Bmatrix} a_1 \\ a_2 \end{Bmatrix} \qquad (3.121)$$

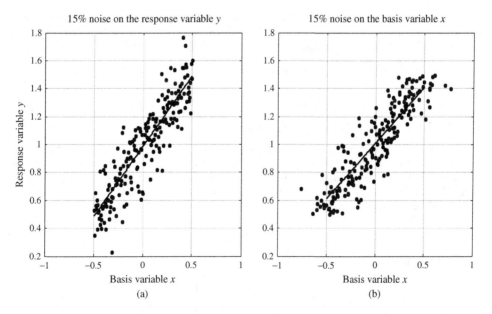

Figure 3.1 Results of fitting data with a straight line using Eq. (3.98). (a) Shows the case where the noise is on the response as it is assumed in the modeling. (b) Shows the results when the noise is on the base variable instead; as it appears a clear bias is present introducing a reduced slope on the regression estimate

For a typical case using 200 points in the data pool, the estimates were found to be

$$\left\{ \begin{array}{c} \hat{a}_1 \\ \hat{a}_2 \end{array} \right\} = \left\{ \begin{array}{c} 1.0041 \\ 0.9868 \end{array} \right\} \tag{3.122}$$

and the corresponding standard deviations for the parameter estimates were found from Eq. (3.112)

$$\left\{ \begin{array}{c} \hat{\sigma}_1 \\ \hat{\sigma}_2 \end{array} \right\} = \left\{ \begin{array}{c} 0.0359 \\ 0.0104 \end{array} \right\} \tag{3.123}$$

Figure 3.1b shows results for the case where no noise is introduced on the response variable, but noise is introduced on the base variable. As it is clearly visible in the plot, the slope now becomes too small, which means that a significant bias is introduced. For the considered case, the estimates were found to be

$$\left\{ \begin{array}{c} \hat{a}_1 \\ \hat{a}_2 \end{array} \right\} = \left\{ \begin{array}{c} 0.7938 \\ 1.0105 \end{array} \right\} \tag{3.124}$$

As we have already observed from the plot, the bias is strong on the slope estimate. The bias is mainly due to statistical errors at the ends of the interval for the base variable. By introducing less weight on these data, the bias can be reduced. In this case, the base variable was sorted in ascending order and a Hanning window was used to suppress the influence from the data at the ends. This reduced the bias as it appears from the following estimates:

$$\left\{ \begin{array}{c} \hat{a}_1 \\ \hat{a}_2 \end{array} \right\} = \left\{ \begin{array}{c} 0.8801 \\ 1.0028 \end{array} \right\} \tag{3.125}$$

However, as indicated, a significant bias is still present. In the case of large errors, the modeling should be carefully considered, for instance by using total least squares that take both the uncertainty on the basis variable and the response variable into account; see Golub and Van Loan [3].

Example 3.3   Fitting a second-order polynomial

In this example, we take a fitting function that is slightly more complicated, a second-order polynomial is fitted to a set of data, thus three parameters are to be estimated, and we illustrate the ability of the LS estimation to provide an estimate of the covariance matrix for the parameters and then further the uncertainty on the fitting estimate. Following the notation used in the Matlab polyfit function, fitting a second-order polynomial takes the form

$$y(x) = a_1 x^2 + a_2 x + a_3 \tag{3.126}$$

and the design matrix is

$$\mathbf{X} = \begin{bmatrix} x_1^2 & x_1 & 1 \\ x_2^2 & x_2 & 1 \\ \vdots & \vdots & \vdots \end{bmatrix} \tag{3.127}$$

In this case, the data is simulated using a second-order polynomial

$$y(x) = x^2 + \varepsilon \tag{3.128}$$

where ε are normal distributed zero mean variables with a certain variance. In this case these have a standard deviation set to 50. The parameters in the Eq. (3.124) were chosen to be $a_1 = 1$, $a_2 = a_3 = 0$. The fitting was performed using the lspoly that is a Matlab function in the OMA toolbox. The function lspoly is working similar to the Matlab polyfit function, but it does also provide estimates of the parameter covariance and the standard deviation on the fitting function according to Eqs. (3.114) and (3.117). For a typical case using 50 points in the data pool, the resulting data together with the results of the LS fitting is shown in Figure 3.2 and the parameters were found to

$$\begin{Bmatrix} \hat{a}_1 \\ \hat{a}_2 \\ \hat{a}_3 \end{Bmatrix} = \begin{Bmatrix} 2.32 \\ -13.72 \\ 34.66 \end{Bmatrix} \tag{3.129}$$

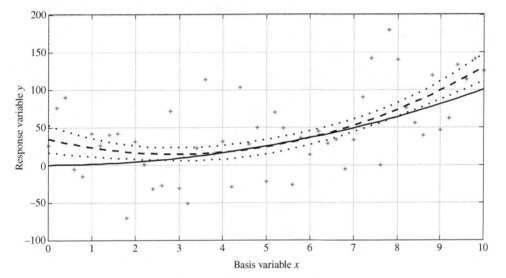

Figure 3.2 Results of fitting data with a second-order polynomial. The asterisks show the data points generated from Eq. (3.128) with a rather high simulated noise, the solid line shows the exact function, the thick dotted line shows the LS estimate, and finally the thin dotted line shows the LS estimate plus/minus the standard deviation on the LS estimate calculated by Eq. (3.117)

The covariance matrix of the parameters was found to

$$\mathbf{C}_{aa} = \begin{bmatrix} 0.66 & -6.65 & 10.86 \\ -6.65 & 71.09 & -131.6 \\ 10.86 & -131.6 & 332 \end{bmatrix} \tag{3.130}$$

As it appears, the large noise clearly reflects on the parameter covariance and on the parameter estimates. The parameter estimates are way off the true values and the covariance estimates are large. This case can be seen as an example of overfitting, because with the large added noise, it might not make sense to fit more than a straight line as we did in the previous example. A better way of doing a fit with a second-order polynomial would have been to force $\hat{a}_2, \hat{a}_3$ to zero; this could have been done by using only the first column in the design matrix given by Eq. (3.127). The example stresses the need for proper model structure selection and the limitation the presence of noise often have on this choice.

It is seen that the true curve in some intervals lies outside of the band indicated by plus/minus the standard deviation on the LS estimate. However, the example illustrates that the standard deviation estimated by Eq. (3.117) gives a good indication of the uncertainty on the fitting estimate.

References

[1] Press, W.H., Flannery, B.B., Teukolsky, S.A. and Vetterling, W.T.: *Numerical recipes in C. The art of scientific computing.*, Cambridge, Cambridge University Press, 1988.
[2] Pandit, S.M. and Wu, S.M.: *Time series and system analysis with applications.* New York, John Wiley and Sons, 1983.
[3] Golub, G.H. and Van Loan, C.F.: *Matrix computations.* Maryland, The John Hopkins University Press. 1996.
[4] Lawson, C.L. and Hanson, R.J.: *Solving least squares problems.* New Jersey, Prentice-Hall, Inc. 1974.
[5] Lancaster, P. and Salkauskas, K.: *Curve and surface fitting.* New York, Academic press. 1986.
[6] Simon, D.: *Optimal state estimation, Kalman, H[infinity] and nonlinear approaches.* New Jersey, Wiley-Interscience, John Wiley & Sons, 2006.

4

Transforms

"What then is time? If no one asks me, I know; if I wish to explain it to one that ask, I know not"

– St. Augustine

Transforms are valuable tools that are extensively used in methods of OMA, in dynamic modeling, in signal processing, as well as in system identification using modal analysis. Without the use of transform tools, modern treatment of these subjects would not be possible.

In this chapter, we discuss the classical theory of the transforms, such as the Fourier transform in continuous time and the discrete Fourier transform, and their generalized counterparts, such as the Laplace transform and the Z-transform. The theory of the Fourier and the Laplace transforms is well covered in undergraduate and graduate engineering courses. The material presented in this chapter follows the treatment of the subject as presented in widely used books dealing with the numerical treatment of the topic. Examples of such books include the "Numerical Recipes" by Press et al. [1] and "The Fast Fourier Transform" by Brigham [2]. Older books such as Doetch [3] and Irwin and Mullineux [4] have been helpful in collecting this material. The readers should have their preferred textbooks covering these transforms in more detail at hand in case they would like to review the fundamentals and details of implementation of these transforms. The Z-transform theory is not a normal reading for students in civil or mechanical engineering, so the subject would be less challenging if the reader gets more familiar with the concepts by reading about it in books that cover the subject in more detail. Suitable references to study are James [5] and Proakis and Manolakis [6].

The material presented in this chapter is fundamental to a clear understanding of the key elements of OMA. Although it should be recognized that a good understanding of the underlying theory behind OMA is a must, practice and experience in the effective use of the OMA tools are necessary for a successful identification of the dynamic properties of a structure. Putting in practice, the theory discussed here using real data also is a must in order to have a better feel for what all the concepts discussed in this book mean in practice.

4.1 Continuous Time Fourier Transforms

The Fourier transform is the basis of all transforms that we use in OMA. Nearly all signal processing techniques make use of this transform in some form or another to modify the characteristics of recorded signals in order to make these signals suitable for further analysis. It is therefore very important to have a good understanding of the properties of the Fourier transform, its application to signal processing, and the basis that it forms for the other transforms.

Introduction to Operational Modal Analysis, First Edition. Rune Brincker and Carlos Ventura.
© 2015 John Wiley & Sons, Ltd. Published 2015 by John Wiley & Sons, Ltd.
Companion Website: www.wiley.com/go/brincker

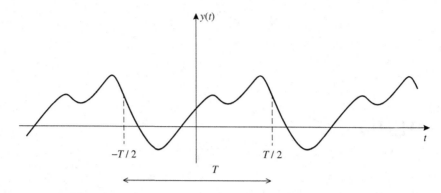

Figure 4.1   General periodic function $y(t)$ with period T to be expressed in terms of an infinite series of harmonics

4.1.1   Real Fourier Series

The basic idea of a Fourier series is to represent a periodic signal in term of its harmonic components. The process is reversible, in the sense that synthesis of all these components restores the original signal. It is commonly accepted that any signal $y(t)$ that is periodical with a period T – see Figure 4.1 – can be expressed as an infinite series of all the harmonic components that are also periodic with the same period T

$$y(t) = a_0 + a_1 \cos \frac{2\pi t}{T} + a_2 \cos \frac{4\pi t}{T} + \cdots$$
$$+ b_1 \sin \frac{2\pi t}{T} + b_1 \sin \frac{4\pi kt}{T} \cdots$$

(4.1)

or in short form

$$y(t) = a_0 + \sum_{k=1}^{\infty} \left(a_k \cos \frac{2\pi kt}{T} + b_k \sin \frac{2\pi kt}{T} \right)$$

(4.2)

The periodic harmonic functions satisfy the following orthogonality conditions

$$\frac{1}{T} \int_{-T/2}^{T/2} \cos \frac{2\pi nt}{T} \cos \frac{2\pi mt}{T} dt = \begin{cases} 0, & n \neq m \\ 1/2, & n = m \end{cases}$$

$$\frac{1}{T} \int_{-T/2}^{T/2} \cos \frac{2\pi nt}{T} \sin \frac{2\pi mt}{T} dt = 0, \quad \text{for all } n, m$$

(4.3)

Using the orthogonality conditions of the periodic harmonic functions and the fact that all the periodic harmonics have a mean value of zero, the Fourier coefficients of each harmonic component of the series can be obtained from the following expressions:

$$a_0 = \frac{1}{T} \int_{-T/2}^{T/2} y(t) \, dt$$

$$a_k = \frac{2}{T} \int_{-T/2}^{T/2} y(t) \cos \left(\omega_k t \right) dt \quad k = 1, 2, 3, \ldots$$

(4.4)

$$b_k = \frac{2}{T} \int_{-T/2}^{T/2} y(t) \sin\left(\omega_k t\right) dt \qquad k = 1, 2, 3, \ldots$$

in which the subscript "k" represents the kth harmonic component and the quantity

$$\omega_k = \frac{2\pi k}{T} \tag{4.5}$$

is the frequency of the kth harmonic. The increment of frequency from one harmonic component to the next is a constant amount, which is given by

$$\Delta\omega = \frac{2\pi}{T} \tag{4.6}$$

A graphical representation of the coefficients of the harmonic components and the spacing between them is given in Figure 4.2.

The aforementioned expressions indicate that the discrete representation in the frequency domain of a continuous periodic signal is not due to a discrete time representation of this signal, but solely due to the fact that the signal is periodic.

4.1.2   Complex Fourier Series

It is common practice to represent the summation in Eq. (4.2) as a summation of terms from minus infinity to plus infinity

$$y(t) = \sum_{k=-\infty}^{\infty} \left(A_k \cos \frac{2\pi kt}{T} + B_k \sin \frac{2\pi kt}{T} \right) \tag{4.7}$$

This is only done in order to simplify the representation of the series in a more amenable mathematical formulation. However, this introduces the "unrealistic" concept of negative frequencies for the harmonic terms associated with the negative value of k.

If the coefficients associated to the negative frequencies are defined as

$$\begin{aligned} A_{-k} &= A_k \\ B_{-k} &= -B_k \end{aligned} \tag{4.8}$$

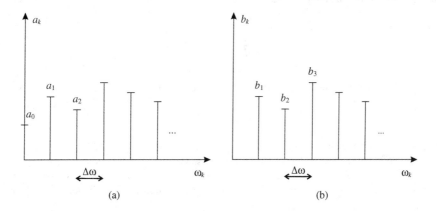

(a)   (b)

Figure 4.2   Graphical representation of the Fourier coefficients at the discrete frequencies ω_k

then, the following simplified set of equations that are valid for all values of k can be used to determine the coefficients A_k and B_k

$$A_k = \frac{1}{T} \int_{-T/2}^{T/2} y(t) \cos\left(\omega_k t\right) dt$$

$$B_k = \frac{1}{T} \int_{-T/2}^{T/2} y(t) \sin\left(\omega_k t\right) dt$$

(4.9)

If we now represent the cosine coefficients A_k and sine coefficients B_k by a complex-valued quantity Y_k, in which the real part is the coefficient A_k and the imaginary part is the coefficient $-B_k$, we can then define the complex Fourier coefficients Y_k as

$$Y_k = A_k - iB_k$$

(4.10)

It is not difficult to demonstrate that by using Eq. (4.10) the infinite series given by Eq. (4.7) can be written in a compact form as

$$y(t) = \sum_{k=-\infty}^{\infty} Y_k e^{i\Delta\omega k t}$$

(4.11)

It is worth noticing that since the signal in Eq. (4.7) is assumed to be real valued, it makes sense to define the real and imaginary parts of the complex Fourier coefficient Y_k like we have done in Eq. (4.8) because this will force the imaginary part of the right side of Eq. (4.7) to be zero as assumed.

If the signal $y(t)$ is an even function, that is if $y(-t) = y(t)$, then all the sine coefficients $B_k = 0$, and the complex Fourier coefficients Y_k are all real. On the other hand, if the signal $y(t)$ is an odd function, that is if $y(-t) = -y(t)$, then all the cosine coefficients $A_k = 0$, and the complex Fourier coefficients Y_k are all imaginary. In the first case, this means that the harmonic components of the signal $y(t)$ are all cosine functions, while in the second case all the harmonic components of $y(t)$ are sine functions.

The complex Fourier coefficients can be found from the orthogonality properties of the complex exponential function

$$\frac{1}{T} \int_{-T/2}^{T/2} e^{i\Delta\omega n t} e^{-i\Delta\omega m t} dt = \frac{1}{T} \int_{-T/2}^{T/2} e^{i\Delta\omega(n-m)t} dt = \delta_{nm}$$

(4.12)

where δ_{nm} is the Kronecker delta defined as $\delta_{nm} = 1$ for $n = m$ and otherwise equal to zero. This property is then used to isolate each Y_k as

$$Y_k = \frac{1}{T} \int_{-T/2}^{T/2} y(t) e^{-i\Delta\omega k t} dt$$

(4.13)

From this point on, we will use Y_k to define the Fourier transform of $y(t)$, and the Fourier transform pair will be denoted as

$$y(t) \leftrightarrow Y_k$$

(4.14)

Here we have used the classical convention that the time function is denoted by a lower case variable, whereas the Fourier transform is denoted by an upper case variable. However, since the terminology must be generalized to vectors and matrices, in some cases we will not always follow the classical convention and use instead a tilde over the variable to identify the Fourier transform of a signal. In such cases, we might also write

$$Y(t) \leftrightarrow \tilde{Y}_k$$

(4.15)

or

$$y(t) \leftrightarrow \widetilde{y}_k \tag{4.16}$$

It is commonly done in textbooks dealing with this subject to leave the case of continuous periodic relatively quickly and move on to the more practical case of discrete time functions. However, since the origin of many of the problems encountered when applying the concepts described here to practical cases are due to misinterpretation of the basic concepts, we will dwell a little longer on the discussion of continuous periodic functions.

First, let us note that the transform is a linear operation. For any set of real-valued periodic functions, say $y_1(t), y_2(t)$, and for any real or complex-valued numbers c_1, c_2, we can write

$$c_1 y_1(t) + c_2 y_2(t) \leftrightarrow c_1 Y_{1k} + c_2 Y_{2k} \tag{4.17}$$

where $y_1(t) \leftrightarrow Y_{1k}$ and $y_2(t) \leftrightarrow Y_{2k}$. Therefore, the Fourier transform concept can be extended to complex signals. For instance, for a general harmonic signal, $y(t) = c \cos(\omega t + \phi)$, we can write this as a complex exponential of the form

$$y(t) = ce^{i(\omega t + \phi)} \tag{4.18}$$

This makes sense because the application of Euler's identity leads to

$$ce^{i(\omega t + \phi)} = ce^{i\phi} e^{i\omega t}$$

$$= c(\cos\phi + i\sin\phi)(\cos\omega t + i\sin\omega t) \tag{4.19}$$

Taking the real value component of the aforementioned expression brings us back to the general harmonic expression

$$\operatorname{Re}\left(ce^{i(\omega t + \phi)}\right) = c(\cos\omega t \cos\phi - \sin\omega t \sin\phi)$$

$$= c\cos(\omega t + \phi) \tag{4.20}$$

The reader is encouraged to review the references mentioned at the beginning of this chapter or can verify as an exercise that the following statements are true.

It can be shown that the following time reversal property holds

$$y(-t) \leftrightarrow Y_{-k} = Y_k^* \tag{4.21}$$

Note that the term to the right side of the equality is only true for real time functions $y(t)$.

The time shift property is

$$y(t - t_0) \leftrightarrow Y_k e^{-i\Delta\omega k t_0} \tag{4.22}$$

This means that shifting a function in the time domain by t_0 has a Fourier transform equal to the Fourier transform of the original function multiplied by a complex exponential introducing a linearly varying phase $\phi_k = -\Delta\omega k t_0$, the slope of the phase being equal to t_0.

If we define convolution for the periodic functions $h(t), g(t)$ over the period T as

$$h(t) * g(t) = \frac{1}{T} \int_{-T/2}^{T/2} h(t - \tau) g(\tau) d\tau = \frac{1}{T} \int_{-T/2}^{T/2} g(t - \tau) h(\tau) d\tau \tag{4.23}$$

then the important convolution theorem holds

$$h(t) * g(t) \leftrightarrow H_k G_k \tag{4.24}$$

which means that the convolution of two functions in the time domain has a Fourier transform that is equal to the product of the Fourier transform of each function.

It can be demonstrated that the corresponding theorem for convolution in the frequency domain

$$H_k * G_k = \sum_{n=-\infty}^{\infty} H_{k-n}G_n = \sum_{n=-\infty}^{\infty} G_{k-n}H_n \qquad (4.25)$$

results in

$$h(t) g(t) \leftrightarrow H_k * G_k \qquad (4.26)$$

In this case, the convolution of the Fourier transform of two functions is the Fourier transform of the product of the two functions.

From Eq. (4.11), the differentiation theorem states that

$$\dot{y}(t) \leftrightarrow Y_k i\omega_k \qquad (4.27)$$

and similarly, for integration

$$\int y(t) \, dt \leftrightarrow Y_k / (i\omega_k) \qquad (4.28)$$

We see that the Fourier transform of the derivative of a function is equal to the Fourier transform of the original function multiplied by the scaling factor $i\omega_k$. Similarly, the Fourier transform of the integral of a function is equal to the Fourier transform of the original function divided by the same scaling factor.

4.1.3   The Fourier Integral

The Fourier integral is the basis for the transform of a continuous time function defined over an infinite time domain without any consideration for periodicity. The transform can be seen as the result of a transition where we assume that the period T extends to infinity. Substituting Eq. (4.13) into (4.11) leads to

$$y(t) = \sum_{k=-\infty}^{\infty} \left\{ \frac{1}{T} \int_{-T/2}^{T/2} y(t) e^{-i\Delta\omega kt} dt \right\} e^{i\Delta\omega kt} \qquad (4.29)$$

Noting from Eq. (4.6) that $1/T = \Delta\omega / 2\pi$ and letting $T \to \infty$, we can see that $\Delta\omega$ approaches zero, and in the limit we replace it with $d\omega$. The discrete variable $\omega_k = \Delta\omega k$ is then replaced by the continuous variable ω and the sum is replaced by an integral of the form

$$y(t) = \frac{1}{2\pi} \int_{-\infty}^{\infty} \left\{ \int_{-\infty}^{\infty} y(t) e^{-i\omega t} dt \right\} e^{i\omega t} d\omega \qquad (4.30)$$

We see that the Fourier integral can be naturally defined as

$$y(t) = \int_{-\infty}^{\infty} Y(\omega) e^{i\omega t} d\omega \qquad (4.31)$$

where the Fourier transform (the continuous coefficient) is given by

$$Y(\omega) = \frac{1}{2\pi} \int_{-\infty}^{\infty} y(t) e^{-i\omega t} dt \qquad (4.32)$$

As it appears, the placement of the factor $1/2\pi$ on the Fourier transform is a matter of convenience. Some authors prefer to place this term on the Fourier integral, while some prefer to use the factor $1/\sqrt{2\pi}$ placed in both equations. We could also have defined the Fourier integral from Eq. (4.30) using the cyclic frequency f instead of the angular frequency ω and that would have removed the 2π from the equations.

The linearity of the Fourier integral is expressed as

$$c_1 y_1(t) + c_2 y_2(t) \leftrightarrow c_1 Y_1(\omega) + c_2 Y_2(\omega) \tag{4.33}$$

and the time reversal property is

$$y(-t) \leftrightarrow Y(-\omega) = Y^*(\omega) \tag{4.34}$$

where the term on right side of the equal sign is only true for real functions.

The time shift property is given by

$$y(t - t_0) \leftrightarrow Y(\omega) e^{-i\omega t_0} \tag{4.35}$$

We define the convolution of signals of infinite duration as

$$h(t) * g(t) = \int_{-\infty}^{\infty} h(t - \tau) g(\tau) d\tau = \int_{-\infty}^{\infty} g(t - \tau) h(\tau) d\tau \tag{4.36}$$

and it can be easily demonstrated that the convolution properties are

$$h(t) * g(t) \leftrightarrow H(\omega) G(\omega) \tag{4.37}$$

$$h(t) g(t) \leftrightarrow H(\omega) * G(\omega) \tag{4.38}$$

The differentiation and integration theorems are

$$\dot{y}(t) \leftrightarrow Y(\omega) i\omega \tag{4.39}$$

$$\int_{-\infty}^{t} y(\tau) d\tau \leftrightarrow Y(\omega) / i\omega \tag{4.40}$$

4.2 Discrete Time Fourier Transforms

Since all real data is represented by discrete samples, it is obvious that the discrete version of the Fourier transform is needed for signal processing of real data.

4.2.1 Discrete Time Representation

As discussed before, a periodic continuous time function has a discrete infinite frequency-domain function. Because of the symmetry between time domain and frequency domain in the Fourier integral/transform, see Eqs. (4.31/4.32), continuous periodic frequency functions with period B_ω (the frequency band measured in terms of angular frequency) will result in discrete time functions with a spacing (the sampling time step) between the time points given by

$$\Delta t = \frac{2\pi}{B_\omega} \tag{4.41}$$

When working with discrete signals, it is common practice to change from angular frequency ω measured in rad/s to frequency f measured in Hz using the well-known relationship

$$\omega = 2\pi f \tag{4.42}$$

Eqs. (4.41–4.42) can be combined to express the sampling time step in terms of the frequency bandwidth B_f measured in terms if Hz as follows

$$\Delta t = \frac{1}{B_f} \tag{4.43}$$

The bandwidth B_f is also called the sampling frequency f_s. If we now decide to place the bandwidth B_f symmetrically on the frequency axis (like we did for the periodic time functions), then the periodic frequency function will span from $-B_f/2 = -f_s/2$ to $B_f/2 = f_s/2$. This defines the Nyquist frequency

$$f_v = B_f/2 = f_s/2 = \frac{1}{2\Delta t} \tag{4.44}$$

The relationship between the Nyquist frequency and the sampling frequency is then given by

$$f_v = \frac{1}{2\Delta t} = \frac{f_s}{2} \tag{4.45}$$

If the time function is periodic with period T, then the corresponding frequency-domain function will be discrete with the spacing given by Eq. (4.6), but it can now be expressed in terms of the frequency measured in Hz, and the discrete values of the frequency will be sampled at a frequency increment of

$$\Delta f = \frac{1}{T} \tag{4.46}$$

In practice, the number of sample points in the time domain is taken to be the same as the number of sample points in the frequency domain. One may notice that the signal in the time domain is normally real, whereas the frequency-domain signal is normally complex, thus it might seem like the frequency domain carries the double information (in both the real and imaginary parts). However, there is no extra information to be carried by the complex representation, because the information related to negative frequencies is essentially the same as that included for the positive frequency values up to the Nyquist frequency, according to the symmetry relations given by Eq. (4.8).

If the number of sample points is N, then

$$T = N\Delta t \tag{4.47}$$

$$f_v = N\Delta f/2 \tag{4.48}$$

Taking the finite number of sample points into account and restricting time to the sample points $t = n\Delta t$ Eq. (4.11) becomes

$$y_n = y(n\Delta t) = \sum_{k=-N/2-1}^{N/2} Y_k e^{i\Delta \omega k n \Delta t} \tag{4.49}$$

The negative values of the discrete frequency are nothing more than the complex conjugates of the corresponding positive values, and this is why the end result of the summation on the right-hand side of the aforementioned equation results always in a real number.

The argument in the exponential function can be written

$$i\Delta\omega kn\Delta t = i\frac{2\pi}{T}kn\Delta t = i2\pi kn/N \qquad (4.50)$$

and the discrete Fourier series can then be expressed as

$$y_n = y(n\Delta t) = \sum_{k=-N/2+1}^{N/2} Y_k e^{i2\pi kn/N} \qquad (4.51)$$

Similarly, in Eq. (4.13) for the discrete values of frequency, the integral becomes a sum, dt becomes Δt and we have the discrete representation of the Fourier transform

$$Y_k = \frac{1}{T}\sum_{n=-N/2+1}^{N/2} y_n e^{-i2\pi kn/N}\Delta t = \frac{1}{N}\sum_{n=-N/2+1}^{N/2} y_n e^{-i2\pi kn/N} \qquad (4.52)$$

As mentioned before, since we are dealing with periodic functions in both time and frequency domains, we can arrange the function interval in relation to any position of the vertical axis along the frequency axis. Until now, we have arranged the periodic functions symmetrically around the vertical axis, but this is an arbitrary choice and this leads to a slightly complicated upper and lower bound for the summations in Eqs. (4.51/4.52). Since we can shift the functions as we wish, for instance, we can use a summation that goes from 1 to N by letting the frequency and time start at zero (the only important thing is that we keep the number of summations points to be N) and we get the following alternative representation of the discrete Fourier series

$$y_n = \sum_{k=1}^{N} Y_k e^{i2\pi(k-1)(n-1)/N} \qquad (4.53)$$

and the corresponding discrete Fourier transform

$$Y_k = \frac{1}{N}\sum_{n=1}^{N} y_n e^{-i2\pi(k-1)(n-1)/N} \qquad (4.54)$$

These are the formulas used in Matlab. However, Matlab has placed the factor $1/N$ on the Fourier series in Eq. (4.53) instead of on the Fourier coefficient in Eq. (4.54).

As observed from Eq. (4.4), the Fourier coefficient at frequency zero – often denoted as DC[1] – is real and equal to the mean value of the signal. However, the Fourier coefficient $Y_{N/2+1}$ corresponding to the Nyquist frequency is also real. This can be explained by the fact that the imaginary part of the Fourier coefficient is an odd function, Eq. (4.8), see Figure 4.3.

Therefore, in the frequency domain, we have two real values, one at DC ($n = 1$) and the other at the Nyquist frequency ($n = N/2 + 1$), and then in between from $n = 2$ to $n = N/2$ we have $(N/2 - 1)2 = N - 2$ real numbers (counting both real and imaginary parts). It should be noted that the Fourier coefficients for negative frequencies are given by Eq. (4.21) and therefore the total number of real parameters to describe the data in the frequency domain is exactly equal to N, which is the same as in the time domain. The discrete Fourier transform provides a one–one relationship between the data in the time domain and in the frequency domain.

In discrete time representations, we might use either the true discrete time $t = (n-1)\Delta t$, $n = 1, 2, \ldots$ or the normalized discrete time $n = 1, 2, \ldots$

---

[1] DC stands for "direct current" referring to a constant value current and might seem like an awkward expression to use in a mechanical discipline like this. However, since it is common practice to use this term instead of "frequency zero", we will also adopt this practice.

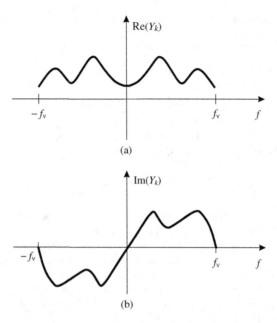

Figure 4.3 Schematic of the real and imaginary parts of the Fourier coefficients. The real part is an even function of frequency whereas the imaginary part is an odd function of frequency forcing the imaginary part to be zero at DC and at the Nyquist frequency

4.2.2 The Sampling Theorem

The sampling theorem is due to Shannon [7] and thus is also called the Shannon sampling theorem. The theorem states that

> Any limited discrete time signal can be brought back to continuous time if the corresponding Fourier transform (the Fourier coefficient) vanishes outside of the Nyquist band

The theorem can be explained as follows. Any limited time signal can be treated as a periodic signal. If the Fourier coefficients of this signal vanish outside of the Nyquist band, and if we restrict the summation to the Nyquist frequency band, then Eq. (4.11) becomes

$$y(t) = \sum_{k=-N/2+1}^{N/2} Y_k e^{i\Delta\omega k t} \qquad (4.55)$$

This equation is similar to Eq. (4.46) for the discrete Fourier series except that the present equation is in terms of continuous time t. When we introduced the discrete time, we sampled this equation at $t = n\Delta t$, but in the present case Eq. (4.55) indicates that we can also obtain the single values at any other time different than $t = n\Delta t$. One can also say that the sampling theorem holds because the harmonic functions used in the Fourier series work as exact spline functions. This equation also indicates that the value of the function at any time t can be recovered exactly if all the Fourier coefficients up to the Nyquist frequency are available and those beyond the Nyquist frequency are zero.

The assumption that the Fourier coefficients vanish outside of the Nyquist band relates to removing all energy in the signal that is outside of the Nyquist band. This process is called antialiasing filtering; see Sections 7.2.6 and 8.3.3.

Determining the series coefficients at a higher sampling rate is called up-sampling. In practice, time series can be up-sampled by zero padding the Fourier coefficients and then transforming the zero padded set of coefficients back to time domain.

Sampling can be considered a data reduction process that does not remove any information. Important implications of the sampling theorem are that the differentiation and the integration theorems still hold. For instance, the differentiation theorem can be proved by differentiating Eq. (4.55) so that

$$\dot{y}(t) = \sum_{k=-N/2+1}^{N/2} Y_k i\Delta\omega k e^{i\Delta\omega kt} \tag{4.56}$$

and then restricting it to discrete time.

Example 4.1   Differentiation and up-sampling

Let us consider one of the signals from the first data set of the Heritage Court Tower data. We consider the signal in channel 1, a small sequence of that data is shown in the top plot of Figure 4.4.

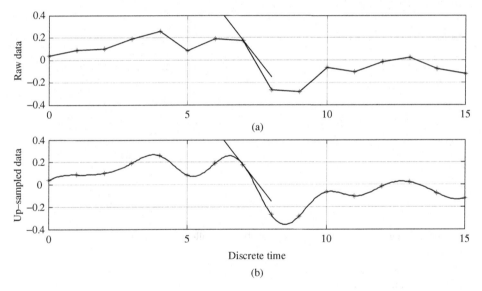

Figure 4.4   (a) A sequence of data from the Heritage Court Tower case in raw form, each data point indicated by an asterisk "*." The slope of the mid-point is calculated and indicated by the tangent. (b) The similar data up-sampled by a factor of 16

We can now use Eq. (4.56) to numerically calculate an estimate of the derivative of the corresponding continuous data at the midpoint of the data sequence. The resulting tangent is also plotted in the top plot of Figure 4.4. Notice, however, from this plot it is difficult to judge if the estimate makes much sense.

We can also up-sample the raw data sequence as described in the preceding section. If we up-sample the data by a factor of 16 and make the corresponding plot as shown in the bottom plot of Figure 4.4, then it becomes more clear that the estimated slope is indeed a good estimate of the derivate of the continuous curve.

It should be noted that using this procedure assumes the data sequence to be periodic – which is in fact not the case. Therefore, errors are introduced near the ends of the up-sampled data segment. These errors are dependent on the discontinuities at the ends of the data segment and are further illustrated in the next example.

According to Eq. (4.56) the derivative is found by multiplying in the frequency domain by $i\omega_k$ and the integral is found by dividing in the frequency domain by $i\omega_k$. In general, performing n times integration can be considered a filtering process with the frequency response function (FRF) $H\left(\omega_k\right) = \left(i\omega_k\right)^{-n}$, and the derivative is found setting $n = -1$. Since numerical integration and differentiation might create errors that are mainly concentrated close to DC and close to the Nyquist frequency, it is practical also to multiply the series by a window in the frequency domain that only allows the mid part of the information in the Nyquist band to remain and the rest is be discarded (band pass filtering). If this frequency-domain window is $W\left(\omega_k\right)$ and the signal to be differentiated is $x(t)$ then the resulting signal $y(t)$ can be found by these simple steps:

1. Define filtering window $W\left(\omega_k\right)$.
2. Define the resulting FRF $H\left(\omega_k\right) = W\left(\omega_k\right)\left(i\omega_k\right)^{-n}$.
3. Calculate the Fourier transform $X\left(\omega_k\right)$.
4. Multiply with the FRF to find the Fourier transform $Y\left(\omega_k\right)$ of $y(t)$.
5. Calculate $y(t)$ by inverse Fourier transform.

The Matlab code used in the OMAtools function fftint1.m includes these steps:

```
[W, f] = Wint(fs, df, f1, f2, m);                        %Step 1, define filtering
                                                         %window
H1 = zeros(Nf, 1);
H1(2:Nf-1) = W(2:Nf-1).*((i*2*pi*f(2:Nf-1)).^(-n));     %Step 2, define one sided FRF
H = [H1; conj(flipud(H1(2:Nf-1)))];                      %Step 2, define double
                                                         %sided FRF
X = fft(x.');                                            %Step 3, Calculate
                                                         %FFT to get X
for c = 1:Nc,
Y(:,c) = X(:, c).*H;                                     %Step 4, Multiply X and H
end
y = ifft(Y).';                                           %Step 5, Take inverse FFT
```

The up-sampling is even simpler and only involves the following main steps:

1. Take the Fourier transform.
2. Pad zeroes in frequency domain.
3. Transform back to the time domain.

The Matlab code used in the OMAtools function fftups1.m to perform these main steps is

```
[N, Nc] = size(y);
Z = zeros(N*(Nups-1),Nc);
Y = fft(y);                           %Step 1, take the Fourier transform
Y1 = [Y(1:N/2+1,:); Z; Y(N/2+2:N);];  %Step 2, pad zeroes in the frequency domain
y1 = Nups*real(ifft(Y1));             %Step 3, Transform back
```

Since Matlab defines the frequency-domain (FD) data from DC to $2f_v$ the FD data must be padded with zeroes by splitting the data at the middle as shown in step 2 of the Matlab code. Furthermore, since

Matlab has the factor $1/N$ on Eq. (4.53) instead of Eq. (4.54) we must remove this factor by multiplying with the up-sampling factor in step 3 of the Matlab code.

Example 4.2   Errors due to discontinuities

The errors due to the discontinuities mentioned in the previous example are further illustrated in this example. Consider a sequence with N points, a cosine with N_c number of cycles

$$y(t) = \cos\left(2\pi \frac{N_c}{N} t + \phi\right) \tag{4.57}$$

where $t = k\Delta t, k = 0, 1, \ldots$ is the discrete time and where the phase ϕ is arbitrary. In this case, the exact solution to the derivative is given by

$$\dot{y}(t) = -2\pi \frac{N_c}{N} \sin\left(2\pi \frac{N_c}{N} t + \phi\right) \tag{4.58}$$

We can numerically calculate the derivative using Eq. (4.56). The results are shown in Figure 4.5 for a cosine sequence with $N = 16$ number of points and $N_c = 3.5$ number of cycles. When the number of cycles N_c is an integer, then Eq. (4.56) is exact and Eq. (4.56) and Eq. (4.58) give the same result (within the round-off errors of the computer program). If the number of cycles N_c is not an integer error due to the discontinuities at the ends of the periodic sequence will appear. As it is seen from the bottom plot, the errors are largest toward the end of the data sequence. The errors can be reduced by windowing the data sequence (see Section 8.1.3).

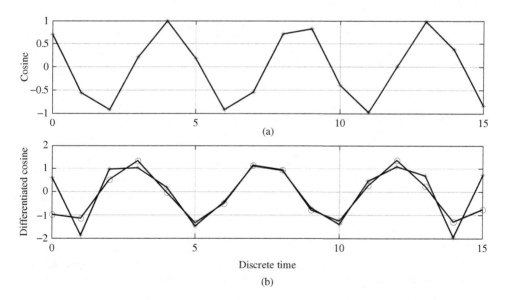

Figure 4.5   (a) A random phase cosine sequence with 16 data points and 3.5 cycles. (b) The numerical differentiation of the cosine sequence (points marked with "*") together with the exact solution given by Eq. (4.59) (marked with "o")

4.3   The Laplace Transform

The Laplace transform is of great importance for the description and solution of analytical models. A good example is the solution of differential equations of motion in structural dynamics, which can be easily obtained with the Laplace transform.

4.3.1   The Laplace Transform as a generalization of the Fourier Transform

The Laplace transform can be seen as a generalization of the Fourier transform. Signals being represented from minus infinity to plus infinity in the time domain are of no practical interest. As a consequence, in the Laplace transform theory, it is normally[2] assumed that the signal start at some finite time, which is taken as the zero point, and it is then observed from there for a long period theoretically until infinity. Introducing this concept of one-sided infinite time functions, the Fourier integral and transform formulas Eqs. (4.31/4.32) can be written as

$$y(t) = \frac{1}{2\pi} \int_{-\infty}^{\infty} Y(\omega) e^{i\omega t} d\omega \qquad (4.59)$$

$$Y(\omega) = \int_{0}^{\infty} y(t) e^{-i\omega t} dt \qquad (4.60)$$

where we have placed the $1/2\pi$ on the Fourier integral instead of on the Fourier transform.

One of the problems with the Fourier integral is that it might not converge. This problem can be avoided if we consider instead a "damped" function[3] of the form $y'(t) = e^{-\sigma t} y(t)$. The corresponding transform of this function is

$$Y'(\omega) = \int_{0}^{\infty} y(t) e^{-\sigma t} e^{-i\omega t} dt = \int_{0}^{\infty} y(t) e^{-(\sigma + i\omega)t} dt \qquad (4.61)$$

Introducing the complex variable, s, defined as

$$s = \sigma + i\omega \qquad (4.62)$$

we can rewrite Eq. (4.61) as follows

$$Y'(s) = \int_{0}^{\infty} y(t) e^{-st} dt \qquad (4.63)$$

This expression is the Laplace transform. From the aforementioned derivation, it is clear that the Fourier transform $Y(\omega)$ can be obtained from the Laplace transform as (except for the constant 2π)

$$Y(\omega) = Y'(i\omega) \qquad (4.64)$$

The inverse Laplace transform can also be obtained from the generalized Fourier integral. We do that by introducing the general complex variable s instead of the real variable ω in Eq. (4.59). When ω varies

---

[2] The Laplace transform can also be defined as the bilateral transform extending the integration over the entire time axis, however we will stick to the traditional one-sided Laplace transform.
[3] The mark on the function only indicate that the function has been modified by the damping term. See Example 4.4 for comments on the value of σ that secures stable solutions.

from $-\infty$ to ∞ while σ is constant, then s varies from $\sigma - i\infty$ to $\sigma + i\infty$. From Eq. (4.62), we see that $ds = id\omega$ and that the Fourier integral becomes

$$e^{-\sigma t} y(t) = \frac{1}{2\pi} \int\limits_{-\infty}^{\infty} Y'(\sigma + i\omega) e^{i\omega t} d\omega$$

$$\Downarrow \qquad\qquad\qquad\qquad (4.65)$$

$$y'(t) = \frac{1}{2\pi i} \int\limits_{\sigma - i\infty}^{\sigma + i\infty} Y'(s) e^{st} ds$$

The last form is the inversion formula of the Laplace transform. The functions $y'(t)$ and $Y'(s)$ constitute a transform pair and we use the same notation as for the other transforms

$$y'(t) \leftrightarrow Y'(s) \qquad\qquad (4.66)$$

We can omit the mark on the Laplace transform to indicate that we are dealing the "damped" version and just write

$$y(t) \leftrightarrow Y(s) \qquad\qquad (4.67)$$

In order not to confuse the Laplace transform with the corresponding Fourier transform, we will normally use the variable s to indicate Laplace and the variable ω to indicate the Fourier transform, thus $F(s)$ is the Laplace transform and $F(\omega)$ is the Fourier transform.

4.3.2 Laplace Transform Properties

For the Laplace transform similar properties exist as for the Fourier transform. It is easy to verify the linearity property

$$c_1 y_1(t) + c_2 y_2(t) \leftrightarrow c_1 Y_1(s) + c_2 Y_2(s) \qquad\qquad (4.68)$$

The time reversal property

$$y(-t) \leftrightarrow Y(-s) \qquad\qquad (4.69)$$

and the time shift property

$$y(t - t_0) \leftrightarrow Y(s) e^{-st_0} \qquad\qquad (4.70)$$

for $t_0 > 0$ where the resulting function must vanish for negative arguments, thus $y(t - t_0) = 0$ for $t < t_0$. Since we are only dealing with the positive time axis, we define the convolution as

$$h(t) * g(t) = \int\limits_{0}^{\infty} h(t - \tau) g(\tau) d\tau = \int\limits_{0}^{\infty} g(t - \tau) h(\tau) d\tau \qquad\qquad (4.71)$$

and the convolution property can be shown to be given by

$$h(t) * g(t) \leftrightarrow H(s) G(s) \qquad\qquad (4.72)$$

There exists also a convolution theorem in the complex plane corresponding to Eq. (4.38), but we shall not need this more complicated convolution property. From the integration theorems, it follows directly the integration theorem for the Laplace transform

$$\int y(t) dt \leftrightarrow Y(s)/s \qquad\qquad (4.73)$$

However, the differentiation theorem is a little more complicated than that for the Fourier transform

$$\dot{y}(t) \leftrightarrow Y(s)s - y(0-) \tag{4.74}$$

where the initial value $y(0-)$ is calculated as the limit of the time function as t tends toward 0 from the left. The reason for having the constant term $y(0-)$ as a limit from the left comes from the fact that we will normally include the starting point $t = 0$ in the integration interval in the Laplace transform. Normally we will use the exact definition

$$Y(s) = \int_{0-}^{\infty} y(t)e^{-st}dt \tag{4.75}$$

Using integration by parts, we obtain

$$Y(s) = \left[\frac{y(t)e^{-st}}{-s}\right]_{0-}^{\infty} - \int_{0-}^{\infty} \dot{y}(t)\frac{e^{-st}}{-s}dt$$

$$\Downarrow \tag{4.76}$$

$$Y(s)s - y(0-) \quad = \int_{0-}^{\infty} \dot{y}(t)e^{-st}dt$$

which proves the differentiation theorem (4.74) and also explains "the limit from the left." Since we will assume that the systems we are considering are at rest at times $t < 0$, we will assume that all considered functions and their derivatives have vanishing limits values from the left. In practice, this means that $y(0-) = 0$, $\dot{y}(0-) = 0$ and we will simplify the differentiation theorem to

$$\dot{y}(t) \leftrightarrow Y(s)s \tag{4.77}$$

4.3.3   Some Laplace Transforms

For the Dirac delta function $\delta(t)$ we have the problem that the meaning of the Laplace transform

$$\Delta(s) = \int_{0}^{\infty} \delta(t)e^{-st}dt \tag{4.78}$$

is uncertain since we have a singularity from the Dirac delta function right at the start of the integration. However, if we specifically include the singularity and integrate from 0− as in Eq. (4.75) we can obtain

$$\Delta(s) = \int_{0-}^{\infty} \delta(t)e^{-st}dt = e^0 = 1 \tag{4.79}$$

If we calculate the shifted delta function $\delta\left(t - t_0\right)$, we obtain

$$\Delta(s) = \int_{0}^{\infty} \delta\left(t - t_0\right)e^{-st}dt = e^{-st_0} \tag{4.80}$$

Table 4.1 Laplace transform table

Time function $y(t)$	Transform $Y(s)$
$\delta(t)$	1
$u(t)$	$\dfrac{1}{s}$
$e^{\lambda t}$	$\dfrac{1}{s-\lambda}$
$\sin at$	$\dfrac{a}{s^2+a^2}$
$\cos at$	$\dfrac{s}{s^2+a^2}$
t	$\dfrac{1}{s^2}$
$te^{\lambda t}$	$\dfrac{1}{(s-\lambda)^2}$
$\dfrac{e^{\lambda_1 t} - e^{\lambda_2 t}}{\lambda_1 - \lambda_2}$	$\dfrac{1}{(s-\lambda_1)(s-\lambda_2)}$

corresponding to what is prescribed by the time shift theorem (4.70). The unit step function[4] $u(t)$

$$u(t) = \int_{-\infty}^{t} \delta(\tau)\,d\tau = \begin{cases} 1 & \text{for } t > 0 \\ 0 & \text{for } t < 0 \end{cases} \tag{4.81}$$

has the Laplace transform

$$U(s) = \int_{0}^{\infty} u(t)\, e^{-st} dt = \int_{0}^{\infty} e^{-st} dt = \left[\frac{e^{-st}}{-s} \right]_{0}^{-\infty} = \frac{1}{s} \tag{4.82}$$

The Laplace transform of the function $e^{\lambda t}$ that is essential for describing free decays of dynamic systems is found in an equally simple manner as

$$\int_{0}^{\infty} e^{\lambda t} e^{-st} dt = \int_{0}^{\infty} e^{-(s-\lambda)t} dt = \left[\frac{e^{-(s-\lambda)t}}{-(s-\lambda)} \right]_{0}^{-\infty} = \frac{1}{s-\lambda} \tag{4.83}$$

Table 4.1 shows a list of typically used Laplace transforms.

Example 4.3   The derivative of the unit step function

We know that the unit step function is the integral of the Dirac delta function, Eq. (4.81), and the derivative of the unit step function is the Dirac delta function. Therefore, it must be true that

$$\dot{u}(t) \leftrightarrow 1 \tag{4.84}$$

---

[4] The unit step function is also often called the Heaviside function and denoted by $H(t)$, but since we reserve this symbol for the FRF's, we will use symbol $u(t)$ for the unit step function.

Let us verify this by using the differentiation theorem. From the Laplace transform table 4.1 and using the differentiation theorem, we obtain

$$\dot{u}(t) \leftrightarrow s\frac{1}{s} - u(0-) = 1 - u(0-) \tag{4.85}$$

We see clearly that it is important to distinguish between the limit from the left and the value at $t = 0$. The unit step function has an indeterminate value at $t = 0$, whereas the limit from the left for sure is equal to zero, thus using the right interpretation of how to calculate the initial values using the differentiation theorem is essential, and doing it rights leads to the important result given by Eq. (4.84).

Example 4.4   Impulse response function of a SDOF system

The impulse response function $h(t)$ is equal to the response $y(t)$ to the differential equation of a single degree-of-freedom (SDOF) system where the input $x(t)$ (the right-hand side) is equal to the Dirac delta function $\delta(t)$

$$m\ddot{y}(t) + c\dot{y}(t) + ky(t) = \delta(t) \tag{4.86}$$

see Section 5.1.3. Using the differential theorem and the transform of the Dirac delta function, we get the transformed equation

$$ms^2 Y(s) + csY(s) + kY(s) = 1 \tag{4.87}$$

that defines the transfer function that is the Laplace transform of the impulse response function

$$H(s) = Y(s) = \frac{1}{m(s - \lambda_1)(s - \lambda_2)} \tag{4.88}$$

where λ_1, λ_2 are the continuous time poles of the system and the roots to the second-order equation

$$ms^2 + cs + k = 0$$

$$\Downarrow \tag{4.89}$$

$$\lambda = \frac{-c}{2m} \pm \frac{\sqrt{c^2 - 4mk}}{2m} = -\varsigma\omega_0 \pm i\omega_d$$

where ω_0, ω_d is the undamped and the damped natural frequency respectively, see Section 5.1.2. As it appears from Eq. (4.89), the real part of the poles is $\sigma = -\varsigma\omega_0$. Positive damping results in a negative real part, and the stable part of the Laplace domain is the negative half plane $\sigma < 0$.

From the Laplace transform table, we see that the corresponding time function – the impulse response function – is given by

$$h(t) = \frac{1}{m}\frac{e^{\lambda_1 t} - e^{\lambda_2 t}}{\lambda_1 - \lambda_2} \tag{4.90}$$

It is easy to verify that this function is real. Omitting the scaling constant $1/(m(\lambda_1 - \lambda_2))$ we see that the impulse response function has the simple form

$$h(t) \propto e^{\lambda t} - e^{\lambda^* t} \tag{4.91}$$

The scaling constant is important in order to make the function fit the physical units of the involved quantities. See Section 5.1.3 for more details on this subject and on the impulse response function in general.

4.4   The Z-Transform

The Z-transform establishes a relationship between exact analytical solutions in the frequency domain and difference equations in the time domain, and, therefore, this transform plays a central role in the formulation and solution of difference equations in the time domain. Difference equations are equations involving terms of the quantity $y(t)$ at different times like $y(n\Delta t), y((n+1)\Delta t), \ldots$ rather than differentials of the quantity at the same time like $\dot{y}(t), \ddot{y}(t), \ldots$

4.4.1   The Z-Transform as a generalization of the Fourier Series

The Z-transform can be seen as a generalization of the Fourier transform for periodic functions – specifically the complex Fourier series considered in Section 4.1.2. In order to introduce the Z-transform, we take the complex Fourier series in Eq. (4.11) and make a swap between frequency and time so that now we write it for the case where the spectrum is periodic[5]

$$Y(\omega) = \sum_{n=-\infty}^{\infty} y_n e^{-i\omega n \Delta t} \tag{4.92}$$

In this equation, we have changed the sign of the exponent so that it corresponds to the other formulas where we are going from the time domain to frequency domain. The time function is given by its discrete values $y_n = y(n\Delta t)$, where Δt is the sampling time. We denote the discrete time values as $y(n)$ or y_n. The expressions (4.41–4.45) still hold with the same arguments for the sampling time, Nyquist frequency, and so on. Considering again the "damped" version of the discrete time function as we did when introducing the Laplace transform in Section 4.3.1, we have the damped function[6] $y_n'' = e^{-\sigma n \Delta t} y_n$. The corresponding transform is

$$Y''(\omega) = \sum_{n=-\infty}^{\infty} y_n e^{-\sigma n \Delta t} e^{-i\omega n \Delta t} = \sum_{n=-\infty}^{\infty} y_n e^{-(\sigma+i\omega)n\Delta t}$$

$$= \sum_{n=-\infty}^{\infty} y_n e^{-sn\Delta t} \tag{4.93}$$

where $s = \sigma + i\omega$ as before. Realizing that $e^{-sn\Delta t} = \left(e^{s\Delta t}\right)^{-n}$ the transform is now defined as a function of the variable z

$$Y''(z) = \sum_{n=-\infty}^{\infty} y_n z^{-n} \tag{4.94}$$

where

$$z = e^{s\Delta t} \tag{4.95}$$

This is called the bilateral or two-sided Z-transform. If we assume that $y_n = 0$ for $n < 0$, then we obtain the unilateral or one-sided Z-transform

$$Y''(z) = \sum_{n=0}^{\infty} y_n z^{-n} \tag{4.96}$$

When dealing with the Z-transform, it is common to use the dimensionless angular frequency

$$w = \omega \Delta t \tag{4.97}$$

---

[5] Just like a periodic time function results in a discrete spectrum, a periodic spectrum results in a discrete time function.
[6] The double mark just indicate that the function has been modified by the damping term.

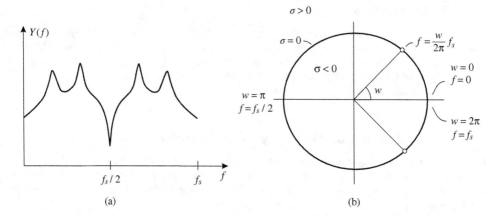

Figure 4.6 The Z-transform can be considered a result of letting the frequency axis (that normally goes from $-f_s/2$ to $f_s/2$ in Fourier transform theory) go from 0 to f_s (a) and then map the frequency axis onto the unit circle, (corresponding to $\sigma = 0$), (b). This automatically introduces the periodic spectral density that is a result of the discrete time function; the stable solutions $\sigma < 0$ are mapped into the unit circle and the unstable solutions $\sigma > 0$ are mapped outside of the unit circle

and to use

$$r = e^{\sigma \Delta t} \tag{4.98}$$

Thus

$$z = e^{s \Delta t} = r e^{iw} \tag{4.99}$$

Let us now consider the frequency variables of the Laplace transform $s = \sigma + i\omega$, and the corresponding Z-transform $z = re^{iw}$. As mentioned earlier (Example 4.4), the region in the s-plane for stable signals is the half plane $\sigma < 0$. For the Z-transform variable $z = re^{iw}$, the damping influence is omitted and mapped onto the unit circle $z = e^{i\omega}$ and the stable region in the s-plane $\sigma < 0$ is mapped into the unit circle. On the unit circle, the Z-transform becomes the one-sided Fourier transform, from Eq. (4.92)

$$Y(\omega) = \sum_{n=0}^{\infty} y_n e^{-i\omega n \Delta t} = \sum_{n=0}^{\infty} y_n e^{-iwn} \tag{4.100}$$

The value of the dimensionless variable $w = 0$ corresponds to $\omega = f = 0$. In this case taking the Nyquist bandwidth going from 0 to f_s (instead of from $-f_s/2$ to $f_s/2$ as we did before), we can imagine that the frequency axis is mapped onto the unit circle in the z-plane, see Figure 4.6, where the angle $w = \pi$ corresponds to the angular frequency $\omega = \pi/\Delta t = 2\pi f_s/2$, thus, the oscillation frequency is $f = f_s/2$, similarly, the total angle $w = 2\pi$ corresponds to $\omega = 2\pi f_s, f = f_s$ and finally an arbitrary angle w corresponds to

$$\omega = w f_s$$
$$f = \frac{w}{2\pi} f_s \tag{4.101}$$

As for the Fourier transform and Laplace transform, we say that the functions y_n and $Y''(z)$ constitute a transform pair, and we use the same notation as for the other transforms

$$y_n \leftrightarrow Y''(z) \tag{4.102}$$

Or we will omit the double mark on the Z-transform to indicate that we are dealing are the "damped" version and just write

$$y_n \leftrightarrow Y(z) \tag{4.103}$$

In order not to confuse the Z-transform with the corresponding Fourier or Laplace transform, we will normally always use the variable z to indicate the Z-transform, the variable ω to indicate the Fourier transform, and the variable s to indicate the Laplace transform. Thus, $Y(z)$ is the Z-transform, $Y(s)$ is the Laplace transform, and $Y(\omega)$ is the Fourier transform.

4.4.2   Z-Transform Properties

For the Z-transform, similar properties exist as for the Fourier and Laplace transforms. It is easy to verify the linearity property

$$c_1 y_1(n) + c_2 y_2(n) \leftrightarrow c_1 Y_1(z) + c_2 Y_2(z) \tag{4.104}$$

The time reversal property

$$y(-n) \leftrightarrow Y(1/z) \tag{4.105}$$

and the time shift property

$$y(n-m) \leftrightarrow z^{-m} Y(z) \tag{4.106}$$

for $m > 0$ where the resulting function must vanish for negative arguments, thus $y(n-m) = 0$ for $n < m$. In the Z-transform theory since we are only dealing with the positive time axis, we define the convolution as

$$h(n) * g(n) = \sum_{m=0}^{\infty} h(n-m) g(m) = \sum_{m=0}^{\infty} g(n-m) h(m) \tag{4.107}$$

and the convolution property can be proved

$$h(n) * g(n) \leftrightarrow H(z) G(z) \tag{4.108}$$

Since we are dealing with discrete time the integration and differentiation theorems do not exist in the Z-transform theory. There exists a theorem for differentiation in the Z-domain, but we shall not need this property. As we shall see in the subsequent example, the differentiation theorem that was central for the Laplace transform is now replaced by the time shift property given by Eq. (4.106).

4.4.3   Some Z-Transforms

As it appears directly from the definition of the Z-transform (4.96), the discrete delta function

$$\delta(n) = \begin{cases} 1, n = 0 \\ 0, n \neq 0 \end{cases} \tag{4.109}$$

has the simple Z-transform

$$\delta(n) \leftrightarrow \sum_{n=0}^{\infty} \delta(n) z^{-n} = 1 \tag{4.110}$$

The discrete unit step function

$$u(n) = \begin{cases} 1, n \geq 0 \\ 0, n < 0 \end{cases} \tag{4.111}$$

is a good example of the fact that usually the infinite series given by the Z-transform can be expressed as a simple rational function. Using Eq. (4.96) we get

$$U(z) = \sum_{n=0}^{\infty} z^{-n} \tag{4.112}$$

which is a geometric series. Considering the same series with the upper limit $N - 1$

$$U(z) = \sum_{n=0}^{N-1} z^{-n} = 1 + z^{-1} + z^{-2} + \cdots + z^{-(N-1)} \Rightarrow$$

$$\frac{1}{z} U(z) = z^{-1} + z^{-2} + \cdots + z^{-(N-1)} + z^{-N} \Rightarrow$$

$$\left(1 - \frac{1}{z}\right) U(z) = 1 - z^{-N} \Rightarrow \tag{4.113}$$

$$U(z) = \frac{1 - z^{-N}}{1 - \frac{1}{z}}$$

And now assuming that z is inside the unit circle, that is $|z| < 1$ so that $z^{-N} \to 0$ for $N \to \infty$ then for the Z-transform (4.112) we obtain

$$U(z) = \frac{1}{1 - \frac{1}{z}} = \frac{z}{z - 1} \tag{4.114}$$

Finally let us consider the exponential decay

$$y(n) = e^{\lambda n \Delta t} = \left(e^{\lambda \Delta t}\right)^n = \mu^n; \quad \mu = e^{\lambda \Delta t} \tag{4.115}$$

Then using the same idea as for the aforementioned geometric series

$$Y(z) = \sum_{n=0}^{\infty} \mu^n z^{-n} = \sum_{n=0}^{\infty} (z/\mu)^{-n}$$

$$= \frac{z/\mu}{z/\mu - 1} = \frac{z}{z - \mu} \tag{4.116}$$

The Z-transform of a sine signal can be found by specializing the general continuous pole to the case where the real part (the damping) is zero $\lambda = i\omega$ and then using the linearity property and Euler's formula

$$\sin(\omega n \Delta t) = \frac{1}{2i} \left(e^{i\omega n \Delta t} - e^{-i\omega n \Delta t}\right)$$

$$= \frac{1}{2i} (\mu^n - \mu^{*n}) \leftrightarrow \frac{1}{2i} \left(\frac{z}{z - \mu} - \frac{z}{z - \mu^*}\right) \tag{4.117}$$

Similarly we find for a cosine signal

$$\cos(\omega n \Delta t) = \frac{1}{2} \left(e^{i\omega n \Delta t} + e^{-i\omega n \Delta t}\right)$$

$$= \frac{1}{2} (\mu^n + \mu^{*n}) \leftrightarrow \frac{1}{2} \left(\frac{z}{z - \mu} + \frac{z}{z - \mu^*}\right) \tag{4.118}$$

Table 4.2 gathers the above derived Z-transforms, often used in practice.

Table 4.2 Z-transform table

Time function $y(t)$	Transform $Y(z)$
$\delta(n)$	1
$u(n)$	$\dfrac{z}{z-1}$
$e^{\lambda n \Delta t} = \mu^n; \quad \mu = e^{\lambda \Delta t}$	$\dfrac{z}{z-\mu}$
$\sin(\omega n \Delta t)$	$\dfrac{1}{2i}\left(\dfrac{z}{z-\mu} - \dfrac{z}{z-\mu^*}\right); \quad \mu = e^{i\omega\Delta t}$
$\cos(\omega n \Delta t)$	$\dfrac{1}{2}\left(\dfrac{z}{z-\mu} + \dfrac{z}{z-\mu^*}\right); \quad \mu = e^{i\omega\Delta t}$

4.4.4 Difference Equations and Transfer Function

In signal processing and modal identification, it is common to work with difference equations of the kind

$$y(n) = \sum_{k=1}^{N} a_k y(n-k) + \sum_{k=0}^{M} b_k x(n-k) \tag{4.119}$$

where $x(n)$ is the system input and $y(n)$ is the system response. The coefficients a_k are called the autoregressive (AR) part of the model since the response at time n is modeled as a regression on the past response; the coefficients b_k are called the moving average (MA) part of the model. A difference equation like this is therefore often called an ARMA model. In some presentations of this subject, there is often a minus on the AR part of the right-hand side of Eq. (4.119) because it is common to write the equation in the form where the AR is placed on the left-hand side of the equation. Also some authors prefer to include $y(n)$ in the AR sum.

By taking the Z-transform of the difference equation using the time shift property (4.106), we obtain

$$Y(z) = \sum_{k=1}^{N} a_k z^{-k} Y(z) + \sum_{k=0}^{M} b_k z^{-k} X(z) \Rightarrow$$

$$\left(1 - \sum_{k=1}^{N} a_k z^{-k}\right) Y(z) = \sum_{k=0}^{M} b_k z^{-k} X(z) \tag{4.120}$$

and the transfer function can then be found as

$$H''(z) = \frac{Y(z)}{X(z)} = \frac{\sum_{k=0}^{M} b_k z^{-k}}{1 - \sum_{k=1}^{N} a_k z^{-k}} \tag{4.121}$$

As it appears the MA part turn into the numerator polynomial and the AR part turn into the denominator polynomial of the transfer function. The corresponding frequency response function is then found by evaluating the transfer function on the unit circle $z = e^{iw} = e^{i\omega\Delta t}$

$$H(\omega) = H''\left(e^{i\omega\Delta t}\right) = \frac{\sum_{k=0}^{M} b_k e^{-i\omega\Delta t k}}{1 - \sum_{k=1}^{N} a_k e^{-i\omega\Delta t k}} \tag{4.122}$$

4.4.5 Poles and Zeros

One of the useful aspects of the use of the Z-transform for signal analysis is the description of a system in terms of the so-called poles and zeros of the system. The zeros of a transfer function $H(z)$ are the values of the variable z for which the numerator polynomial is equal to zero. Similarly, the poles of $H(z)$ are the roots of the polynomial on the denominator. When the denominator is equal to zero, then the transfer function has a singularity. The poles and zeros are found by rewriting (4.121) in order to avoid negative powers of the variable z in the numerator and denominator polynomials

$$H''(z) = \frac{b_0 + b_1 z^{-1} + b_2 z^{-2} + \cdots + b_M z^{-M}}{1 - a_1 z^{-1} - a_2 z^{-2} - \cdots - a_N z^{-N}}$$

$$= \frac{b_0 z^{-M}}{z^{-N}} \frac{z^M + (b_1/b_0) z^{M-1} + (b_2/b_0) z^{M-2} + \cdots + (b_M/b_0)}{z^N - a_1 z^{N-1} - a_2 z^{N-2} - \cdots - a_N} \tag{4.123}$$

The constant b_0 is called the gain factor of the transfer function. Denoting the zeros and the poles as z_k and p_k, respectively, the numerator and denominator polynomials can be factorized and the resulting expression can be written as

$$H''(z) = b_0 z^{N-M} \frac{(z - z_1)(z - z_2) \cdots (z - z_M)}{(z - p_1)(z - p_2) \cdots (z - p_N)} \tag{4.124}$$

This equation shows that the transfer function has M zeros and N poles. The poles are associated with the AR coefficients of the system equation (physical properties of the system), and the zeros are associated with the MA part of the system equation that is often said to describe the statistical properties of the system (see the description of the ARMA model in Sections 5.3.5 and 9.3).

Example 4.5 Discrete impulse response function of an single degree-of-freedom (SDOF) system

In this example, we will find the discrete representation of the continuous time impulse response function that we found in Example 4.4 and we will establish the corresponding difference equation. Using a sampling time step of Δt, we can sample the continuous time impulse response function from Eq. (4.91) and obtain

$$h(n) = \frac{1}{m} \frac{e^{\lambda n \Delta t} - e^{\lambda^* n \Delta t}}{\lambda - \lambda^*} \tag{4.125}$$

Now, using the discrete pole

$$\mu = e^{\lambda \Delta t} \iff \lambda \Delta t = \ln(\mu) \tag{4.126}$$

we get the sampled impulse response expressed in discrete properties

$$h(n) = \frac{\Delta t}{m} \frac{\mu^n - \mu^{*n}}{\ln \mu - \ln \mu^*} = C(\mu^n - \mu^{*n}) \tag{4.127}$$

where

$$C = \frac{\Delta t}{m(\ln \mu - \ln \mu^*)} \tag{4.128}$$

Using the Z-transform table and the linearity property, we obtain the transfer function

$$H(z) = C\left(\frac{z}{z - \mu} - \frac{z}{z - \mu^*}\right) \tag{4.129}$$

This equation can be re-arranged to give

$$H(z) = C \; \frac{\mu - \mu^*}{z - \mu - \mu^* + \mu\mu^* \frac{1}{z}}$$

(4.130)

which, upon using the properties of the discrete delta function given by Eq. (4.109) defines the difference equation

$$y(n+1) + a_1 y(n) + a_2 y(n-1) = C\delta(n)$$

$$a_1 = -\mu - \mu^*; \quad a_2 = \mu\mu^*$$

(4.131)

This difference equation has exactly the same physical properties as the differential equation considered in Example 4.4. This is easily checked by finding one of the discrete poles of the difference equation

$$p = \frac{a_1 \pm \sqrt{a_1^2 - 4a_2}}{2}$$

(4.132)

and plotting the discrete impulse response function

$$h(n) = C(p^n + p^{*n})$$

(4.133)

Based on the present analysis, we cannot be sure about the constant C because the poles of the difference equation (4.131) are independent of the constant C. However, following the same analysis that was performed when scaling the continuous time impulse response function, it can be shown[7] that the right discrete impulse response function is given by Eq. (4.133) where the constant C is given by Eq. (4.128).

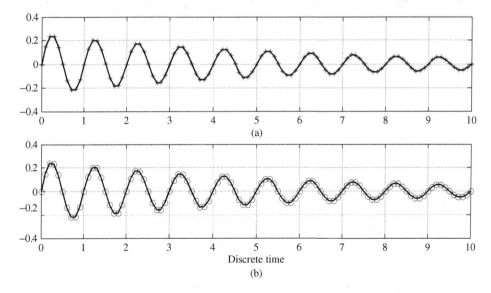

Figure 4.7 Plots of the different versions of the impulse response of the considered SDOF system. (a) The continuous time function in full line together with the sampled version of the continuous time impulse response given by Eq. (4.127) where each sample point is marked by "*." (b) The continuous time function in full line together with the discrete impulse response given by Eq. (4.133) where each sample point is marked by "o"

---

[7] See Section 5.1.3.

The plots are shown in Figure 4.7 for the following physical properties of the continuous Eq. (4.86)

$$m = 1.5\,\text{kg}$$
$$c = 0.2\,\text{Ns/m} \tag{4.134}$$
$$k = 10\,\text{N/m}$$

showing 10 periods with 10 samples per period, and we have $\Delta t = T/10$ where T is the undamped period of the SDOF system.

Example 4.6   Discrete poles of an SDOF system

The continuous time poles of an SDOF system is given by Eq. (4.89)

$$\lambda = \frac{-c}{2m} \pm \frac{\sqrt{c^2 - 4mk}}{2m} = -\varsigma\omega_0 \pm i\omega_d \tag{4.135}$$

The discrete poles are given by Eq. (4.126)

$$\mu = e^{\lambda\Delta t} = e^{-\varsigma\omega_0\Delta t}e^{i\omega_d\Delta t} \tag{4.136}$$

The angle and the absolute value of the discrete pole according to Eq. (4.99) are then given by

$$w = \omega_d\Delta t$$
$$r = e^{-\varsigma\omega_0\Delta t} \tag{4.137}$$

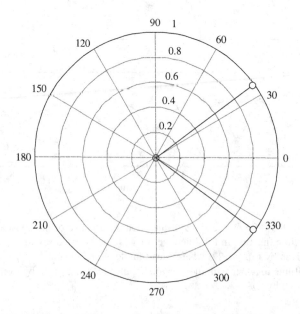

Figure 4.8   Plots of the two discrete poles of the considered SDOF system in the z-plane

For the system considered in the previous example, we obtain

$$\omega_0 = 2.58 \text{ rad/s}$$
$$\varsigma = 0.0258 = 2.58\%$$
$$\omega_d = 2.58 \text{ rad/s} \tag{4.138}$$
$$w = 0.628 \text{ rad} = 36.0 \text{ deg}$$
$$r = 0.984$$

and the discrete poles are

$$\mu_1 = 0.796 + 0.578i$$
$$\mu_2 = 0.796 - 0.578i \tag{4.139}$$

The positions of the two discrete poles in the z-plane are plotted in Figure 4.8. A plot like this is often denoted as a pole plot, and if it contains also the zeroes of the transfer function it is denoted a pole/zero plot.

References

[1] Press, W.H., Flannery, B.P, Teukolsky, S.A. and Vetterling, W.T.: *Numerical recipes in C*. Cambridge, Cambridge University Press, 1989.
[2] Brigham, E.O.: *The fast Fourier transform*. New Jersey, Prentice Hall, Inc. 1974.
[3] Doetch, G.: *Guide to the applications of the Laplace transform*. New Jersey, D. Van Nostrand Company, Ltd., 1961.
[4] Irwin, J. and Mullineux, N.: *Mathematics in physics and engineering*. New York, Academic Press, 1959.
[5] James, G.: *Advanced modern engineering mathematics*. 4th edition. Harlow, Pearson, Prentice-Hall, 2011.
[6] Proakis, G.P. and Manolakis, D.G.: *Digital signal processing. Principles, algorithms, and applications*. 4th edition. Harlow, Pearson, Prentice-Hall, 2007.
[7] Shannon, E.: *Communication in the presence of noise*. Proc. Ins. Radio Eng. V. 37, No. 1, p. 10–21, 1949. Reprint as classic paper in: Proc. IEEE, vol. 86, no. 2, Feb. 1998.

5

Classical Dynamics

"Proportional damping is like suggesting that all men are linear combinations of Robert Redford and George Clooney"

– Sam Ibrahim

In this chapter, we will provide a general overview of the fundamental concepts of deterministic structural dynamics of discrete systems. This constitutes the underlying modeling of all modal identification of discrete linear multi-degree-of-freedom (MDOF) systems.

We will follow the classical way of presenting the deterministic dynamic models. To this end, we will first consider the simplest case, the single-degree-of-freedom system (SDOF) and discuss about its properties and how to evaluate its response in the time and the frequency domain. Then the MDOF case will be considered in relation to the type of damping in the system.

We will limit our discussion to systems with viscous damping. Most of OMA techniques have been developed for the analysis of signals under the assumption that viscous damping is an acceptable representation of the energy dissipation of the measured system, although it should be recognized that the capacity of a physical system to dissipate energy is a more complex process than the simplistic assumption that energy is dissipated through viscous damping. It is expected that future developments of OMA techniques will incorporate the ability to identify other types of damping, but for the present time, we will focus our efforts in understanding the theory behind OMA under the viscous damping assumption.

For the treatment of MDOF systems, we will discuss first the behavior of undamped systems so that concepts such as modal frequency (or modal period), modal mass, modal stiffness, and mode shapes of the undamped system can be introduced and explained. Next, we will extend the treatment to MDOF systems with proportional (or classical) damping and discuss concepts such that modal damping and the relationship between the mode shapes of the undamped system and those of the proportionally damped system. And finally, the case of general damping will be considered where concepts such as the complex mode shapes are introduced by solving the equation of motion using a state space formulation.

The presentation of the dynamic models is inspired by classical texts on the subject, like Newland's book on classical dynamics [1], and the books by Thomson [2], and Chopra [3]. This step-by-step presentation will help the readers comprehend better the basic ideas of modeling and modal identification using a physically motivated approach. The authors recognize that there are more formal mathematical ways to introduce these concepts, and readers who would prefer such type of presentation can read the presentation of the same material in Appendix C, where the dynamic modeling is presented using a state space formulation from the very beginning.

Introduction to Operational Modal Analysis, First Edition. Rune Brincker and Carlos Ventura.
© 2015 John Wiley & Sons, Ltd. Published 2015 by John Wiley & Sons, Ltd.
Companion Website: www.wiley.com/go/brincker

At the end of the chapter, some topics of importance for the application of OMA are presented. These include the mathematical concepts related to structural modification theory, sensitivity equations, especially the sensitivity of closely spaced modes, model reduction, and discrete time representations.

5.1 Single Degree of Freedom System

5.1.1 Basic Equation

We will start with the formulation of the dynamic equation of equilibrium of a simple SDOF system subjected to an external force that varies with time. Figure 5.1 shows a simple linear mass–spring–damper system subjected to an external force that changes with time. The mass of the block is m, the spring has a stiffness constant k, the damper has a damping constant c, and the amplitude of the force at any time t is F_x. In order to formulate the equation of dynamic equilibrium for this SDOF system, we need to account for the inertia force resulting from the motion of the mass of the SDOF, the restoring force associated with the spring that prevents this mass from moving freely, and the force associated with the dissipation of energy provided by the damper in this SDOF system.

Let us start first with the treatment of the inertia force. One of the most important contributions to structural dynamics was made by Isaac Newton (1643–1727) when he formulated his Second Law of Motion. The original Latin text reads "*Lex II: Mutationem motus proportionalem esse vi motrici impressae, et fieri secundum lineam rectam qua vis illa imprimatur*," which has been translated into English as "Law II: The change in momentum of a body is proportional to the impulse impressed on the body and happens along the straight line on which that impulse is impressed." If the mass of the body remains constant, which is the case for the type of systems that we are dealing with in this book, then the change of momentum in nothing more than the product of the mass times the acceleration of the body, and a corollary of this Law is the well-known relationship between the total force, F_{tot}, the mass, m, and the acceleration of the body ($\ddot{y}$)

$$F_{tot} = ma = m\ddot{y} \tag{5.1}$$

where y is the coordinate describing the location of the mass; this coordinate also defines the positive direction for the translational problem, see Figure 5.1. An equally important contribution was made by Robert Hooke (1635–1703) formulating the empirical "law" of elasticity as "*Ut tensio, sic vis*," meaning "as the extension, so is the force." When expressed in mathematical terms, this law indicates that the restoring force in a spring is proportional to the extension (displacement) of that spring according to this relationship

$$F_s = ky \tag{5.2}$$

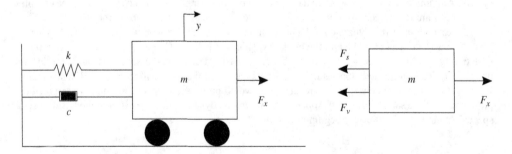

Figure 5.1 Mechanical representation of a SDOF system and corresponding free-body-diagram

where the force F_s is acting opposite the direction of motion. All bodies experiencing vibration dissipate energy in some manner. Dissipation of energy is a complex process and incorporating this term in the equation of dynamic equilibrium of a simple system is not an easy task. So we will resort to a simple formulation of the dissipation of energy in the system by making the rough assumption that the dissipation of energy can be represented in terms of a viscous damping force that restrains the motion

$$F_v = cv = c\dot{y} \tag{5.3}$$

where also the force F_v is acting opposite the direction of motion. The free body diagram in Figure 5.1b shows that the acting forces on the body are F_s, F_v and the external force F_x so that $F_{tot} = F_x - F_s - F_v$. By gathering all terms in Eq. (5.1), we arrive at the well-known basic equation in structural dynamics for a single degree-of-freedom system

$$m\ddot{y} + c\dot{y} + ky = F_x \tag{5.4}$$

in which it is understood that F_x and y and its derivatives are functions of the time.

We can rewrite this equation in a more generic form where $x(t)$ is the input and $y(t)$ is the output

$$m\ddot{y}(t) + c\dot{y}(t) + ky(t) = x(t) \tag{5.5}$$

In the following sections, we explore some important aspects of this equation and explain how they are related to the OMA concepts.

5.1.2 Free Decays

The free decays represent the possible solutions to Eq. (5.5) for the free vibration case, where the right-hand side is set to zero $x(t) = 0$. There are many ways to derive the expression for the free decays, but we will use here the Laplace transform approach. So, if we take the Laplace transform of both sides of Eq. (5.5) for the case of $x(t) = 0$, one obtains the equation

$$\left(ms^2 + cs + k\right) Y(s) = 0 \tag{5.6}$$

We are not interested in the trivial solutions corresponding to $Y(s) = 0$ but we are looking for the values of s that make the second-order polynomial equation equal to zero; that is

$$ms^2 + cs + k = 0 \tag{5.7}$$

The solution to this equation for the case of lightly damped systems has complex conjugate roots

$$\lambda = \frac{-c \pm i\sqrt{4mk - c^2}}{2m} \tag{5.8}$$

The roots – also denoted the continuous time poles of the system – describe the physical properties of the system. A further simplification of the aforementioned expression can be achieved by combining the mass, stiffness, and damping values into two parameters that also describe the physical characteristics of the system, but in terms of the ratios between the mass, stiffness, and damping. To this end, we define, respectively, the undamped angular natural frequency of vibration ω_0 (or simply the undamped natural frequency) and the damping ratio ς of the system as

$$\omega_0 = \sqrt{\frac{k}{m}}; \quad \varsigma = \frac{c}{2\sqrt{mk}} \tag{5.9}$$

The undamped natural frequency depends only on the ratio between the mass and the stiffness of the system, and is independent of the damping of the system, while the damping ratio depends on the ratio between the damping constant of the system and the corresponding mass and stiffness.

The solution presented in Eq. (5.8) reduces to

$$\lambda = -\varsigma\omega_0 + i\omega_0\sqrt{1-\varsigma^2}$$
$$\lambda^* = -\varsigma\omega_0 - i\omega_0\sqrt{1-\varsigma^2} \tag{5.10}$$

The quantity in the imaginary term

$$\omega_d = \omega_0\sqrt{1-\varsigma^2} \tag{5.11}$$

is called the damped natural frequency. Using this terminology, the pole equations are rewritten as

$$\lambda = -\varsigma\omega_0 + i\omega_d$$
$$\lambda^* = -\varsigma\omega_0 - i\omega_d \tag{5.12}$$

If we know the poles of the system, then the angular frequency and the damping ratio can be found from

$$\omega_0 = \sqrt{\lambda\lambda^*} = |\lambda|$$
$$\varsigma = -\frac{\text{Re}\,(\lambda)}{\omega_0} \tag{5.13}$$

Any free decay of the system can be found as the complete solution to the homogenous equation

$$y(t) = C_1 e^{\lambda t} + C_2 e^{\lambda^* t} \tag{5.14}$$

where the general complex constants C_1, C_2 can be determined from the initial conditions of motion of the SDOF system. However, in order to ensure that the solution given by Eq. (5.14) is real, we must limit the constants to be a complex conjugate pair, which is the case for undamped and underdamped systems; that is, $\varsigma < 1$

$$C_1 = C_2^* \tag{5.15}$$

which correspond to

$$C_1 = Re^{i\phi}; \quad C_2 = Re^{-i\phi} \tag{5.16}$$

where the real-valued quantity R is the magnitude of the complex numbers C_1, C_2 and the real-valued quantity ϕ is the corresponding phase angle between the real and imaginary parts of C_1. In a complex plane representation, C_1, C_2 are two vectors, and R will represent the distance from the origin of the plane to the end of each vector, and ϕ is the angle from the real axis to the complex number C_1 (the phase). Now, combining Eq. (5.14) with Eqs. (5.11/5.16), we obtain

$$y(t) = Re^{-\varsigma\omega_0 t}\left(e^{i\phi}e^{i\omega_d t} + e^{-i\phi}e^{-i\omega_d t}\right)$$
$$= Re^{-\varsigma\omega_0 t}\left(e^{i(\omega_d t+\phi)} + e^{-i(\omega_d t+\phi)}\right) \tag{5.17}$$
$$= 2Re^{-\varsigma\omega_0 t}\cos\left(\omega_d t + \phi\right)$$

The midterm $e^{-\varsigma\omega_0 t}$ is the damping term describing how fast the oscillatory motion dies out, often called the envelope function, and the last term is a harmonic term describing frequency and phase of the oscillation. The initial displacement of the system at time $t = 0$ is $2R\cos\phi$.

The damped period T_d is defined as the time it takes the system to complete one full oscillation, and it is related to the damped natural frequency of the system by the following relationship:

$$\omega_d T_d = 2\pi \quad \Rightarrow \quad T_d = \frac{2\pi}{\omega_d} \quad \Rightarrow \quad \omega_d = \frac{2\pi}{T_d} \tag{5.18}$$

The units of the angular natural frequency (undamped or damped) of the system are rad/sec, but it is customary to express this quantity in another unit. The alternative undamped frequency f_0 and damped frequency f_d are given by

$$f_0 = \frac{1}{T_0} = \frac{\omega_0}{2\pi}; \quad f_d = \frac{1}{T_d} = \frac{\omega_d}{2\pi} \tag{5.19}$$

that define the frequency in terms of cycles per second (cps) – the unit that we denote Hz. For each single oscillation, using Eq. (5.17), we obtain that the amplitude of motion is reduced by the factor

$$\frac{y(t+T_d)}{y(t)} = \frac{2R\ e^{-\varsigma\omega_0(t+T_d)}\cos\left(\omega_d t + \phi\right)}{2R\ e^{-\varsigma\omega_0 t}\cos\left(\omega_d t + \phi\right)} = e^{-\omega\varsigma T_d} \tag{5.20}$$

where the damping quantity

$$\delta = \omega\varsigma T_d = 2\pi\frac{\varsigma}{\sqrt{1-\varsigma^2}} \tag{5.21}$$

is known as the logarithmic decrement. This way of presenting the rate of decay of motion from one cycle to the next can be shown to correspond to the natural logarithm of the ratio between the amplitude of motion of the system at any time (t) and the amplitude after one full cycle of response $(t+T_d)$.

Example 5.1   Free decay with zero initial velocity

We consider the case with the initial conditions

$$y(0) = y_0; \quad \dot{y}(0) = 0 \tag{5.22}$$

From Eq. (5.14), we obtain

$$y_0 = C_1 + C_2; \quad 0 = C_1\lambda + C_2\lambda^* \tag{5.23}$$

That gives the coefficients

$$C_1 = -\frac{y_0\lambda^*}{\lambda - \lambda^*}$$
$$C_2 = \frac{y_0\lambda}{\lambda - \lambda^*} \tag{5.24}$$

and we have the solution

$$y(t) = \frac{y_0}{\lambda - \lambda^*}\left(\lambda e^{\lambda^* t} - \lambda^* e^{\lambda t}\right) \tag{5.25}$$

Using Eq. (5.17), it can be shown that the free decay for the case of zero velocity can be reduced to

$$y(t) = \frac{y_0}{2i\omega_d}e^{-\varsigma\omega_0 t}\left(\left(-\varsigma\omega_0 + i\omega_d\right)e^{-i\omega_d t} - \left(-\varsigma\omega_0 - i\omega_d\right)e^{i\omega_d t}\right)$$
$$= y_0 e^{-\varsigma\omega_0 t}\cos\left(\omega_d t + \phi\right)/\cos\left(\phi\right) \tag{5.26}$$

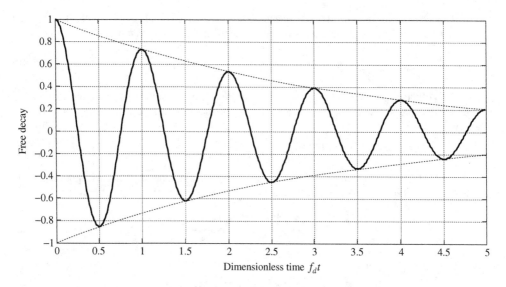

Figure 5.2   Free decay with zero initial velocity corresponding to Eqs. (5.25/5.26). The plot is made for unit initial amplitude, damping ratio of 5% and is plotted in dimensionless time $f_d t$ equal to the number of periods. The envelope function $\pm e^{-\varsigma \omega_0 t}$ is indicated by the dashed line

where the phase angle is approximately equal to $\phi = -\varsigma$. For the case of zero velocity, the phase is small and the response of the system as a function of time normalized with respect to the period of the system is presented in Figure 5.2.

Example 5.2   Free decay with zero initial displacement

We consider the case with the initial conditions

$$y(0) = 0; \quad \dot{y}(0) = v_0 \tag{5.27}$$

From Eq. (5.14), we obtain

$$0 = C_1 + C_2; \quad v_0 = C_1 \lambda + C_2 \lambda^* \tag{5.28}$$

That gives the coefficients

$$C_1 = \frac{v_0}{\lambda - \lambda^*}$$

$$C_2 = -\frac{v_0}{\lambda - \lambda^*} \tag{5.29}$$

and the solution for this case is

$$y(t) = v_0 \frac{e^{\lambda t} - e^{\lambda^* t}}{\lambda - \lambda^*} \tag{5.30}$$

Using Eq. (5.17), we see that the free decay for zero initial displacement reduces to

$$y(t) = v_0 \left(e^{(-\varsigma \omega_0 + i \omega_d)t} - e^{(-\varsigma \omega_0 - i \omega_d)t} \right) / (2i \omega_d)$$

$$= v_0 e^{-\varsigma \omega t} \left(\cos \omega_d t + i \sin \omega_d t - \cos \omega_d t + i \sin \omega_d t \right) / \left(2i\omega_d \right) \qquad (5.31)$$

$$= \frac{v_0}{\omega_d} e^{-\varsigma \omega_0 t} \sin \omega_d t = \frac{v_0}{\omega_d} e^{-\varsigma \omega_0 t} \cos \left(\omega_d t + \pi/2 \right)$$

For the case of zero initial displacement, the initial amplitude is v_0/ω_d and the phase is $\pi/2$ the response time history is shown in Figure 5.3.

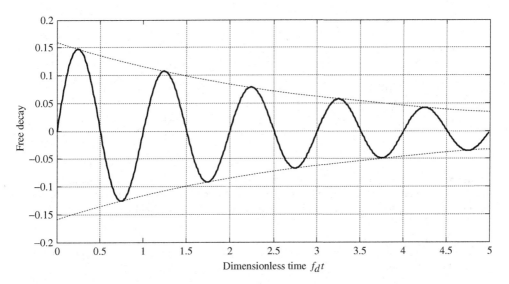

Figure 5.3 Free decay with zero initial displacement corresponding to Eqs. (5.30/5.31). The plot is made for unit initial velocity, a damping ratio of 5% and is plotted in dimensionless time $f_d t$ equal to the number of periods. The envelope function $\pm e^{-\varsigma \omega_0 t}$ is indicated by the dashed line

5.1.3   Impulse Response Function

Consider a SDOF system initially at rest, that is, the displacement and velocity of the system at $t = 0$ are both equal to zero. The impulse response function (IRF) of the system, generally denoted as $h(t)$ is the response when the system, is disturbed by an impulsive excitation of very short duration at time $t = 0$. This very short impulsive excitation is represented in terms of the Dirac delta function $\delta(t)$; see Section 4.3.3. We note that since the system is time invariant, if we shift the time of the impulse by the time shift τ, we just get the response $y(t)$ shifted accordingly, and a pulse $\delta(t - \tau)$ produces the response $h(t - \tau)$. This is illustrated on the left side plots of Figure 5.4, where the top figure represents the instant in which the impulsive force is applied and the bottom figure shows how the system responds to this impulse.

By definition, an impulse is the result of the product between the intensity of the force and the time this force is acting on the system. As the duration of the pulse tends toward zero, we use the Dirac delta function to describe the effect of finite force acting over a very short duration of time.

If we consider the continuous input $x(t)$ as a superposition of a train of impulses of various amplitudes but of very short duration each, the impulse of the force from the time τ to $\tau + d\tau$ is then $x(\tau) d\tau$. The response due to this pulse is $h(t - \tau)x(\tau) d\tau$ (see the right-hand plots of Figure 5.4). Since the system is linear, the principle of superposition applies and the total response can be expressed as the sum (the

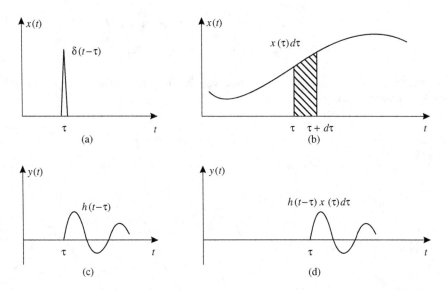

Figure 5.4 (a) Shows a Dirac delta impulse applied at $t = \tau$ and (c) shows the corresponding response, given by the shifted IRF $h\,(t - \tau)$. (b) Shows the impulse of a continuously acting force in the infinitesimal time interval from τ to $\tau + d\tau$ and (d) shows the corresponding scaled impulse response

integral) of all individual responses. The response to the continuous input $x\,(t)$ is then written in the form of Duhamel's integral as

$$y\,(t) = \int_{-\infty}^{\infty} h\,(t - \tau)\,x\,(\tau)\,d\tau \qquad (5.32)$$

This integral is nothing more than the convolution $y\,(t) = h\,(t) * x\,(t)$ between the IRF $h\,(t)$ and the input $x\,(t)$. The convolution operation is commutative, so

$$y\,(t) = h\,(t) * x\,(t) = x\,(t) * h\,(t) \qquad (5.33)$$

The integral in Eq. (5.32) can be expressed in different ways without changing the result

$$
\begin{aligned}
y\,(t) &= \int_{-\infty}^{\infty} h\,(t - \tau)\,x\,(\tau)\,d\tau \\
&= \int_{0}^{\infty} h\,(t - \tau)\,x\,(\tau)\,d\tau \qquad (5.34) \\
&= \int_{-\infty}^{t} h\,(t - \tau)\,x\,(\tau)\,d\tau
\end{aligned}
$$

The reason for the second form of the convolution integral is that we normally assume that the input is zero for negative values of time, thus $x\,(t) = 0$ for $t < 0$, and the reason for the last form of the convolution integral is that $h\,(t) = 0$ for $t < 0$. Using the calculation rules of the Dirac delta function, it is easy to check that if we use $x\,(t) = \delta\,(t)$ in Eq. (5.32) then in fact we obtain the response $y\,(t) = h\,(t)$.

Since the IRF is the response to a Dirac delta input, we can rewrite the differential equation Eq. (5.5) for an impulse of unit amplitude as

$$m\ddot{h}(t) + c\dot{h}(t) + kh(t) = \delta(t) \tag{5.35}$$

Taking the Laplace transform of both sides of the equation, $h(t) \leftrightarrow H(s)$ and using the properties of the delta function from Table 4.1, we get

$$\left(ms^2 + cs + k\right) H(s) = 1 \tag{5.36}$$

If we use the poles of the system described in the previous section, we get the following expression:

$$H(s) = \frac{1}{m(s - \lambda)(s - \lambda^*)} \tag{5.37}$$

We can also show that by using the results from the Laplace transform in Table 4.1 we obtain the corresponding time function

$$h(t) = \frac{1}{m} \frac{e^{\lambda t} - e^{\lambda^* t}}{\lambda - \lambda^*} \tag{5.38}$$

Using the results from Example 5.2, we can show that the aforementioned expression reduces to

$$h(t) = \frac{v_0}{\omega_d} e^{-\varsigma \omega_0 t} \sin \omega_d t \tag{5.39}$$

where the initial velocity v_0 is yet to be determined. We can find the unknown velocity considering the equilibrium of motion at the time when the impulse is applied. We note that just before the impulse is applied the system is at rest, thus all the terms in Eq. (5.35) are zero. For a system initially at rest, which accelerates to a certain velocity in a very short period of time, we can use the impulse-momentum relationship to determine the velocity of the system immediately after the action of the impulse. For an impulse of unit magnitude, we can see that

$$mv_0 = 1 \quad \Rightarrow \quad v_0 = \frac{1}{m} \tag{5.40}$$

Upon substituting Eq. (5.40) into Eq. (5.39), one obtains

$$h(t) = \frac{1}{m\omega_d} e^{-\varsigma \omega_0 t} \sin \omega_d t \tag{5.41}$$

which is the traditional form of expressing the IRF of an underdamped SDOF system.

Equation (5.41) shows that the physical units of the IRF are given in terms of s/kg in the SI system, thus the physical units of the right-hand side of Eq. (5.32) are (s/kg) × N × s. Since N = kg × m/s² the units of the right side of Eq. (5.32) are in fact (s/kg) × kg × m/s² × s = m, which makes physical sense since the response $y(t)$ is indeed given in units of displacement.

5.1.4   Transfer Function

The transfer function is defined as the Laplace transform of the IRF. We will rewrite the transfer function given by Eq. (5.37) as

$$H'(s) = \frac{1}{m} \frac{1}{(s - \lambda)(s - \lambda^*)} \tag{5.42}$$

Since we will use the symbol H for both the transfer function and the frequency response function (FRF) (to be defined in the following section), we will put an apostrophe mark on the transfer function like we did in Section 4.3.1 in order to make a clear distinction between the two functions. Sometimes it might be convenient to use the "unscaled transfer function," which is simply the transfer function given by Eq. (5.42) but without the factor $1/m$.

Taking the Laplace transform of both sides of Eq. (5.5) yields

$$\left(ms^2 + cs + k\right) Y(s) = X(s) \tag{5.43}$$

The solution to this equation can be expressed in terms of the mass-scaled transfer function. Using the pole formulas Eq. (5.8), we obtain

$$m(s - \lambda)(s - \lambda^*) Y(s) = X(s)$$

$$\Downarrow \tag{5.44}$$

$$Y(s) = \frac{1}{m(s - \lambda)(s - \lambda^*)} X(s) = H'(s) X(s)$$

where $y(t) \leftrightarrow Y(s)$ and $x(t) \leftrightarrow X(s)$ are the Laplace transform pairs. This result states that in the Laplace domain the solution is just the product of the force and the transfer function. This result can also be derived from the convolution equation Eq. (5.33) using the convolution property of the Laplace transform (see Section 4.3.2).

5.1.5   Frequency Response Function

The FRF $H(\omega)$ can be defined as the transfer function calculated at the imaginary axis $s = i\omega$

$$H(\omega) = H'(i\omega) = \frac{1}{m} \frac{1}{(i\omega - \lambda)(i\omega - \lambda^*)} \tag{5.45}$$

In physical terms, the FRF represents the amplitude and phase of the steady-state response of a viscously damped SDOF system subjected to a harmonic force of unit amplitude and frequency ω. We can plot the FRF as a function of the angular frequency ω. Since the function is complex, the function is normally plotted in two separate plots, one plot showing the absolute value of the FRF as a function of frequency (also called the amplification function), and another plot showing the phase as a function of frequency.

The solution to the equilibrium Eq. (5.5) can be expressed in terms of the mass-scaled FRF. By taking the Fourier transform of both sides of Eq. (5.5) and using the differentiation theorem given by Eq. (4.39) yields

$$\left(m(i\omega)^2 + ci\omega + k\right) Y(\omega) = X(\omega) \tag{5.46}$$

Using the pole formulas in Eq. (5.8), we obtain

$$m(i\omega - \lambda)(i\omega - \lambda^*) Y(\omega) = X(\omega)$$

$$\Downarrow$$

$$Y(\omega) = \frac{1}{m(i\omega - \lambda)(i\omega - \lambda^*)} X(\omega) = H(\omega) X(\omega) \tag{5.47}$$

where $y(t) \leftrightarrow Y(\omega)$ and $x(t) \leftrightarrow X(\omega)$ are now the Fourier transform pairs. This result states that in the frequency domain, the solution of the differential equation of motion is the product of the force and the FRF. This result can again be derived from the convolution equation Eq. (5.33) using the convolution property of the Fourier transform (see Section 4.1.3).

From Eq. (5.47), it is clear that the FRF has the simple symmetry property such that $H^*(-\omega) = H(\omega)$. We note that for the case of damping values smaller than 10% or so, the undamped and damped frequencies of the system are, for practical purposes, the same ($\omega_d \cong \omega_0$). For the positive values of ω, the amplitude of the response given by Eq. (5.47) is most sensitive to the values of $i\omega - \lambda$ than the rest of the terms in this equation. At the resonance condition, when $\omega = \omega_0$, it can be shown that the product $(i\omega - \lambda)(i\omega - \lambda^*)$ is close to minimum, and since this value is in the denominator of Eq. (5.47), it is obvious that the amplitude of the FRF will be the largest in the vicinity of the resonant frequency. Similarly, for negative values of ω the FRF amplitude will be the largest when $\omega = -\omega_0$. This explains the role of the two complex conjugate poles in controlling the location of the peaks in the FRF plots. The pole $\lambda = -\varsigma\omega_0 + i\omega_d$ controls the peak in the FRF for positive frequencies, whereas its complex conjugate controls the peak in the FRF for negative frequencies.

Example 5.3   Comparing the mass-scaled FRF with the traditional formulation

In the literature, it is common to write the FRF in a form that might appear to be different from Eq. (5.45). For instance, in Bendat and Piersol [4], the FRF is written as a function of the oscillation frequency in the following form

$$H(f) = \frac{1}{k}\frac{1}{\left(1 - (f/f_0)^2 + i2\varsigma f/f_0\right)} \tag{5.48}$$

where f_0 is the natural frequency of the system, Eq. (5.19). This equation gives exactly the same results as Eq. (5.45). The plots of the two functions are shown in Figure 5.5.

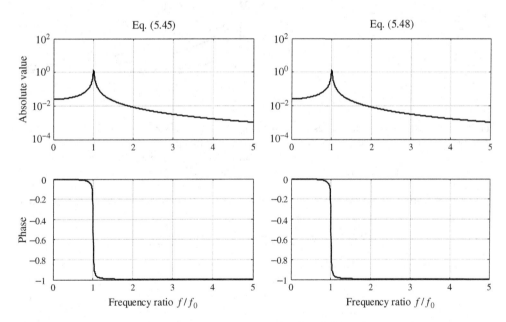

Figure 5.5   Comparison of the mass normalized FRF (left plot) given by Eq. (5.45) with the traditional FRF (right plot) given by Eq. (5.48). Top plots show the absolute value of the FRF's, whereas bottom plots show the phase. The frequency used at the frequency axis is the frequency ratio f/f_0 and damping ratio is equal to 1%. The plot follows the tradition of plotting the absolute value using a logarithmic vertical axis, and using a natural frequency axis

Example 5.4   Plotting the FRF for different damping ratios

It is common to illustrate how the FRF varies for different damping ratios in a plot where a logarithmic axis is used for the frequency and where the amplification function is also shown in a logarithmic plot. The result is shown in Figure 5.6.

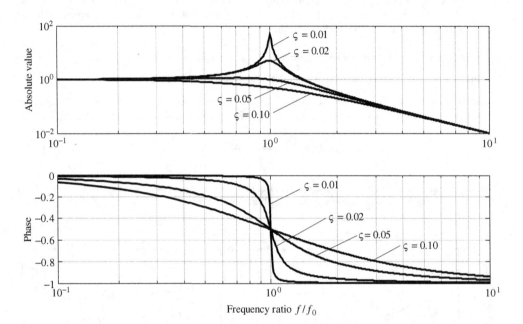

Figure 5.6 FRF for different values of the damping ratio $\varsigma = 0.01, 0.1, 0.5, 1$. The plots show the amplification function in the top plot and the phase in the bottom plot. The frequency used for the frequency axis is the frequency ratio f/f_0. This plot follows the tradition of a double logarithmic plot for the absolute value, and a natural vertical axis and logarithmic horizontal axis for the phase

The amplification function is plotted normalized to unity at DC (when $\omega = 0$), thus the plotted function is

$$kH(\omega) = \frac{k}{m} \frac{1}{(i\omega - \lambda)(i\omega - \lambda^*)} = \frac{\omega_0^2}{(i\omega - \lambda)(i\omega - \lambda^*)} \tag{5.49}$$

Using Eq. (5.13), it can be verified that

$$kH(0) = \frac{\omega_0^2}{\lambda\lambda^*} = 1 \tag{5.50}$$

5.2   Multiple Degree of Freedom Systems

For an N-degree of freedom viscously damped system, Eq. (5.5) is generalized to a matrix equation of the form

$$\mathbf{M\ddot{y}}(t) + \mathbf{C\dot{y}}(t) + \mathbf{Ky}(t) = \mathbf{x}(t) \tag{5.51}$$

where the input force $\mathbf{x}(t)$ and the response $\mathbf{y}(t)$ are vectors with N elements, and the mass matrix $\mathbf{M}$, the damping matrix $\mathbf{C}$, and the stiffness matrix $\mathbf{K}$ are all symmetric real $N \times N$ matrices with constant coefficients. The mass matrix is positive definite, but the stiffness and damping matrices should be at least positive semi-definite. In some presentations, a formulation such as Eq. (5.51) will assume that the response vector only consists of components belonging to one single direction and including other directions will define a new vector. We shall not assume any of such limitations, but assume that all components of the general external 3-dimensional forces and all components of the corresponding 3-dimensional responses are included in the vectors $\mathbf{x}(t)$ and $\mathbf{y}(t)$. For OMA purposes, solutions to Eq. (5.51) are generally found assuming that the damping forces are small with respect to the inertia and restoring forces and neglecting the effect of damping on the spatial solution (the mode shapes).

5.2.1   Free Responses for Undamped Systems

Let us consider the simple case without external forces and where the damping forces are so small that they can be neglected, thus we assume $\mathbf{x}(t) = 0$ and $\mathbf{C} = 0$. The solutions can be found by the method of separation of variables assuming that the free response is the product of some column vector $\mathbf{b}$ (the spatial solution) times a harmonic function $e^{i\omega t}$ (the time solution), that is,

$$\mathbf{y}(t) = \mathbf{b}e^{i\omega t} \tag{5.52}$$

By substituting Eq. (5.52) into Eq. (5.51), we obtain the equation of what is known as the extended eigenvalue problem

$$-\omega^2 \mathbf{M}\mathbf{b} + \mathbf{K}\mathbf{b} = 0 \tag{5.53}$$

Premultiplying both sides of this equation by the inverse of the mass matrix leads to the classical eigenvalue problem with the nonsymmetrical dynamic matrix $\mathbf{D} = \mathbf{M}^{-1}\mathbf{K}$ and eigenvalue ω^2

$$\mathbf{D}\mathbf{b} = \mathbf{M}^{-1}\mathbf{K}\mathbf{b} = \omega^2 \mathbf{b} \tag{5.54}$$

We can express the dynamic properties of Eq. (5.51) for the undamped system by the eigenvalue decomposition of the dynamic matrix $\mathbf{D}$ as

$$\mathbf{M}^{-1}\mathbf{K} = \left[\mathbf{b}_n\right]\left[\omega_n^2\right]\left[\mathbf{b}_n\right]^{-1} \tag{5.55}$$

For the type of systems being considered here, the N eigenvectors $\mathbf{b}_1, \mathbf{b}_2 \ldots \mathbf{b}_N$ are called the mode shapes of the system and are arranged as columns in the mode shape matrix $\left[\mathbf{b}_n\right]$ and the corresponding set of eigenvalues are the angular natural frequencies squared and are arranged as the elements of the diagonal matrix $\left[\omega_n^2\right]$.

It is common that symbols such as $\boldsymbol{\varphi}$ or $\boldsymbol{\psi}$ are used to represent the mode shapes. In this book, we need to make a distinction between mode shapes obtained analytically and experimentally, so we will use the symbol $\mathbf{b}$ to denote mode shapes obtained from the analysis of a finite element model and the symbol $\mathbf{a}$ to denote mode shapes obtained from experimental data.

For the undamped case, it can be shown that all eigenvalues are real positive numbers. Even though the dynamic matrix $\mathbf{D}$ is in general nonsymmetric, the eigenvectors $\mathbf{b}_n$ can be expressed as real-valued vectors. Although the numerical process to obtain eigenvectors may result in complex valued vectors, one can always reduce them to real-valued vectors because the phase between the real and imaginary parts of each nodal coordinates of each eigenvector is the same. In other words, the shape of the real part of a complex eigenvector is the same as the shape of the imaginary part of the same vector. The real modes for the undamped system are often referred to as normal modes.

The mode shapes of an undamped system satisfy some important orthogonality properties that are easily verified. We can write Eq. (5.53) for two different modes $\mathbf{b}_n, \mathbf{b}_m$ with corresponding natural frequencies ω_n, ω_m. If we premultiply each of the resulting equations by the other transposed mode shape, we get the following two equations:

$$-\omega_m^2 \mathbf{b}_n^T \mathbf{M} \mathbf{b}_m + \mathbf{b}_n^T \mathbf{K} \mathbf{b}_m = 0$$
$$-\omega_n^2 \mathbf{b}_m^T \mathbf{M} \mathbf{b}_n + \mathbf{b}_m^T \mathbf{K} \mathbf{b}_n = 0 \tag{5.56}$$

Since the mass matrix and the stiffness matrix are symmetric matrices, the scalar valued inner products $\mathbf{b}_n^T \mathbf{M} \mathbf{b}_m$ and $\mathbf{b}_n^T \mathbf{K} \mathbf{b}_m$ are indifferent to the placement of $\mathbf{b}_n, \mathbf{b}_m$, thus $\mathbf{b}_n^T \mathbf{M} \mathbf{b}_m = \mathbf{b}_m^T \mathbf{M} \mathbf{b}_n$ and $\mathbf{b}_n^T \mathbf{K} \mathbf{b}_m = \mathbf{b}_m^T \mathbf{K} \mathbf{b}_n$. Making use of this fact and subtracting the two equations in Eq. (5.56) yields

$$\left(\omega_n^2 - \omega_m^2 \right) \mathbf{b}_n^T \mathbf{M} \mathbf{b}_m = 0 \tag{5.57}$$

If $\omega_n^2 \neq \omega_m^2$ then the only way to satisfy Eq. (5.57) is by ensuring that the inner matrix product is equal to zero. The orthogonality condition of the normal modes with respect to the mass matrix is then given by

$$\mathbf{b}_n^T \mathbf{M} \mathbf{b}_m = 0 \tag{5.58}$$

And from Eq. (5.53), it follows that the corresponding orthogonality condition of the normal modes with respect to the stiffness matrix is given by

$$\mathbf{b}_n^T \mathbf{K} \mathbf{b}_m = 0 \tag{5.59}$$

The length of the eigenvector is not determined by the eigenvalue problem (5.53), and it can be arbitrarily chosen. The vectors $\mathbf{b}_n$ are often chosen to have unit length so that

$$\mathbf{b}_n^T \mathbf{b}_n = 1 \tag{5.60}$$

We can define the so-called modal mass m_n that is dependent on the scaling of the eigenvectors

$$\mathbf{b}_n^T \mathbf{M} \mathbf{b}_n = m_n \tag{5.61}$$

and similarly, the inner product over the stiffness matrix defines the modal stiffness k_n

$$\mathbf{b}_n^T \mathbf{K} \mathbf{b}_n = k_n \tag{5.62}$$

It is common practice to work with mass normalized modes, which can be defined as

$$\boldsymbol{\beta}_n = \mathbf{b}_n / \sqrt{m_n} \tag{5.63}$$

So that

$$\boldsymbol{\beta}_n^T \mathbf{M} \boldsymbol{\beta}_n = 1 \tag{5.64}$$

For the mass normalized modes, $\boldsymbol{\beta}_n$ the inner product with the stiffness matrix reduces to the natural frequency squared

$$\boldsymbol{\beta}_n^T \mathbf{K} \boldsymbol{\beta}_n = \frac{k_n}{m_n} = \omega_n^2 \tag{5.65}$$

For the mode shape matrix where mode shapes are arbitrarily scaled, the orthogonality conditions can be expressed as

$$\left[\mathbf{b}_n \right]^T \mathbf{M} \left[\mathbf{b}_n \right] = \left[m_n \right]; \quad \left[\mathbf{b}_n \right]^T \mathbf{K} \left[\mathbf{b}_n \right] = \left[k_n \right] \tag{5.66}$$

For the mass normalized mode shapes, $\left[\boldsymbol{\beta}_n \right]$ the equations simplifies to

$$\left[\boldsymbol{\beta}_n \right]^T \mathbf{M} \left[\boldsymbol{\beta}_n \right] = \mathbf{I}; \quad \left[\boldsymbol{\beta}_n \right]^T \mathbf{K} \left[\boldsymbol{\beta}_n \right] = \left[\omega_n^2 \right] \tag{5.67}$$

5.2.2   Free Responses for Proportional Damping

When the viscous damping matrix satisfies certain properties such that orthogonality of the normal modes of the undamped system with respect to the damping matrix is also satisfied, the system is said to have proportional (or classical) damping. Caughey and O'Kelly, [5] showed in 1965 that if the damping matrix satisfies the identity $\mathbf{CM}^{-1}\mathbf{K} = \mathbf{KM}^{-1}\mathbf{C}$ then the normal modes of vibration of the system are real-valued and identical to those of the associated undamped system. The well-known Rayleigh damping is one case that satisfies this condition. In this case, the damping matrix is expressed as a linear combination of the mass and stiffness matrix

$$\mathbf{C} = \alpha\mathbf{M} + \beta\mathbf{K} \tag{5.68}$$

It is not difficult to show that in this case the orthogonality conditions of the normal modes of the undamped system also apply to the damping matrix and that the damping matrix can be diagonalized using Eqs. (5.61) and (5.62). The orthogonality of the normal modes with respect to the damping matrix can be expressed as

$$\mathbf{b}_n^T\mathbf{Cb}_n = \alpha m_n + \beta k_n = c_n \tag{5.69}$$

We will define the term c_n as the modal damping coefficient.

Let us assume that a suitable solution to the free vibration form of Eq. (5.51) can be given in terms of the mode shape $\mathbf{b}_n$

$$\mathbf{y}(t) = \mathbf{b}_n e^{\lambda t} \tag{5.70}$$

Replacing this expression and its corresponding derivatives into Eq. (5.51), setting the right side equal to zero, premultiplying both sides of the resulting expression by the transpose of the nth mode, and carrying out the inner products given by Eqs. (5.61/5.62/5.69) we obtain the characteristic equation for mode n

$$m_n\lambda^2 + c_n\lambda + k_n = 0 \tag{5.71}$$

The solutions to this equation are of the form

$$\lambda_n = \frac{-c_n \pm i\sqrt{c_n^2 - 4m_n k_n}}{2m_n} \tag{5.72}$$

and the angular undamped and damped natural frequencies and damping ratio of mode n can be expressed in terms of the modal mass, modal stiffness, and modal damping coefficient as

$$\omega_{0n} = \sqrt{\frac{k_n}{m_n}}; \quad \varsigma_n = \frac{c_n}{2\sqrt{m_n k_n}}; \quad \omega_{dn} = \omega_{0n}\sqrt{1 - \varsigma_n^2} \tag{5.73}$$

These relations are identically the same as those for the case for SDOF systems (see Eqs. (5.9) and (5.11)). This means that a general free decay can be generalized from Eq. (5.14) as a linear combination of all possible free decays of the system, and we can express this as

$$\mathbf{y}(t) = \sum_{n=1}^N \mathbf{b}_n(c_{n1}e^{\lambda_n t} + c_{n2}e^{\lambda_n^* t}) \tag{5.74}$$

where the constants c_{n1}, c_{n2} describe the initial conditions of each modal free decay.

5.2.3   General Solutions for Proportional Damping

Under the assumption of proportional damping, the normal modes defined by Eq. (5.54) form an eigen basis for all possible solutions to the N-degree of freedom system given by Eq. (5.51). Any general response $\mathbf{y}$ can be expressed as a linear combination of the eigenvectors

$$\mathbf{y} = \sum_{n=1}^{N} \mathbf{b}_n q_n = \left[\mathbf{b}_n \right] \mathbf{q} \tag{5.75}$$

where the vector $\mathbf{q}$ is a column vector that includes the so-defined modal coordinates of the general response $\mathbf{y}$. Since the eigenvectors are constant vectors and the general response is a function of time, a more general expression of the response of the system is given by

$$\mathbf{y}(t) = \sum_{n=1}^{N} \mathbf{b}_n q_n(t) = \left[\mathbf{b}_n \right] \mathbf{q}(t) \tag{5.76}$$

This formulation is a modal decomposition of the response $\mathbf{y}(t)$. Inserting the modal decomposition into the general equation of motion Eq. (5.51), premultiplying both sides of this equation by the transpose of eigenvector $\mathbf{b}_n$ and using the orthogonality conditions of the normal modes, we can show that the system of simultaneous second-order differential equations given by Eq. (5.51) is decomposed into a set of independent SDOF equations of the form

$$m_n \ddot{q}(t) + c_n \dot{q}(t) + k_n q(t) = p_n(t) \tag{5.77}$$

where the right-hand side is called the modal load (or modal force) given by

$$p_n(t) = \mathbf{b}_n^T \mathbf{x}(t) \tag{5.78}$$

The general solution for proportional damping can be found simply by determining the modal loads using Eq. (5.78), for each mode solving the SDOF equation of motion given by Eq. (5.77) and then adding the modal solutions together using Eq. (5.76).

As for the SDOF solution, the modal coordinates can be found by taking the Laplace transform. In parallel to Eq. (5.44), we get

$$Q_n(s) = \frac{1}{m_n \left(s - \lambda_n \right) (s - \lambda_n^*)} P_n(s) = H_n(s) P_n(s) \tag{5.79}$$

where $q_n(t) \leftrightarrow Q_n(s)$ and $p_n(t) \leftrightarrow P_n(s)$ are Laplace transform pairs, and $H_n(s)$ is the mass-scaled transfer function[1] of mode n

$$H_n(s) = \frac{1}{m_n \left(s - \lambda_n \right) (s - \lambda_n^*)} \tag{5.80}$$

5.2.4 Transfer Function and FRF Matrix for Proportional Damping

The transfer function matrix is readily found by taking the Laplace transform of Eq. (5.76)

$$\widetilde{\mathbf{y}}(s) = \sum_{n=1}^{N} \mathbf{b}_n Q_n(s) = \left[\mathbf{b}_n \right] \widetilde{\mathbf{q}}(s) \tag{5.81}$$

where $\mathbf{y}(t) \leftrightarrow \widetilde{\mathbf{y}}(s)$ and $\mathbf{q}(t) \leftrightarrow \widetilde{\mathbf{q}}(s)$ are Laplace transform pairs. Using Eq. (5.79), we get

$$\widetilde{\mathbf{y}}(s) = \sum_{n=1}^{N} \mathbf{b}_n H_n(s) P_n(s) \tag{5.82}$$

and the Laplace transform of Eq. (5.78)

$$P_n(s) = \mathbf{b}_n^T \widetilde{\mathbf{x}}(s) \tag{5.83}$$

---

[1] Often we might put a mark like $H_n^I(s)$ on this function in order to indicate that we are dealing with the Laplace transform, but for simplicity we omit this indication here.

where $\mathbf{x}(t) \leftrightarrow \tilde{\mathbf{x}}(s)$ is a Laplace transform pair. From this, we obtain

$$\mathbf{y}(s) = \left(\sum_{n=1}^{N} \mathbf{b}_n H_n(s) \mathbf{b}_n^T \right) \tilde{\mathbf{x}}(s) \tag{5.84}$$

Upon noticing that the series of outer products $H_1(s)\mathbf{b}_1\mathbf{b}_1^T + H_2(s)\mathbf{b}_2\mathbf{b}_2^T + \cdots$ can be expressed as the matrix product

$$\sum_{n=1}^{N} \mathbf{b}_n H_n(s) \mathbf{b}_n^T = \left[\mathbf{b}_n\right] \left[H_n(s)\right] \left[\mathbf{b}_n\right]^T \tag{5.85}$$

where $\left[H_n(s)\right]$ is a diagonal matrix holding the modal transfer functions, we see that the Laplace transform of the response is given by

$$\tilde{\mathbf{y}}(s) = \tilde{\mathbf{H}}(s)\tilde{\mathbf{x}}(s) \tag{5.86}$$

where the transfer function matrix is

$$\tilde{\mathbf{H}}(s) = \left[\mathbf{b}_n\right] \left[H_n(s)\right] \left[\mathbf{b}_n\right]^T \tag{5.87}$$

The corresponding FRF matrix is then given by

$$\tilde{\mathbf{H}}(i\omega) = \left[\mathbf{b}_n\right] \left[H_n(i\omega)\right] \left[\mathbf{b}_n\right]^T \tag{5.88}$$

From Eq. (5.87/5.88), we clearly see that the transfer function and FRF function matrices are symmetric. It should be noted that the mass-scaled modal transfer functions are used in Eq. (5.87). The modal mass can instead be distributed on the two mode shapes in the outer products in Eq. (5.84) as given by Eq. (5.63); this leads to the expression for the transfer matrix given by Eq. (5.87). In this case the mode shape matrix contains the mass-scaled mode shapes and the modal transfer function is given without mass scaling (or we can say with the modal masses equal to unity)

$$H_n(s) = \frac{1}{(s - \lambda_n)(s - \lambda_n^*)} \tag{5.89}$$

Note that the modal transfer function of mode n can be found as the solution to the N-degree-of-freedom system Eq. (5.51) where the right-hand side is set equal to the Dirac delta function times the mode shape of mode n

$$\mathbf{M}\ddot{\mathbf{y}}(t) + \mathbf{C}\dot{\mathbf{y}}(t) + \mathbf{K}\mathbf{y}(t) = \delta(t)\mathbf{b}_n \tag{5.90}$$

Taking the Laplace transform of both sides of this equation, using the modal decomposition Eq. (5.81) and finally premultiplying both sides of the equation by the transpose of the eigenvector $\mathbf{b}_n$ leads directly to Eq. (5.80). From the arguments that lead us to Eq. (5.40), we can similarly see that this corresponds to a free decay with zero initial displacement and an initial velocity vector equal to

$$\mathbf{v}_0 = \dot{\mathbf{y}}(0+) = \frac{\mathbf{b}_n}{m_n} \tag{5.91}$$

The aforementioned expressions for the transfer function are very simple. However, they are based on the assumption of proportional damping, which is not necessarily the most realistic representation of how the system dissipates energy when it is vibrating. The only justifications for the assumption of proportional damping are the various mathematical simplifications that follow from this assumption. It simplifies the solution of the equation of motion and takes advantage of the fact that under normal operating conditions of the system, damping forces are small compared to the other forces in Eq. (5.51).

One may argue that the idea of modeling damping forces in real structures by a linear viscous term is unrealistic as much of the damping in real structures is the result of viscoelastic material response, friction between components of the system and other complex nonlinear phenomena. The assumption of proportional damping might seem to be the least of the challenges when modeling an MDOF system. However, recognizing the limitation of the assumption of viscous damping as a way to represent the energy dissipation characteristics of a structure, it is still of practical interest to consider the solution to the linear model given by Eq. (5.51) for cases where the damping is not proportional.

Example 5.5   Comparison with traditional versions of the transfer function matrix

In the literature, the transfer function matrix is often written in the form

$$\widetilde{\mathbf{H}}(s) = \sum_{n=1}^{N} \left(\frac{\mathbf{A}_n}{s - \lambda_n} + \frac{\mathbf{A}_n^*}{s - \lambda_n^*} \right) \tag{5.92}$$

where the residue matrix $\mathbf{A}_n$ for mode n is given by

$$\mathbf{A}_n = \frac{\mathbf{b}_n \mathbf{b}_n^T}{2i\omega_{dn} m_n} \tag{5.93}$$

and where ω_{dn} is the damped natural frequency of mode n. Unit length eigenvectors are assumed in this case. From Eq. (5.87), one obtains

$$\widetilde{\mathbf{H}}(s) = \sum_{n=1}^{N} \mathbf{b}_n H_n(s) \mathbf{b}_n^T \tag{5.94}$$

Realizing the partial fraction expansion[2]

$$\frac{1}{(s - \lambda_1)(s - \lambda_2)} = \frac{1}{(\lambda_1 - \lambda_2)} \left(\frac{1}{s - \lambda_1} - \frac{1}{s - \lambda_2} \right) \tag{5.95}$$

and using Eq. (5.12), we get

$$\lambda_n - \lambda_n^* = 2i\omega_{dn} \tag{5.96}$$

we then have a modal transfer function of the form

$$H_n(s) = \frac{1}{2i\omega_{dn} m_n} \frac{1}{s - \lambda_n} + \frac{1}{-2i\omega_{dn} m_n} \frac{1}{s - \lambda_n^*} \tag{5.97}$$

which leads directly to the traditional Eqs. (5.92/5.93).

Example 5.6   Transfer function of a 2-DOF system

The system in Figure 5.7 with the two DOFs y_1 and y_2 has the mass and stiffness matrices

$$\mathbf{M} = \begin{bmatrix} m + \dfrac{\Delta m}{2} & \dfrac{\Delta m}{2} \\ \dfrac{\Delta m}{2} & m + \dfrac{\Delta m}{2} \end{bmatrix} = m \begin{bmatrix} 1 + \dfrac{\Delta m}{2m} & \dfrac{\Delta m}{m} \\ \dfrac{\Delta m}{m} & 1 + \dfrac{\Delta m}{2m} \end{bmatrix}; \quad \mathbf{K} = \begin{bmatrix} k & 0 \\ 0 & k \end{bmatrix} = k \begin{bmatrix} 1 & 0 \\ 0 & 1 \end{bmatrix} \tag{5.98}$$

---

[2] The result can be found by the Heaviside expansion theorem, or it can be derived using the results from the Laplace transform table in Chapter 4.

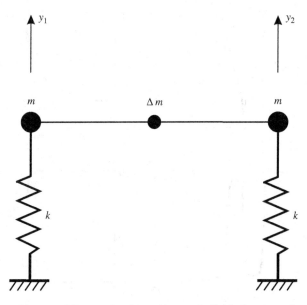

Figure 5.7   System with two DOFs y_1 and y_2 formed by two individual mass-spring systems connected by a weightless rigid bar with an additional mass in the middle

Let $\Delta m = m$. In this case, the normal modes and natural frequencies are

$$\left[\mathbf{b}_n\right] = \begin{bmatrix} 0.7071 & 0.7071 \\ 0.7071 & -0.7071 \end{bmatrix}; \quad \left[\omega_n^2\right] = \frac{k}{m}\begin{bmatrix} 0.5 & 0 \\ 0 & 1.0 \end{bmatrix} \tag{5.99}$$

which correspond to the frequencies $f_1 = 0.1125\sqrt{k/m}$ and $f_2 = 0.1592\sqrt{k/m}$. The modal masses are found to be

$$m_1 = 2.0m; \quad m_2 = 1.0m \tag{5.100}$$

These results make sense because for the first mode shape the delta mass is included in the motion (the two masses are moving in the same direction), whereas for the second mode shape, the delta mass is not moving at all as the two masses are moving in opposite directions. The modal stiffness are both equal to unity for this case.

The FRF's $H_{11}(f)$ and $H_{12}(f)$ can then be plotted for a damping ratio of 1% using Eq. (5.58) as shown in Figure 5.8.

According to Eqs. (5.90/5.91) if we apply two simultaneous impulses scaled as the components in the mode shape then we excite only the corresponding mode. For instance, if we apply two simultaneous pulses with same direction and strength on the two masses, then we excite only mode one, and the other mode is left motionless.

5.2.5   General Damping

When the damping matrix does not satisfy the Caughey and O'Kelly conditions and the normal modes do not diagonalize the damping matrix then other ways of solving the Eq. (5.51) must be used. In this

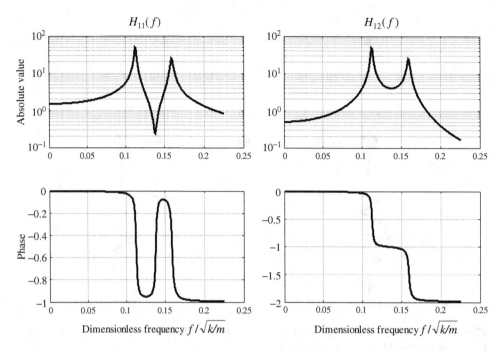

Figure 5.8 The FRF's H_{11} (f) and H_{12} (f) plotted for a damping ratio equal to 1%. The two plots to the left show H_{11} (f) and the two plots to the right show H_{12} (f). The horizontal axis shows the dimensionless frequency $f/\sqrt{k/m}$

case, the equation of motion may be written in a different format that includes both the displacement $\mathbf{y}$ (t) and its derivate $\dot{\mathbf{y}}$ (t) to form a new variable vector as

$$\mathbf{u}\,(t) = \begin{Bmatrix} \dot{\mathbf{y}}\,(t) \\ \mathbf{y}\,(t) \end{Bmatrix} \qquad (5.101)$$

This makes it possible to write the equation of motion Eq. (5.51) in the so-called state space format.[3] We will use here the state space format described in Heylen et al. [8]

$$\mathbf{A}\dot{\mathbf{u}}\,(t) + \mathbf{B}\mathbf{u}\,(t) = \mathbf{f}\,(t)$$
$$\mathbf{y}\,(t) = \mathbf{P}\mathbf{u}\,(t) \qquad (5.102)$$

where

$$\mathbf{A} = \begin{bmatrix} [0] & \mathbf{M} \\ \mathbf{M} & \mathbf{C} \end{bmatrix}; \quad \mathbf{B} = \begin{bmatrix} -\mathbf{M} & [0] \\ [0] & \mathbf{K} \end{bmatrix}; \quad \mathbf{f}\,(t) = \begin{Bmatrix} \{0\} \\ \mathbf{x}\,(t) \end{Bmatrix}; \quad \mathbf{P} = \begin{bmatrix} [0] & \mathbf{I} \end{bmatrix} \qquad (5.103)$$

The matrix [0] is a null matrix and the vector {0} is a null vector. Since all matrices of the system are of order $N \times N$, this formulation expand the new system matrices $\mathbf{A}$ and $\mathbf{B}$ to a $2N \times 2N$ space, but at the same time the problem simplifies to a set of simultaneous first-order differential equations. The matrix $\mathbf{P}$ is called the observation matrix since it includes the last N elements of the state vector $\mathbf{u}$ (t),

---

[3] Other state formulations might be formed, see for instance Hoen [6], Hurty and Rubenstein [7], the derivation in appendix C, or the section about discrete time later in this chapter.

which can be used to obtain the response $\mathbf{y}(t)$. Considering a general free decay, a possible solution of Eq. (5.102) is

$$\mathbf{u}(t) = \boldsymbol{\varphi} e^{\lambda t} \tag{5.104}$$

We then have the eigenvalue problem expressed as

$$\begin{aligned} \lambda \mathbf{A}\boldsymbol{\varphi} + \mathbf{B}\boldsymbol{\varphi} = 0 \Rightarrow \\ -\mathbf{A}^{-1}\mathbf{B}\boldsymbol{\varphi} = \lambda\boldsymbol{\varphi} \end{aligned} \tag{5.105}$$

The free response properties of the general equation are now described by the eigenvalue decomposition of the matrix $-\mathbf{A}^{-1}\mathbf{B}$

$$-\mathbf{A}^{-1}\mathbf{B} = \left[\boldsymbol{\varphi}_n\right] \left[\lambda_n\right] \left[\boldsymbol{\varphi}_n\right]^{-1} \tag{5.106}$$

where the matrix $\left[\boldsymbol{\varphi}_n\right]$ holds the mode shapes in the following manner

$$\boldsymbol{\varphi}_n = \left\{ \begin{array}{c} \lambda \mathbf{b}_n \\ \mathbf{b}_n \end{array} \right\} \tag{5.107}$$

This equation indicates that the lower half of the vector includes the nodal coordinates of the nth mode shape displacements, while the upper half contains the corresponding values of the mode shape velocities. The dimension of the space has doubled in this formulation, and now the mode shape matrix contains additional modal information about the system.

The diagonal matrix $\left[\lambda_n\right]$ holds the eigenvalues. Depending on the characteristics of the damping matrix, the eigenvalues may correspond to underdamped or overdamped modes. If all the values of the eigenvalues are complex-valued, the modes are said to be underdamped and it can be shown that the $2N$ eigenvalues and modes shapes will appear in complex conjugate pairs. In contrast, if the eigenvalues are all real, then there will be a set of $2N$ overdamped modes and all the eigenvectors will appear as real-valued vectors. Of course, it is also possible that the set of eigenvalues and corresponding eigenvectors could be a combination of underdamped and overdamped modes. For more details about this, the reader is referred to Veletsos and Ventura [9] or Chopra [3].

In this book, we will limit our treatment of systems with general damping to the case when all the eigenvalues and corresponding eigenvectors are complex valued and correspond to underdamped modes. From Eqs. (5.51) and (5.101/5.104), it can be shown that the state space eigenvalue problem is equivalent to the equation

$$\left(\mathbf{M}\lambda^2 + \mathbf{C}\lambda + \mathbf{K}\right) \mathbf{b}_n = 0 \tag{5.108}$$

In general, the matrices $\left[\mathbf{b}_n\right]^T \mathbf{M} \left[\mathbf{b}_n\right]$, $\left[\mathbf{b}_n\right]^T \mathbf{C} \left[\mathbf{b}_n\right]$, and $\left[\mathbf{b}_n\right]^T \mathbf{K} \left[\mathbf{b}_n\right]$ are not diagonal. Following a similar approach to prove the orthogonality of the normal modes of the undamped system presented in Section 5.2.1, it can be shown that

$$\left(\lambda_n - \lambda_m\right) \boldsymbol{\varphi}_n^T \mathbf{A}\boldsymbol{\varphi}_m = 0 \tag{5.109}$$

If $\lambda_n \neq \lambda_m$ we can determine that the orthogonality condition of the complex mode shapes with respect to the matrix $\mathbf{A}$ is given by

$$\boldsymbol{\varphi}_n^T \mathbf{A}\boldsymbol{\varphi}_m = 0$$

$$\Downarrow \tag{5.110}$$

$$\left(\lambda_m + \lambda_n\right) \mathbf{b}_n^T \mathbf{M}\mathbf{b}_m + \mathbf{b}_n^T \mathbf{C}\mathbf{b}_m = 0$$

and from Eq. (5.105) the orthogonality of the modes with respect to the matrix $\mathbf{B}$ is

$$\boldsymbol{\varphi}_n^T \mathbf{B} \boldsymbol{\varphi}_m = 0$$

$$\Downarrow \tag{5.111}$$

$$\mathbf{b}_n^T \mathbf{K} \mathbf{b}_m - \left(\lambda_m + \lambda_n \right) \mathbf{b}_n^T \mathbf{M} \mathbf{b}_m = 0$$

This leads us to the generalized complex-modal coefficients for mode n associated to the $\mathbf{A}$ and $\mathbf{B}$ matrices defined as

$$a_n = \boldsymbol{\varphi}_n^T \mathbf{A} \boldsymbol{\varphi}_n = 2\lambda_n \, \mathbf{b}_n^T \mathbf{M} \mathbf{b}_n + \mathbf{b}_n^T \mathbf{C} \mathbf{b}_n \tag{5.112}$$

and

$$b_n = \boldsymbol{\varphi}_n^T \mathbf{B} \boldsymbol{\varphi}_n = \mathbf{b}_n^T \mathbf{K} \mathbf{b}_n - \lambda_n^2 \, \mathbf{b}_n^T \mathbf{M} \mathbf{b}_n \tag{5.113}$$

Evaluating Eq. (5.105) for mode n, premultiplying both sides of the resulting expression by the transpose of the nth mode and carrying out the inner products given by Eqs. (5.112/5.113), we obtain the characteristic equation for mode n

$$\lambda_n^2 \, \mathbf{b}_n^T \mathbf{M} \mathbf{b}_n + \lambda_n \mathbf{b}_n^T \mathbf{C} \mathbf{b}_n + \mathbf{b}_n^T \mathbf{K} \mathbf{b}_n = 0$$

$$\Updownarrow \tag{5.114}$$

$$a_n \lambda_n + b_n = 0$$

It should be noted that in Eq. (5.114) all quantities are complex quantities. Although we can no longer associate the inner products for the mass, damping and stiffness matrices to the real-valued modal mass, modal damping, and modal stiffness, we could carry out the same concepts but with the understanding that these are now complex valued numbers, and that there is always a complex conjugate associated to each of them.

The solution to Eq. (5.114) will result in a set of $2N$ eigenvalues, but half of them are the complex conjugates of the others. Before proceeding any further with the solution of the equation of motion, it is of interest to explore a way to provide a physical meaning to the eigenvalues as it was done for the case of undamped and proportionally damped systems earlier. To this end, we consider a pair of eigenvalues

$$\left. \begin{array}{c} \lambda_n \\ \lambda_n^* \end{array} \right\} = \sigma_n \pm i\omega_{dn} \tag{5.115}$$

where σ_n and ω_n are the real and imaginary parts of the eigenvalues. We can then define the modulus of the eigenvalue

$$\omega_{0n} = \sqrt{\lambda_n \lambda_n^*} = \sqrt{\sigma_n^2 + \omega_{dn}^2} \tag{5.116}$$

and the real-valued quantity

$$\varsigma_n = -\frac{\sigma_n}{\omega_{0n}} \tag{5.117}$$

that is positive as long as the eigenvalue is located in the negative half plane $\sigma_n < 0$. The eigenvalues can then be expressed as

$$\left. \begin{array}{c} \lambda_n \\ \lambda_n^* \end{array} \right\} = -\omega_{0n} \varsigma_n \pm i\omega_{dn} \tag{5.118}$$

and the following relationship between ω_{dn} and the modulus ω_{0n} is determined from Eq. (5.116)

$$\omega_{dn} = \omega_{0n} \sqrt{1 - \varsigma_n^2} \tag{5.119}$$

We see that for the general case of damping we can interpret ω_{0n} as the undamped natural frequency of mode n similarly to what we did for an SDOF system (see Eq. (5.11)), we can interpret ω_{dn} as the damped natural frequency of mode n and finally we can interpret ς_n as the damping ratio of mode n.

It should be noted that the equivalent or pseudo undamped natural frequency ω_{0n} is a function of the damping and hence, differs in that sense from the corresponding frequency of the undamped system. Further, it should be noted that even though some of the quantities defined are similar to the quantities defined for the SDOF system, the quantities do not in all aspects have a similar physical meaning. For instance, using Eqs. (5.61, 5.62 and 5.69) and the complex mode shapes to calculate the undamped natural frequency and the damping ratio will in general lead to complex valued quantities with a different meaning than what is defined here.

Similarly to Eq. (5.75), we express the general solution as a linear combination of the mode shapes

$$\mathbf{u}(t) = \sum_{n=1}^{2N} \boldsymbol{\varphi}_n q_n(t) = \left[\boldsymbol{\varphi}_n\right] \mathbf{q}(t) \tag{5.120}$$

where the modal coordinates are now also double in size, and again they appear in complex conjugate pairs. If the sum is restricted to the number of modes and each mode only is associated with one mode $\boldsymbol{\varphi}_n$ and one modal coordinate $q_n(t)$, then

$$\mathbf{u}(t) = \sum_{n=1}^{N} \left(\boldsymbol{\varphi}_n q_n(t) + \boldsymbol{\varphi}_n^* q_n^*(t)\right) = 2\mathrm{Re}\left[\sum_{n=1}^{N} \left(\boldsymbol{\varphi}_n q_n(t)\right)\right] \tag{5.121}$$

where Re means "the real part of" the expression that follows. Applying this solution in Eq. (5.102), using the orthogonality properties given by Eqs. (5.110) and (5.111) and premultiplying both sides of the equation by the transpose of the eigenvector $\boldsymbol{\varphi}_n$ we arrive at the following two equations for mode n

$$a_n \dot{q}_n(t) + b_n q_n(t) = p_n(t)$$
$$a_n^* \dot{q}_n^*(t) + b_n^* q_n^*(t) = p_n^*(t) \tag{5.122}$$

where the modal load is now given by

$$p_n(t) = \boldsymbol{\varphi}_n^T \mathbf{f}(t) \tag{5.123}$$

The two equations in Eq. (5.122) are equivalent because they are each other's complex conjugate, so we can use only one equation for each mode to determine the modal response. But we have to keep in mind that for each complex modal response, there will always be a complex conjugate. To this end, we can express the modal response as

$$a_n \dot{q}_n(t) + b_n q_n(t) = p_n(t) \tag{5.124}$$

Finally taking the Laplace transform of both sides of the equation, we obtain

$$\left(a_n s + b_n\right) Q_n(s) = P_n(s) \tag{5.125}$$

The eigenvalue for mode n is found from Eq. (5.114)

$$\lambda_n = -\frac{b_n}{a_n} \tag{5.126}$$

and the "half component" of the modal transfer function is

$$H_n(s) = \frac{1}{a_n\left(s - \lambda_n\right)} \tag{5.127}$$

Therefore, the Laplace transformed general response is

$$\tilde{\mathbf{u}}(s) = \sum_{n=1}^{N} \left(\boldsymbol{\varphi}_n \frac{P_n(s)}{a_n\left(s - \lambda_n\right)} + \boldsymbol{\varphi}_n^* \frac{P_n^*(s)}{a_n^*\left(s - \lambda_n^*\right)}\right) \tag{5.128}$$

Now using the Laplace transformed version of (5.123), we obtain

$$\tilde{\mathbf{u}}(s) = \sum_{n=1}^{N} \left(\frac{\boldsymbol{\varphi}_n \boldsymbol{\varphi}_n^T}{a_n \left(s - \lambda_n\right)} + \frac{\boldsymbol{\varphi}_n^* \boldsymbol{\varphi}_n^{*T}}{a_n^* \left(s - \lambda_n^*\right)} \right) \tilde{\mathbf{f}}(s) \tag{5.129}$$

which defines the state space transfer function matrix

$$\tilde{\mathbf{H}}_{ss}(s) = \sum_{n=1}^{N} \left(\frac{\boldsymbol{\varphi}_n \boldsymbol{\varphi}_n^T}{a_n \left(s - \lambda_n\right)} + \frac{\boldsymbol{\varphi}_n^* \boldsymbol{\varphi}_n^{*T}}{a_n^* \left(s - \lambda_n^*\right)} \right) \tag{5.130}$$

And the corresponding FRF matrix is given by

$$\tilde{\mathbf{H}}_{ss}(i\omega) = \sum_{n=1}^{N} \left(\frac{\boldsymbol{\varphi}_n \boldsymbol{\varphi}_n^T}{a_n \left(i\omega - \lambda_n\right)} + \frac{\boldsymbol{\varphi}_n^* \boldsymbol{\varphi}_n^{*T}}{a_n^* \left(i\omega - \lambda_n^*\right)} \right) \tag{5.131}$$

We see that again the transfer function and FRF matrices are symmetric. Both transfer function and FRF matrix are easily reduced to the displacement by including only the last N elements in the mode shapes, thus the normal size transfer function is

$$\tilde{\mathbf{H}}(s) = \sum_{n=1}^{N} \left(\frac{\mathbf{b}_n \mathbf{b}_n^T}{a_n \left(s - \lambda_n\right)} + \frac{\mathbf{b}_n^* \mathbf{b}_n^{*T}}{a_n^* \left(s - \lambda_n^*\right)} \right) = \sum_{n=1}^{N} \left(\frac{\mathbf{A}_n}{s - \lambda_n} + \frac{\mathbf{A}_n^*}{s - \lambda_n^*} \right) \tag{5.132}$$

and the corresponding FRF matrix is

$$\tilde{\mathbf{H}}(i\omega) = \sum_{n=1}^{N} \left(\frac{\mathbf{b}_n \mathbf{b}_n^T}{a_n \left(i\omega - \lambda_n\right)} + \frac{\mathbf{b}_n^* \mathbf{b}_n^{*T}}{a_n^* \left(i\omega - \lambda_n^*\right)} \right) = \sum_{n=1}^{N} \left(\frac{\mathbf{A}_n}{i\omega - \lambda_n} + \frac{\mathbf{A}_n^*}{i\omega - \lambda_n^*} \right) \tag{5.133}$$

The residue matrices $\mathbf{A}_n = \mathbf{b}_n \mathbf{b}_n^T / a_n$ and $\mathbf{A}_n^* = \mathbf{b}_n^* \mathbf{b}_n^{*T} / a_n^*$ of rank one should not be confused with the system matrix $\mathbf{A}$. Also in this reduced form, we can notice that the transfer function and FRF matrices are symmetric. We also see that the FRF has the symmetry property

$$\left(\tilde{\mathbf{H}}(-i\omega) \right)^* = \sum_{n=1}^{N} \left(\frac{\mathbf{A}_n}{-i\omega - \lambda_n} + \frac{\mathbf{A}_n^*}{-i\omega - \lambda_n^*} \right)^* = \sum_{n=1}^{N} \left(\frac{\mathbf{A}_n}{i\omega - \lambda_n} + \frac{\mathbf{A}_n^*}{i\omega - \lambda_n^*} \right) = \tilde{\mathbf{H}}(i\omega) \tag{5.134}$$

In simple words, the negative part of any scalar FRF can be established by flipping the positive part and taking the complex conjugate. This symmetry property is due to the fact that the IRFs are real-valued functions.

The traditional format given by (5.92/93) is in fact a general format that is valid for both proportional damping and general damping. We shall often assume that the mode shapes $\mathbf{b}_n$ are scaled to unity, which in this case means that

$$\mathbf{b}_n^H \mathbf{b}_n = 1 \tag{5.135}$$

For each term in the linear combination $e^{\lambda_n t} \leftrightarrow 1 / \left(s - \lambda_n\right)$, we get[4] the corresponding IRF matrix

$$\mathbf{H}(t) = \sum_{n=1}^{N} \left(\frac{\mathbf{b}_n \mathbf{b}_n^T}{a_n} e^{\lambda_n t} + \frac{\mathbf{b}_n^* \mathbf{b}_n^{*T}}{a_n^*} e^{\lambda_n^* t} \right) = \sum_{n=1}^{N} \left(\mathbf{A}_n e^{\lambda_n t} + \mathbf{A}_n^* e^{\lambda_n^* t} \right) \tag{5.136}$$

---

[4] See the section on Laplace transforms in Chapter 4.

where $\mathbf{H}(t) \leftrightarrow \tilde{\mathbf{H}}(s)$ is the Laplace transform pair. The transfer function Eq. (5.132) can also be written in matrix form. If we define the scaling of the mode shapes such that

$$a_n = 1 \tag{5.137}$$

then we can represent each of the modal contributions in Eq. (5.132) as $\mathbf{b}_n \mathbf{b}_n^T (s - \lambda_n)^{-1} + \mathbf{b}_n^* \mathbf{b}_n^H (s - \lambda_n^*)^{-1}$. Gathering the mode shape from the mode shape matrix $\mathbf{B}$ as it was done before and the terms $(s - \lambda_n)$ from the diagonal matrix $\mathbf{I}s - [\lambda_n] = [s - \lambda_n]$, we have

$$\tilde{\mathbf{H}}(s) = \mathbf{B} \ [s - \lambda_n]^{-1} \mathbf{B}^T + \mathbf{B}^* [s - \lambda_n^*]^{-1} \mathbf{B}^H \tag{5.138}$$

Before we close the matter of classical dynamics, we will consider two issues of practical interest. One is the question if for general damping it is still possible to excite only one single mode (which it is), and the other issue is why we need to introduce, for instance, complex-valued modal mass.

One would think that since for general damping the equation of motion does not decouple and thus the different modal coordinates are coupled through the damping matrix, it might not be possible in general to excite the system in such a way that only one mode is excited. It is, however, easy to prove that it is possible to excite only one mode.

From the matrix formulation of the response $\tilde{\mathbf{y}}(s) = \tilde{\mathbf{H}}(s) \tilde{\mathbf{x}}(s)$ where the transfer function $\tilde{\mathbf{H}}(s)$ is given by Eq. (5.138), it is clear that the external force vector $\tilde{\mathbf{x}}(s)$ in the Laplace domain that only excites mode n must satisfy the condition

$$\mathbf{B}^T \tilde{\mathbf{x}}(s) = \boldsymbol{\delta}_n \tag{5.139}$$

where $\boldsymbol{\delta}_n$ is a zero vector except for the element nth element, which is equal to unity. In the Laplace domain, this force vector is a constant vector

$$\tilde{\mathbf{x}}(s) = \mathbf{x}_{0n} = \left(\mathbf{B}^T\right)^{-1} \boldsymbol{\delta}_n \tag{5.140}$$

We see that if this condition is fulfilled, both terms in Eq. (5.138) reduce to a sole contribution from mode n. Therefore, in the time domain the impulse

$$\mathbf{x}(t) = \mathbf{x}_{0n} \delta(t) \tag{5.141}$$

will only excite mode n. It is seen from Eq. (5.121) that a general free decay must be a linear combination of the mode shapes. The coefficients in the linear combination are free decays of the corresponding modal coordinates, each modal coordinate having an initial amplitude and velocity as shown for the 1-DOF case in Section 5.1.2. This is shown in detail in Appendix C and further formulated in discrete time in Section 5.3.5.

Before we leave this section, some comments should be made about the complexity of the mode shape, modal mass, modal stiffness, and modal damping.

The general damping has been taken into account by allowing the mode shapes to become complex. This seems reasonable since a complex mode shape means that the different parts of the mode shape are not moving "in phase" or "out of phase" as for normal modes. But the phase is able to describe a traveling wave behavior of the mode, which means that energy is transported throughout the structure. The oscillation energy must be transported to the place where it can be dissipated by the dampers. Proportional damping means that the energy can be dissipated "where it is," therefore, no transport of the kinetic energy is needed, and thus, the mode shapes are real. This is essentially the concept of a standing wave, and we can say that the modes of a system with proportional damping are nothing more than standing waves.

It might seem unnecessary that the modal masses become complex, but in order to understand why this is in fact necessary, we have to remember that a mode shape **b** is determined by the eigenvalue problem except for the normalizing constant, and what comes out of an eigenvalue algorithm is the eigenvector scaled by some constant. This constant might be complex, thus not only the length of the vector – that is the real number $\sqrt{\mathbf{b}^H \mathbf{b}}$ – has to be taken into account, but also the more or less random phase must be taken into account. Let us assume that a random phase ϕ is introduced to the real mode shape **b**. Instead of the real mode shape **b**, we now have the mode shape

$$\mathbf{b}' = \mathbf{b}e^{i\phi} \tag{5.142}$$

If we define the modal mass as a real number, such as $m = \mathbf{b}^H \mathbf{M} \mathbf{b}$, and $m' = \mathbf{b}'^H \mathbf{M} \mathbf{b}'$, then in fact the modal mass is the same in both cases, but we have the problem that the influence of the random phase is not removed when we compute the transfer function in Eq. (5.133) using Eq. (5.114). If, on the other hand, we define the modal mass as in Eq. (5.61), then we have

$$m' = me^{i2\phi} \tag{5.143}$$

but in this case the influence of the phase is removed in the transfer function because the term

$$\frac{\mathbf{b}'\mathbf{b}'^T}{a'} = \frac{\mathbf{b}'\mathbf{b}'^T}{2im'\omega_d} = \frac{\mathbf{b}\mathbf{b}^T}{2im\omega_d} \tag{5.144}$$

is independent of the phase ϕ. The reader might verify that the damped frequency ω_d is the same in both cases using the classical definition of the modal quantities given by Eqs. (5.61/5.62/5.69) together with Eq. (5.114).

In conclusion, even though it might seem strange that the modal mass, the modal stiffness and also the modal damping according to Eqs. (5.61/5.62/5.69) should be defined as complex quantities, these are the only definitions that make sense in the end because they ensure that the transfer function is independent of a random phase in the mode shapes.

Example 5.7   Transfer function of a 2-DOF system with nonproportional damping

We are considering the same case as in Example 5.6, but now using the damping matrix

$$\mathbf{C} = \begin{bmatrix} c & 0 \\ 0 & 0 \end{bmatrix} = c \begin{bmatrix} 1 & 0 \\ 0 & 0 \end{bmatrix} \tag{5.145}$$

Using $k = 1\,N/m$, $m = dm = 1\,kg$, and $c = 0.5Ns/m$, we get a system with a relatively strong nonproportional damping. The eigenvectors of the state space system scaled to unit length such that $\boldsymbol{\varphi}_n^H \boldsymbol{\varphi}_n = 1$ are

$$[\boldsymbol{\varphi}_n] = \begin{bmatrix} 0.066 - 0.506i & 0.066 + 0.506i & -0.246 - 0.317i & -0.246 + 0.317i \\ 0.187 + 0.433i & 0.187 - 0.433i & 0.037 - 0.432i & 0.037 + 0.432i \\ -0.528 & -0.528 & -0.402 + 0.372i & -0.402 - 0.372i \\ 0.420 - 0.250i & 0.420 + 0.250i & -0.593 & -0.593 \end{bmatrix} \tag{5.146}$$

The frequency and damping are

$$\begin{bmatrix} f_1 & f_2 \\ \varsigma_1 & \varsigma_2 \end{bmatrix} = \begin{bmatrix} 0.117\,\text{Hz} & 0.154\;\text{Hz} \\ 8.43\% & 13.0\% \end{bmatrix} \tag{5.147}$$

and the generalized complex modal masses are

$$\begin{bmatrix} m_1 \\ m_2 \end{bmatrix} = \begin{bmatrix} 0.889 + 1.101i \\ 0.397 + 0.749i \end{bmatrix} \text{ kg} \tag{5.148}$$

The response $y_1(t)$ of DOF 1 (the degree of freedom where the damper is located) is now calculated for an impulse vector $\mathbf{x}(t) = \mathbf{x}_0\delta(t)$ exciting only the first mode as given by Eq. (5.141). The result is found by multiplying $\mathbf{x}_0$ on the IRF matrix (5.136). The free response is the result of the contribution from the first mode $y_{11}(t)$ and the contribution from the second mode $y_{12}(t)$. The results are shown in Figure 5.9. As it appears from the figure, only the first mode is excited, and the contribution from the second mode is zero when accounting for the computational accuracy.

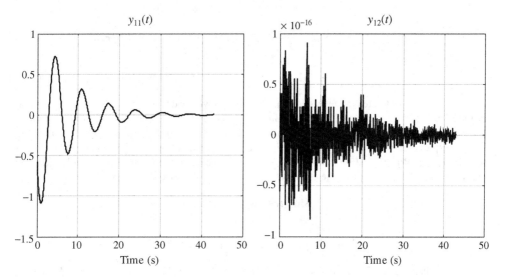

Figure 5.9 Free response of the DOF where the damper is placed when an impulse is applied such that only the first mode is contributing to the response. The plot to the left shows the contribution from the first mode and the plot to the right the contribution from the second mode. The response to the right is for all practical purposes zero

5.3 Special Topics

The main concepts of the classical dynamics of discrete linear systems have been covered earlier in this chapter. However, there are other topics of importance to the theory of OMA that need to be addressed before we leave this chapter.

5.3.1 Structural Modification Theory

The presentation in this section follows the ideas discussed by Sestieri [10]. Consider two structural systems with N DOF's. An undisturbed system "B" has a mass matrix $\mathbf{M}$, and a stiffness matrix $\mathbf{K}$. The undamped equation of motion for this system is

$$\mathbf{M}\ddot{\mathbf{y}}_b(t) + \mathbf{K}\mathbf{y}_b(t) = \mathbf{x}(t) \tag{5.149}$$

The solution to the corresponding eigenvalue problem is given by the mode shape matrix $\mathbf{B} = \begin{bmatrix} \mathbf{b}_1 \cdots \mathbf{b}_N \end{bmatrix}$ containing the mass-scaled mode shapes and the matrix $\begin{bmatrix} \omega_{bn}^2 \end{bmatrix}$ holding the eigenvalues. The orthogonality conditions lead to

$$\mathbf{B}^T\mathbf{MB} = \mathbf{I}; \quad \mathbf{B}^T\mathbf{KB} = \begin{bmatrix} \omega_{bn}^2 \end{bmatrix} \tag{5.150}$$

and expanding the response in terms of the eigenvectors of the system one obtains

$$\mathbf{y}_b(t) = \mathbf{B}\mathbf{q}_b(t) \tag{5.151}$$

The corresponding decoupled equations of motion are

$$\ddot{\mathbf{q}}_b(t) + \begin{bmatrix} \omega_{bn}^2 \end{bmatrix}\mathbf{q}_b(t) = \mathbf{B}^T\mathbf{x}(t) \tag{5.152}$$

A second system "A" has a mass matrix $\mathbf{M} + \Delta\mathbf{M}$, and a stiffness matrix $\mathbf{K} + \Delta\mathbf{K}$. The undamped equation of motion is of the form

$$(\mathbf{M} + \Delta\mathbf{M})\ddot{\mathbf{y}}_a(t) + (\mathbf{K} + \Delta\mathbf{K})\mathbf{y}_a(t) = \mathbf{x}(t) \tag{5.153}$$

This system can be treated in the usual form. This process will lead to the mode shapes $\mathbf{A} = \begin{bmatrix} \mathbf{a}_1 \cdots \mathbf{a}_N \end{bmatrix}$ of the system together with their corresponding eigenvalues. However, we will follow an alternative approach that allows us to formulate a relation between the sets of mode shapes for systems $\mathbf{A}$ and $\mathbf{B}$.

If we express the response of system A using the mode shapes of system B similarly to Eq. (5.151), such as $\mathbf{y}_a(t) = \mathbf{B}\mathbf{q}_b(t)$, and premultiply Eq. (5.153) by the transpose of the undisturbed mode shape matrix (we make a projection of the equation on the undisturbed mode shapes), we get

$$\widetilde{\mathbf{M}}\ddot{\mathbf{q}}_b(t) + \widetilde{\mathbf{K}}\mathbf{q}_b(t) = \mathbf{B}^T\mathbf{x}(t) \tag{5.154}$$

where

$$\widetilde{\mathbf{M}} = \mathbf{I} + \mathbf{B}^T\Delta\mathbf{MB}; \quad \widetilde{\mathbf{K}} = \begin{bmatrix} \omega_{bn}^2 \end{bmatrix} + \mathbf{B}^T\Delta\mathbf{KB} \tag{5.155}$$

We note that Eq. (5.154) is not decoupled unless we have the special cases when $\Delta\mathbf{M} \propto \mathbf{M}$ and $\Delta\mathbf{K} \propto \mathbf{K}$. Solving the new eigenvalue problem defined by Eq. (5.154) results in a new set of eigenvectors $\mathbf{t}_1$ with a mode shape matrix $\mathbf{T} = \begin{bmatrix} \mathbf{t}_1 \cdots \mathbf{t}_N \end{bmatrix}$ and corresponding eigenvalues ω_{an}^2 defining the diagonal matrix $\begin{bmatrix} \omega_{an}^2 \end{bmatrix}$. The solution to Eq. (5.154) can then be expanded using this new set of eigenvectors

$$\mathbf{q}_b(t) = \mathbf{T}\mathbf{q}_a(t) \tag{5.156}$$

The new eigenvectors satisfy their own orthogonality equations and we can use these to decouple Eq. (5.154) to obtain the following equation of motion

$$\ddot{\mathbf{q}}_a(t) + \begin{bmatrix} \omega_{an}^2 \end{bmatrix}\mathbf{q}_a(t) = \mathbf{T}^T\mathbf{B}^T\mathbf{x}(t) \tag{5.157}$$

Since the right-hand side of this equation is the projection of the equation of motion on the transpose of the matrix of mode shapes of the disturbed system, see Eq. (5.78), we can conclude that the mode shapes of the disturbed system are given by the transpose of the matrix product $\mathbf{T}^T\mathbf{B}^T$ that is,

$$\mathbf{A} = \mathbf{BT} \tag{5.158}$$

From this analysis, we can see that the mode shapes resulting from any change of either the mass or stiffness, or both, the mode shapes of the changed system can be obtained as a linear combination of the mode shapes of the original system.

5.3.2   Sensitivity Equations

In the preceding section, we considered arbitrary large changes of the system physical properties. In this section, we discuss the sensitivity of the mode shapes due to small changes of the system. This problem was initially discussed by Fox and Kapoor [11] and Nelson [12]. We will make use of the results from the approach to deal with this problem as presented by Heylen et al. [8]. In Heylen's book, it is shown that the sensitivity of the mode shapes to changes of a parameter u in the dynamic system can be expressed as

$$\frac{\partial \mathbf{b}_i}{\partial u} = -\frac{1}{2m_i} \mathbf{b}_i^T \frac{\partial \mathbf{M}}{\partial u} \mathbf{b}_i \mathbf{b}_i + \sum_{r=1, r \neq i}^{N} \frac{1}{\omega_i^2 - \omega_r^2} \frac{1}{m_r} \mathbf{b}_r^T \left(-\omega_i^2 \frac{\partial \mathbf{M}}{\partial u} + \frac{\partial \mathbf{K}}{\partial u} \right) \mathbf{b}_i \mathbf{b}_r \qquad (5.159)$$

where N is the number of modes in the model. Consider now a finite but small mass and stiffness changes $\Delta \mathbf{M}, \Delta \mathbf{K}$. Because all terms of the form $\mathbf{b}_r^T \Delta \mathbf{M} \mathbf{b}_i$ are inner products (scalars), we can rearrange the terms in this equation and move the last vector in the products $\mathbf{b}_r^T \Delta \mathbf{M} \mathbf{b}_i \mathbf{b}_r$ to the front. For this case, Eq. (5.159) leads to the following approximation

$$\Delta \mathbf{b}_i \cong -\frac{1}{2m_i} \mathbf{b}_i \mathbf{b}_i^T \Delta \mathbf{M} \mathbf{b}_i + \sum_{r=1, r \neq i}^{N} \frac{1}{m_r} \left(-\frac{\omega_i^2}{\omega_i^2 - \omega_r^2} \mathbf{b}_r \mathbf{b}_r^T \Delta \mathbf{M} \mathbf{b}_i + \frac{1}{\omega_i^2 - \omega_r^2} \mathbf{b}_r \mathbf{b}_r^T \Delta \mathbf{K} \mathbf{b}_i \right) \qquad (5.160)$$

If the changes are infinitesimally small, this expression is equivalent to the exact solution given by Eq. (5.159). According to Heylen et al. [8], the corresponding changes of the natural frequencies can be expressed as

$$\Delta \omega_i = \frac{\omega_i}{2m_i} \mathbf{b}_i^T \left(-\Delta \mathbf{M} + \frac{1}{\omega_i^2} \Delta \mathbf{K} \right) \mathbf{b}_i \qquad (5.161)$$

Realizing that all terms in Eq. (5.160) are proportional to the outer vector products $\mathbf{b}_r \mathbf{b}_r^T$, we can write this equation in the following form

$$\Delta \mathbf{b}_i \cong \left(\gamma_{M,1} \mathbf{b}_1 \mathbf{b}_1^T + \cdots + \gamma_{M,N} \mathbf{b}_N \mathbf{b}_N^T \right) \Delta \mathbf{M} \mathbf{b}_i + \left(\gamma_{K,1} \mathbf{b}_1 \mathbf{b}_1^T + \cdots + \gamma_{K,N} \mathbf{b}_N \mathbf{b}_N^T \right) \Delta \mathbf{K} \mathbf{b}_i \qquad (5.162)$$

A series of outer products such as $\gamma_1 \mathbf{b}_1 \mathbf{b}_1^T + \cdots + \gamma_N \mathbf{b}_N \mathbf{b}_N^T$ can always be written in matrix form as $\mathbf{B} \left[\gamma_n \right] \mathbf{B}^T$ where $\boldsymbol{\Gamma} = \left[\gamma_n \right]$ is a diagonal matrix holding the coefficients γ_n. Thus we can rewrite Eq. (5.160) as

$$\Delta \mathbf{b}_i \cong \mathbf{B} \boldsymbol{\Gamma}_{M,i} \mathbf{B}^T \Delta \mathbf{M} \mathbf{b}_i + \mathbf{B} \boldsymbol{\Gamma}_{K,i} \mathbf{B}^T \Delta \mathbf{K} \mathbf{b}_i \qquad (5.163)$$

A careful inspection of the coefficients in Eq. (5.159) shows that the diagonal matrices are given by

$$\boldsymbol{\Gamma}_{M,i} = \left[\gamma_r \right] = \begin{cases} \dfrac{-\omega_i^2}{m_r \left(\omega_i^2 - \omega_r^2 \right)} & \text{for } r \neq i \\ -\dfrac{1}{2m_i} & \text{for } r = i \end{cases} \quad ; \quad \boldsymbol{\Gamma}_{K,i} = \left[\gamma_r \right] = \begin{cases} \dfrac{1}{m_r \left(\omega_i^2 - \omega_r^2 \right)} & \text{for } r \neq i \\ 0 & \text{for } r = i \end{cases} \qquad (5.164)$$

Following the principle of local correspondence (LC) by Brincker et al. [13], it is possible to show that this leads to the following approximate solution for the relation between a perturbed mode shape cluster $\mathbf{A} = \left[\mathbf{a}_n, \mathbf{a}_{n+1} \cdots \right]$, and the corresponding mode shape cluster $\mathbf{B}$

$$\mathbf{A} \cong \mathbf{B} \left(\mathbf{I}' + \mathbf{T}' \right) \qquad (5.165)$$

where $\mathbf{I}'$ is the identity matrix, but truncated in cases where the unperturbed mode shape cluster $\mathbf{B}$ needs to be bigger than the perturbed mode shape cluster $\mathbf{A}$. Each of the column vectors in the transformation vector $\mathbf{T}'$ is given by

$$\mathbf{t}'_i = \left(\mathbf{\Gamma}_{M,i} \mathbf{B}_i^T \Delta \mathbf{M} + \mathbf{\Gamma}_{K,i} \mathbf{B}_i^T \Delta \mathbf{K} \right) \mathbf{b}_i \qquad (5.166)$$

Letting $\mathbf{T} = \mathbf{I}' + \mathbf{T}'$ proves the existence of the approximate transformation between two corresponding mode shapes clusters

$$\mathbf{A} \cong \mathbf{BT} \qquad (5.167)$$

This is the same transformation that we showed in the preceding section as an exact relation for a complete set of modes. We have now shown that for the case of a limited number of modes the transformation still holds approximately if the changes of the system are reasonably small, and if the set of unperturbed mode shapes in the matrix is large enough to give an appropriate approximation according to Eq. (5.167).

5.3.3 Closely Spaced Modes

Identification of closely spaced modes always represents a challenge. In order to understand why this is the case, we will briefly review how the basic theory of structural dynamics deals with the subject. As a first step, it is useful to consider the case where the eigenvalues are identical, also denoted as the case of repeated poles.

We are considering here a case where two eigenfrequencies, say ω_1 and ω_2, with corresponding normal modes $\mathbf{b}_1$ and $\mathbf{b}_2$ are identical, that is, $\omega_1 = \omega_2 = \omega$. The first important thing to note is that in this case the orthogonality between the two mode shapes $\mathbf{b}_1, \mathbf{b}_2$ is no longer assured. For the two corresponding mode shapes, we have in general that

$$\mathbf{b}_1^T \mathbf{M} \mathbf{b}_2 \neq 0 \qquad (5.168)$$

However, any linear combination of the two mode shapes

$$\mathbf{b} = t_1 \mathbf{b}_1 + t_2 \mathbf{b}_2 = [\mathbf{b}_1, \mathbf{b}_2] \begin{Bmatrix} t_1 \\ t_2 \end{Bmatrix} = \mathbf{Bt} \qquad (5.169)$$

is also an eigenvector. We can verify this by reference to the eigenvalue problem stated by Eq. (5.54)

$$\mathbf{M}^{-1}\mathbf{K} \left(t_1 \mathbf{b}_1 + t_2 \mathbf{b}_2 \right) = \omega^2 t_1 \mathbf{b}_1 + \omega^2 t_2 \mathbf{b}_2$$

$$= \omega^2 \mathbf{b} \qquad (5.170)$$

We can always define a new set of eigenvectors such that the new set satisfies the orthogonality conditions. For instance, we can choose the new set of eigenvectors as $\mathbf{b}_1$ and $\mathbf{b}'_2$ where the vector $\mathbf{b}'_2 = [\mathbf{b}_1, \mathbf{b}_2] \mathbf{t}$ is orthogonal to $\mathbf{b}_1$,

$$\mathbf{b}_1^T \mathbf{M} \mathbf{B} \mathbf{t} = 0 \qquad (5.171)$$

We can define the transformation vector with unit length by the angle θ if we let $\mathbf{t}^T = \{\cos \theta, \sin \theta\}$. And we can rotate the vector $\mathbf{b}_2$ in the subspace defined by $\mathbf{b}_1$ and $\mathbf{b}_2$ until $\mathbf{b}_2$ is perpendicular to $\mathbf{b}_1$, see Figure 5.10. Calculating the row vector

$$\{m_1, m_2\} = \mathbf{b}_1^T \mathbf{M} \mathbf{B} \qquad (5.172)$$

using Eq. (5.169) and recognizing that the new vector $\mathbf{b}'_2$ is orthogonal to $\mathbf{b}_1$ with respect to the mass matrix we get as a possible solution

$$\theta = \arctan \left(-m_1 / m_2 \right) \qquad (5.173)$$

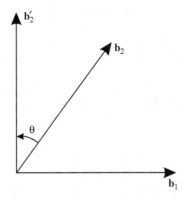

Figure 5.10 In case of repeated poles, the corresponding mode shapes are not necessarily orthogonal, but one of the vectors might be chosen for the basis – in this case $\mathbf{b}_1$ – and the other one – in this case $\mathbf{b}_2$ – is then rotated in the subspace defined by $\mathbf{b}_1$ and $\mathbf{b}_2$ until it is perpendicular to $\mathbf{b}_1$

It should be noted that this procedure does not ensure that the length (scaling) of the rotated vector is kept unchanged. If, for instance, the eigenvector $\mathbf{b}_1$ and $\mathbf{b}_2$ are scaled to unit length, then by performing this rotation will result in a new vector whose length can be reduced to

$$\sqrt{\mathbf{b'}_2^T \mathbf{b'}_2} = \sqrt{1 + 2\cos\theta \sin\theta\, \mathbf{b}_1^T \mathbf{b}_2} \tag{5.174}$$

The transformation that keeps the unit scaling is given by

$$\mathbf{t}^T = \{\cos\theta, \sin\theta\} / \sqrt{1 + 2\cos\theta \sin\theta\, \mathbf{b}_1^T \mathbf{b}_2} \tag{5.175}$$

It is important to note that in the case of repeated poles, the corresponding mode shapes do not exist as individual vectors. Rather, they only define a subspace, and if individual mode shapes are going to be used as a basis for this subspace, the choice of basis must be made by the user.

We will consider now the case where the frequencies of modes are close but not identical to each other as considered earlier. We are now considering two modes with the frequencies $\omega_1 = \omega$, $\omega_2 = \omega + \Delta\omega$ and we will assume that the frequency difference $\Delta\omega = \omega_2 - \omega_1$ is small compared to ω, say, not more than 5% difference, see Brincker and Lopez-Aenlle [14]. Using the results from the preceding section, we see that if the frequency separation is also small compared to the distance to all other modes, only the weighting terms from the two closely spaced modes will be significant, see Eq. (5.164). In this case, the approximate transformation between the perturbed and the unperturbed mode shapes given by Eq. (5.167) consists of the mode shapes of two closely spaced modes. The weighting matrices given by Eq. (5.164) reduce to

$$\Gamma_{M,1} = \frac{1}{2m}\begin{bmatrix} -1 & 0 \\ 0 & \omega/\Delta\omega \end{bmatrix} ; \quad \Gamma_{M,2} = \frac{1}{2m}\begin{bmatrix} -\omega/\Delta\omega & 0 \\ 0 & -1 \end{bmatrix}$$

$$\Gamma_{K,1} = \frac{1}{2m\omega^2}\begin{bmatrix} 0 & 0 \\ 0 & -\omega/\Delta\omega \end{bmatrix} ; \quad \Gamma_{K,2} = \frac{1}{2m\omega^2}\begin{bmatrix} \omega/\Delta\omega & 0 \\ 0 & 0 \end{bmatrix} \tag{5.176}$$

For this discussion, we have used the approximation $\omega_2^2 / \left(\omega_2^2 - \omega_1^2\right) \cong \omega / (2\Delta\omega)$, $1/\left(\omega_2^2 - \omega_1^2\right) \cong 1/(2\omega\Delta\omega)$ and we have assumed that the modal masses of the two closely spaced modes are the same.

The corresponding transformation vectors given by Eq. (5.166) are

$$\mathbf{t}_1' = \frac{1}{2m}\begin{bmatrix} -1 & 0 \\ 0 & \omega/\Delta\omega \end{bmatrix}\begin{bmatrix} \mathbf{b}_1^T \\ \mathbf{b}_2^T \end{bmatrix}\Delta\mathbf{Mb}_1 + \frac{1}{2m\omega^2}\begin{bmatrix} 0 & 0 \\ 0 & -\omega/\Delta\omega \end{bmatrix}\begin{bmatrix} \mathbf{b}_1^T \\ \mathbf{b}_2^T \end{bmatrix}\Delta\mathbf{Kb}_1$$

$$\mathbf{t}_2' = \frac{1}{2m}\begin{bmatrix} -\omega/\Delta\omega & 0 \\ 0 & -1 \end{bmatrix}\begin{bmatrix} \mathbf{b}_1^T \\ \mathbf{b}_2^T \end{bmatrix}\Delta\mathbf{Mb}_2 + \frac{1}{2m\omega^2}\begin{bmatrix} \omega/\Delta\omega & 0 \\ 0 & 0 \end{bmatrix}\begin{bmatrix} \mathbf{b}_1^T \\ \mathbf{b}_2^T \end{bmatrix}\Delta\mathbf{Kb}_2$$

(5.177)

that reduces to

$$\mathbf{t}_1' = \left\{ \begin{array}{c} -\dfrac{1}{2m}\mathbf{b}_1^T\Delta\mathbf{Mb}_1 \\ \dfrac{1}{2m}\dfrac{\omega}{\Delta\omega}\mathbf{b}_2^T\left(\Delta\mathbf{M} - \dfrac{1}{\omega^2}\Delta\mathbf{K}\right)\mathbf{b}_1 \end{array} \right\} ; \quad \mathbf{t}_2' = \left\{ \begin{array}{c} -\dfrac{1}{2m}\dfrac{\omega}{\Delta\omega}\mathbf{b}_1^T\left(\Delta\mathbf{M} - \dfrac{1}{\omega^2}\Delta\mathbf{K}\right)\mathbf{b}_2 \\ -\dfrac{1}{2m}\mathbf{b}_2^T\Delta\mathbf{Mb}_2 \end{array} \right\}$$

(5.178)

In the terms $\mathbf{b}_1^T\Delta\mathbf{Mb}_1$ and $\mathbf{b}_2^T\Delta\mathbf{Mb}_2$, the mass change is measured in terms of the inner product with the mode shapes, thus we can define the mass changes as

$$\Delta m_1 = \mathbf{b}_1^T\Delta\mathbf{Mb}_1$$

$$\Delta m_2 = \mathbf{b}_2^T\Delta\mathbf{Mb}_2$$

(5.179)

or the dimensionless mass changes as

$$\Delta\theta_1 = \frac{1}{2m}\mathbf{b}_1^T\Delta\mathbf{Mb}_1$$

$$\Delta\theta_2 = \frac{1}{2m}\mathbf{b}_2^T\Delta\mathbf{Mb}_2$$

(5.180)

If we define

$$\theta \cong \frac{\omega}{2m\Delta\omega}\left(\mathbf{b}_1^T\Delta\mathbf{Mb}_2 - \frac{1}{\omega^2}\mathbf{b}_1^T\Delta\mathbf{Kb}_2\right) = \frac{\omega}{2m\Delta\omega}\left(\mathbf{b}_2^T\Delta\mathbf{Mb}_1 - \frac{1}{\omega^2}\mathbf{b}_2^T\Delta\mathbf{Kb}_1\right)$$

(5.181)

the transformation vectors reduce to

$$\mathbf{t}_1' = \left\{ \begin{array}{c} -\Delta\theta_1 \\ \theta \end{array} \right\} ; \quad \mathbf{t}_2' = \left\{ \begin{array}{c} -\theta \\ -\Delta\theta_2 \end{array} \right\}$$

(5.182)

Here we have assumed that the change matrices must be symmetric, thus the two inner products are independent on the placement of the vectors, $\mathbf{b}_1^T\Delta\mathbf{Mb}_2 = \mathbf{b}_2^T\Delta\mathbf{Mb}_1$ and $\mathbf{b}_1^T\Delta\mathbf{Kb}_2 = \mathbf{b}_2^T\Delta\mathbf{Kb}_1$. From Eq. (5.165) and assuming that the mass changes are small, $1 - \Delta\theta_1 \cong 1 - \Delta\theta_2 \cong 1$ the new set of closely spaced modes is given by

$$[\mathbf{a}_1, \mathbf{a}_2] = [\mathbf{b}_1, \mathbf{b}_2]\left(\mathbf{I} + \begin{bmatrix} -\Delta\theta_1 & -\theta \\ \theta & -\Delta\theta_2 \end{bmatrix}\right) \cong [\mathbf{b}_1, \mathbf{b}_2]\begin{bmatrix} 1 & -\theta \\ \theta & 1 \end{bmatrix}$$

(5.183)

The transpose can be written as

$$\begin{bmatrix} \mathbf{a}_1^T \\ \mathbf{a}_2^T \end{bmatrix} = \mathbf{R}\begin{bmatrix} \mathbf{b}_1^T \\ \mathbf{b}_2^T \end{bmatrix}$$

(5.184)

where the matrix

$$\mathbf{R} = \begin{bmatrix} 1 & \theta \\ -\theta & 1 \end{bmatrix}$$

(5.185)

is a rotation matrix. In general, the definition of a rotation matrix defined by the counterclockwise rotation angle θ in a plane is

$$\mathbf{R} = \begin{bmatrix} \cos\theta & \sin\theta \\ -\sin\theta & \cos\theta \end{bmatrix} \tag{5.186}$$

However, for small rotations $\cos\theta \cong 1$, $\sin\theta \cong \theta$ and the general rotation matrix reduces to Eq. (5.185).

We can conclude that the change caused by a perturbation of the system is mainly a rotation of the unperturbed mode shapes in their subspace (see Figure 5.11). This rotation angle is proportional to the change of mass and stiffness measured by the inner products $\mathbf{b}_2^T \Delta \mathbf{M} \mathbf{b}_1$ and $\mathbf{b}_2^T \Delta \mathbf{K} \mathbf{b}_1 / \omega^2$. The rotation angle is also proportional to the frequency ratio, $\omega / \Delta\omega$. Thus the smaller the frequency separation $\Delta\omega$, the larger the sensitivity of the mode shapes to changes in mass and stiffness.

The solution derived here is only useful for "moderate closeness" of the two closely spaced modes because the transformation of vectors given by Eq. (5.178) does not make sense when $\omega / \Delta\omega \to \infty$. Also the rotation angle given by Eq. (5.181) does not make sense in this case. The solution given by these equations is only valid for moderate rotations of the mode shapes.

It can be shown that for very closely spaced modes, the rotation angle and the frequency shift can be obtained from a special eigenvalue problem only involving the mode shape vectors of the closely spaced modes, Brincker and Lopez-Aenlle [14]

$$\mathbf{B}^T \left(-\Delta \mathbf{M} + \frac{1}{\omega^2} \Delta \mathbf{K} \right) \mathbf{B} = \mathbf{T} \mathbf{D} \mathbf{T}^{-1} \tag{5.187}$$

The transformation $\mathbf{T}$ can be obtained in one step as the eigenvectors of the matrix $\mathbf{B}^T \left(-\Delta \mathbf{M} + \Delta \mathbf{K} / \omega^2 \right) \mathbf{B}$. The eigenvalues define the frequency shift $\Delta\omega_1, \Delta\omega_2$

$$\mathbf{D} = \frac{2m}{\omega} \begin{bmatrix} \Delta\omega_1 & 0 \\ 0 & \Delta\omega_2 \end{bmatrix} \tag{5.188}$$

From this equation, we can see that if the matrix $\mathbf{B}^T \left(-\Delta \mathbf{M} + \Delta \mathbf{K} / \omega^2 \right) \mathbf{B}$ is computed using the mass-scaled mode shapes, then the modal mass disappears from the problem, and the corresponding eigenvalues are proportional to two times the relative frequency shifts.

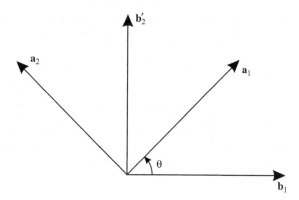

Figure 5.11 In the case of closely spaced modes, the change of mode shapes from the unperturbed set $\mathbf{b}_1$ and $\mathbf{b}_2$ to the perturbed set of mode shapes $\mathbf{a}_1$ and $\mathbf{a}_2$ the change is mainly a rotation of the unperturbed set in the subspace spanned by $\mathbf{b}_1$ and $\mathbf{b}_2$, the rotation angle is defined as the angle θ given by Eq. (5.181)

Example 5.8   Forming an orthogonal set of eigenvectors for a case of repeated poles

A 4-DOF system has the following eigenvector matrix

$$\mathbf{B} = \begin{bmatrix} 0.3670 & -0.1363 & -0.9202 & 0.4140 \\ -0.8817 & -0.2606 & -0.3131 & 0.3533 \\ 0.1022 & -0.9025 & 0.1745 & -0.4181 \\ 0.2781 & -0.3146 & 0.1576 & 0.7273 \end{bmatrix} \tag{5.189}$$

and it has two repeated eigenvalues

$$[\lambda_n] = \begin{bmatrix} 4.2229 & & & \\ & 2.979 & & \\ & & 1.1120 & \\ & & & 1.1120 \end{bmatrix} \tag{5.190}$$

We can then construct the corresponding defective mass and stiffness matrices from the orthogonality conditions in Eq. (5.67), and solve the eigenvalue problem Eq. (5.53). This gives us a new set of eigenvectors that are not orthogonal. If we perform the inner product of the eigenvectors with respect to the mass matrix, we get the result

$$\mathbf{B}^T\mathbf{MB} = \begin{bmatrix} 1.0000 & 0 & 0 & 0 \\ 0 & 1.0000 & 0 & 0 \\ 0 & 0 & 0.9999 & 0.7510 \\ 0 & 0 & 0.7510 & 1.7296 \end{bmatrix} \tag{5.191}$$

As we see, the two eigenvectors corresponding to the repeated eigenvalues are far from orthogonal. However, if we compute the rotation angle as per by Eqs. (5.172/5.173), we obtain

$$\theta = -48.17 \ \text{deg} \tag{5.192}$$

and if we rotate the last eigenvector by this angle, then the final modified set of eigenvectors constitute an orthogonal set. After normalizing the orthogonal set of eigenvectors with the square root of the modal mass, the inner product with respect to the mass matrix produces the identity matrix, and the corresponding inner product with respect to the stiffness matrix produces the eigenvalue matrix according to Eq. (5.67).

The conclusion is that when we have closely spaced modes, the individual mode shapes are sensitive to small changes of the system. This means that since small changes of the system cannot be of any real physical significance, we can conclude that, in the case of closely spaced modes, the individual mode shapes are not of any physical interest. In this case, only the subspace defined by the closely space modes is of physical importance.

5.3.4   *Model Reduction (SEREP)*

When we deal with both analytical models and experimental results, it is important to be able to compare the results of the model with the results of the experiment in detail in a meaningful way. For this purpose, it is often useful to have the analytical model in a reduced form formulated only by reference to the DOF's that were measured in the experiment.

   Different techniques exist for model reduction; however, almost all of them are approximate in one way or the other. This subject is extensive and significant amount of research has been done in the past and we are only going to give a general brief overview of the important issues of model reduction. More information about this can be found in, for instance, Chopra [3]. In this section, we will concentrate on a model reduction technique that is effective in representing the true properties of the full analytical model. This reduction technique is called "system equivalent reduction expansion process" (SEREP) and is due to O'Callahan et al. [15]. In SEREP, the starting point is the modally decomposed response given by Eq. (5.76). The full set of DOF's in the model is divided into two sets of DOF's in the following manner

$$y(t) = Bq(t) = \begin{bmatrix} B_a \\ B_d \end{bmatrix} q(t) \tag{5.193}$$

where B_a is the mode shape matrix expressed in terms of the "active" DOF's (we can think of these DOF's as being the same as the experimental DOF's), and B_d contains the mode shapes in the DOF's that are going to be "deleted" from the comparative analysis. The response in the active DOF's is then given as

$$y_a(t) = B_a q(t) \tag{5.194}$$

   If the number of active DOF's is larger than the number of modes to be used for the analysis, then we can solve this equation as an overdetermined problem to find the modal coordinates

$$\hat{q}(t) = B_a^+ y_a(t) \tag{5.195}$$

where B_a^+ is the pseudo inverse, and $\hat{q}(t)$ is an estimate of the modal coordinates. The total response is given by Eq. (5.193), and by combining this equation with Eq. (5.195) we can obtain the estimate $\hat{y}(t)$ of the total response as

$$\hat{y}(t) = B\,\hat{q}(t) = \begin{bmatrix} B_a \\ B_d \end{bmatrix} B_a^+ y_a(t) = \begin{bmatrix} B_a B_a^+ \\ B_d B_a^+ \end{bmatrix} y_a(t) = Ty_a(t) \tag{5.196}$$

where the transformation matrix T is given by

$$T = \begin{bmatrix} B_a B_a^+ \\ B_d B_a^+ \end{bmatrix} = BB_a^+ \tag{5.197}$$

   Taking the undamped form of Eq. (5.51), using the modal decomposition given by Eq. (5.193) and premultiplying both sides of the resulting expression by the transpose of the full mode shape matrix, we get

$$B^T MB\ddot{q}(t) + B^T KBq(t) = B^T x(t) \qquad \cdot \tag{5.198}$$

where $B^T x(t)$ is the modal load as given by Eq. (5.78). Using Eq. (5.193), the estimate given by Eq. (5.196) and the transformation given by Eq. (5.197), we can obtain the following equation in terms of the reduced set of DOF's

$$B^T MBB_a^+ \ddot{y}_a(t) + B^T KBB_a^+ y_a(t) = B^T x(t) \tag{5.199}$$

   By premultiplying both sides of this equation by the transpose of B_a^+ and using Eq. (5.197), one obtains

$$T^T MT\ddot{y}_a(t) + T^T KTy_a(t) = T^T x(t) \tag{5.200}$$

This is the equation of motion in the reduced set of DOF's with the reduced mass and stiffness matrices and we define these as

$$\mathbf{M}_a = \mathbf{T}^T \mathbf{M} \mathbf{T}; \quad \mathbf{K}_a = \mathbf{T}^T \mathbf{K} \mathbf{T} \tag{5.201}$$

and the corresponding modal load as

$$\mathbf{p}_a(t) = \mathbf{T}^T \mathbf{x}(t) \tag{5.202}$$

If the number of DOF's is larger than the number of modes to be used in the analysis, then the transformation matrix is just an estimate because in Eqs. (5.195–5.199), the inverse of $\mathbf{B}_a$ is an estimate. However, if the number of DOF's is the same as the number of modes, then the inverse of $\mathbf{B}_a$ becomes exact and so does the transformation matrix $\mathbf{T}$. In this case, Eq. (5.200) is the exact equation of motion in the reduced set of DOF's, and for this case, the mass and stiffness matrices given by Eq. (5.201) define the correct mass and stiffness elements for the reduced system.

5.3.5   Discrete Time Representations

By its own nature, most modern experimental data are digitally sampled, and it is often practical to have our models in the same format – that is, in discrete time. We will consider, in this section, two cases of interest related to the discrete time representation of models. One case is when we have estimated a model in discrete time and we would want to know the inherent dynamic properties of the system described by the difference equation; and the other one is when we have a model in continuous time and want to formulate a similar model in discrete time. In both cases, an easy approach to find a solution is to use a state space formulation.

When we have estimated a model in discrete time using an ARMA format; see Eq. (4.119), in which the constants are now matrices[5] the measured response $\mathbf{y}(n)$ vector is expressed as

$$\mathbf{y}(n) = \sum_{k=1}^{N} \mathbf{A}_k \mathbf{y}(n-k) + \sum_{k=0}^{M} \mathbf{B}_k \mathbf{x}(n-k) \iff$$

$$\mathbf{y}(n) - \mathbf{A}_1 \mathbf{y}(n-1) - \cdots - \mathbf{A}_N \mathbf{y}(n-N) = \mathbf{B}_0 \mathbf{x}(n) + \cdots + \mathbf{B}_M \mathbf{x}(n-M) \tag{5.203}$$

where the time step between sample points is Δt, the input to the system is given by the discrete time vector $\mathbf{x}(n)$, $\mathbf{A}_1, \ldots, \mathbf{A}_N$ are the auto regressive matrices (AR part) and $\mathbf{B}_0, \ldots, \mathbf{B}_M$ are the moving average matrices (MA part). As we will discuss later in the chapter about identification in the time domain, an ARMA model might be chosen such that both the physical properties and the statistical properties of the model are reflected in a way that we can trust the results of the estimation.[6] The MA part of this equation represents the statistics and the external forces acting on the system, whereas the AR part represents the system's physical properties. Setting the left-hand side to zero, the equation

$$\mathbf{y}(n) - \mathbf{A}_1 \mathbf{y}(n-1) - \cdots - \mathbf{A}_N \mathbf{y}(n-N) = 0 \tag{5.204}$$

describes the free response properties of the system. A way to solve this equation is to define the discrete state vector $\mathbf{u}_d$ made by stacking N response vectors

$$\mathbf{u}_d(n) = \begin{Bmatrix} \mathbf{y}(n-N+1) \\ \vdots \\ \mathbf{y}(n-1) \\ \mathbf{y}(n) \end{Bmatrix} \tag{5.205}$$

---

[5] The matrices $\mathbf{A}_k$ and $\mathbf{B}_k$ defined here should not be confused with other matrices earlier defined in relation to other subjects, for instance, the $\mathbf{A}_k$ matrices here are not the same as the residue matrices defined in Eqs. (5.132–5.133).
[6] If the ARMA model is covariance equivalent.

and defining the companion matrix for discrete time

$$
\mathbf{A}_C = \begin{bmatrix} 0 & \mathbf{I} & 0 & 0 \\ \vdots & 0 & \ddots & \vdots \\ 0 & \vdots & & \mathbf{I} \\ \mathbf{A}_N & \mathbf{A}_{N-1} & \cdots & \mathbf{A}_1 \end{bmatrix} \tag{5.206}
$$

where $\mathbf{I}$ is an identity matrix of the same size as the autoregressive matrices. Realizing that Eq. (5.204) can be written as

$$
\mathbf{y}(n+1) = \mathbf{A}_1 \mathbf{y}(n) + \cdots + \mathbf{A}_N \mathbf{y}(n-N+1) \tag{5.207}
$$

and using the definitions of the state vector and the companion matrix we have that

$$
\mathbf{A}_C \mathbf{u}_d(n) = \mathbf{u}_d(n+1) \tag{5.208}
$$

We see that the companion matrix advances the state vector one time step. This can be used to form an eigenvalue problem noting that the observation matrix $\mathbf{P}$ picks the mode shape $\mathbf{b}$ from the state space mode shape $\boldsymbol{\varphi}$ as $\mathbf{b} = \mathbf{P}\boldsymbol{\varphi}$ and from Eq. (5.205) we obtain for a response containing only one single mode

$$
\mathbf{u}_d(n) = \begin{Bmatrix} \mathbf{P}\boldsymbol{\varphi} e^{\lambda(n-N+1)\Delta t} \\ \vdots \\ \mathbf{P}\boldsymbol{\varphi} e^{\lambda(n-1)\Delta t} \\ \mathbf{P}\boldsymbol{\varphi} e^{\lambda n \Delta t} \end{Bmatrix} = \begin{Bmatrix} \mathbf{b} e^{\lambda(-N+1)\Delta t} \\ \vdots \\ \mathbf{b} e^{\lambda(-1)\Delta t} \\ \mathbf{b} \end{Bmatrix} e^{\lambda n \Delta t} = \boldsymbol{\varphi}_d \mu^n \tag{5.209}
$$

where λ is the continuous time pole, $\mu = e^{\lambda \Delta t}$ is the discrete time pole and $\boldsymbol{\varphi}_d$ is the discrete eigenvector. Using this result in Eq. (5.208) we get

$$
\mathbf{A}_C \boldsymbol{\varphi}_d = \mu \boldsymbol{\varphi}_d \tag{5.210}
$$

which is the eigenvalue problem that we are looking for. This means that the discrete mode shapes and the discrete poles can be found as the eigenvectors and eigenvalues to the companion matrix $\mathbf{A}_C$. As before for the state space solution, we only use the first part (or the last part) of the eigenvectors corresponding to the size of the data vector $\mathbf{y}$. Once the continuous time eigenvalues and the mode shapes are known, all continuous time properties are defined. This is because the eigenvalues and the mode shapes define the continuous time system as we have seen in the preceding sections.

If we have a continuous time model and want to convert this model to discrete time, we can define the state space formulation in terms of Eq. (5.102). However, in this case, we will use another state space formulation of Eq. (5.51) of the form

$$
\dot{\mathbf{u}}(t) = \mathbf{A}\mathbf{u}(t) + \mathbf{B}\mathbf{x}(t)
$$
$$
\mathbf{y}(t) = \mathbf{P}\mathbf{u}(t) \tag{5.211}
$$

where the state vector is given by Eq. (5.101), but where the system matrix $\mathbf{A}$, the load distribution matrix $\mathbf{B}$ and the observation matrix $\mathbf{P}$ are now given by

$$
\mathbf{A} = \begin{bmatrix} -\mathbf{M}^{-1}\mathbf{C} & -\mathbf{M}^{-1}\mathbf{K} \\ \mathbf{I} & [0] \end{bmatrix}; \quad \mathbf{B} = \begin{bmatrix} \mathbf{M}^{-1} \\ [0] \end{bmatrix}; \quad \mathbf{P} = \begin{bmatrix} [0] & \mathbf{I} \end{bmatrix} \tag{5.212}
$$

Mode shapes and eigenvalues are found by the modal decomposition

$$
\mathbf{A} = \begin{bmatrix} \boldsymbol{\varphi}_n \end{bmatrix} \begin{bmatrix} \lambda_n \end{bmatrix} \begin{bmatrix} \boldsymbol{\varphi}_n \end{bmatrix}^{-1} \tag{5.213}
$$

where the matrix again $\left[\boldsymbol{\varphi}_n\right]$ contains the mode shapes as shown by Eq. (5.107), and the diagonal matrix $\left[\lambda_n\right]$ contains the corresponding eigenvalues.

By using a state space formulation such as the one given by Eqs. (5.211/5.212), the general solution can be readily obtained as, see for instance Kailath [16] or Appendix C,

$$\mathbf{u}(t) = \exp(\mathbf{A}t)\,\mathbf{u}(0) + \int\limits_0^t \exp(\mathbf{A}(t-\tau))\,\mathbf{B}\mathbf{x}(\tau)\,d\tau \tag{5.214}$$

The first term is the solution to the homogenous equation (the transient response due to general initial conditions of the system) and the second term is the particular solution (related to the external forces acting on the system). It should be noted that the exponential function of a matrix is defined by its power series.[7]

To convert this solution to discrete time, we sample all variables, such as $\mathbf{y}(n) = \mathbf{y}(n\Delta t)$, and the solution to the homogenous equation becomes

$$\mathbf{u}(n) = \exp(\mathbf{A}n\Delta t)\,\mathbf{u}_0 = \mathbf{D}^n\mathbf{u}_0 \tag{5.215}$$

where $\mathbf{u}_0 = \mathbf{u}(0)$ is the vector of initial conditions of the system. The discrete time system matrix is given by

$$\mathbf{D} = \exp(\mathbf{A}\Delta t) \tag{5.216}$$

and the free response solution is given by

$$\mathbf{y}(n) = \mathbf{P}\mathbf{D}^n\mathbf{u}_0 \tag{5.217}$$

Forming the discrete state vector for the present case using Eq. (5.205) and stacking two response vectors, we obtain

$$\mathbf{u}_d(n) = \left\{ \begin{array}{c} \mathbf{y}(n-1) \\ \mathbf{y}(n) \end{array} \right\} \tag{5.218}$$

which we can rewrite using Eq. (5.217) as

$$\mathbf{u}_d(n) = \left\{ \begin{array}{c} \mathbf{y}(n-1) \\ \mathbf{y}(n) \end{array} \right\} = \left\{ \begin{array}{c} \mathbf{P}\mathbf{D}^{n-1}\mathbf{u}_0 \\ \mathbf{P}\mathbf{D}^n\mathbf{u}_0 \end{array} \right\} = \begin{bmatrix} \mathbf{P} \\ \mathbf{P}\mathbf{D} \end{bmatrix} \mathbf{D}^{n-1}\mathbf{u}_0 \tag{5.219}$$

Assuming that the square matrix $\begin{bmatrix} \mathbf{P} \\ \mathbf{P}\mathbf{D} \end{bmatrix}$ can be inverted, we can isolate the term $\mathbf{D}^{n-1}\mathbf{u}_0$ as

$$\mathbf{D}^{n-1}\mathbf{u}_0 = \begin{bmatrix} \mathbf{P} \\ \mathbf{P}\mathbf{D} \end{bmatrix}^{-1} \mathbf{u}_d(n) \tag{5.220}$$

Writing Eq. (5.219) for the time step $n+1$ and using Eq. (5.220), we get

$$\mathbf{u}_d(n+1) = \begin{bmatrix} \mathbf{P} \\ \mathbf{P}\mathbf{D} \end{bmatrix} \mathbf{D}^n\mathbf{u}_0$$

$$= \begin{bmatrix} \mathbf{P} \\ \mathbf{P}\mathbf{D} \end{bmatrix} \mathbf{D} \begin{bmatrix} \mathbf{P} \\ \mathbf{P}\mathbf{D} \end{bmatrix}^{-1} \mathbf{u}_d(n) \tag{5.221}$$

---

[7] See Chapter 3 about matrices and estimation or see the specific explanation in Appendix C.

By comparing this equation with Eq. (5.208), we see that the discrete time companion matrix is given by

$$A_C = \begin{bmatrix} P \\ PD \end{bmatrix} D \begin{bmatrix} P \\ PD \end{bmatrix}^{-1} \tag{5.222}$$

The AR coefficient matrices can then be identified from the bottom half part of the companion matrix according to Eq. (5.206).

Example 5.9   Discrete formulation of a 2DOF system

We consider the same case described in Example 5.7. We form the system matrix according to Eq. (5.212) and perform an eigenvalue decomposition to find the mode shapes and the poles. The results are exactly the same as those given in Example 5.7.

To simulate a free response in discrete time, we can calculate the companion matrix according to Eq. (5.222). However, first we need to estimate the discrete system matrix using the matrix function expansion formula given by Eqs. (3.54) and (5.213). This gives

$$D = \exp(A\Delta t) = [\boldsymbol{\varphi}_n] \left[\exp(\lambda_n \Delta t)\right] [\boldsymbol{\varphi}_n]^{-1} \tag{5.223}$$

and then the companion matrix is found to be

$$A_C = \begin{bmatrix} 0 & 0 & 1 & 0 \\ 0 & 0 & 0 & 1 \\ -0.9377 & 0 & 1.916 & 0.0071 \\ -0.0207 & -1.0000 & 0.0279 & 1.9780 \end{bmatrix} \tag{5.224}$$

The AR matrices are then identified in accordance to Eq. (5.206)

$$A_1 = \begin{bmatrix} 1.9163 & 0.0071 \\ 0.0279 & 1.9780 \end{bmatrix} ; \quad A_2 = \begin{bmatrix} -0.9377 & 0 \\ -0.0207 & -1.0000 \end{bmatrix} \tag{5.225}$$

For a given set of initial conditions, say for instance

$$\mathbf{u}_d(0) = \begin{Bmatrix} \mathbf{y}(-1) \\ \mathbf{y}(0) \end{Bmatrix} = \begin{Bmatrix} 0 \\ 1 \\ 0 \\ 1 \end{Bmatrix} \tag{5.226}$$

we can calculate the free response using either the formulation given by Eq. (5.207) or the state space formulation given by Eq. (5.208). The results are the same; both estimates are shown in Figure 5.12.

5.3.6   Simulation of OMA Responses

Performing a simulation in relation to OMA normally means to simulate random responses of a considered dynamic system. The basic idea in doing such a simulation is that we know the exact properties

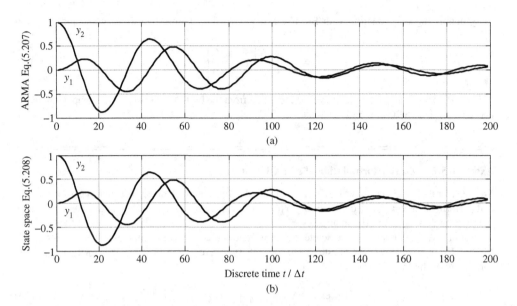

Figure 5.12 Showing two different estimates of a free decay for the same system as considered in Example 5.7. (a) Shows the free response estimated using an ARMA model according to Eq. (5.207). (b) Shows the free response estimated using the state space formulation according to (5.208)

of the underlying physical system, and, therefore, these properties can be compared to the results of the OMA analysis of the simulated data. The possibility for comparing the true physical values with the information obtained from OMA analysis of the simulated data makes the simulation approach important for investigating the applicability and reliability of testing setups and identification procedures.

Since the idea of a simulation is to determine the correct physical properties of the system, it is essential to use simulation techniques that do not introduce any significant bias on the physical properties of the system. For this reason, traditional procedures based on numerical solution of the differential equations, for instance, the Runge–Kutta-based techniques and similar procedures, should be avoided.

Reliable and accurate simulations can be based on covariance equivalent ARMA models of the form given by Eq. (5.203) because when the models are covariance equivalent, the physical properties of the ARMA model is the same as the physical properties of the initial system and thus no bias is present. This approach is to be preferred for cases where the simulated response is of short duration and where the FFT-based techniques are significantly influenced by leakage errors.

However, we shall assume that the simulated response is a typical OMA response that is relatively long compared to the memory of the system, and leakage introduced by FFT methods is minimum and has little effect on the simulation results. Therefore, FFT-based simulations with long duration records can be effectively used to test OMA identification techniques. The basis of FFT simulation is the simple linear relation given by Eq. (5.86), which can also be formulated in the frequency domain as

$$\tilde{\mathbf{y}}(\omega) = \tilde{\mathbf{H}}(i\omega)\,\tilde{\mathbf{x}}(\omega) \tag{5.227}$$

It follows from the derivations in Section 5.2.5 that the equation is always valid and that the FRF function matrix $\mathbf{H}(i\omega)$ should be estimated from Eq. (5.133). The advantage of using this equation for simulation is the simplicity of the procedure to be implemented to generate response data. The procedure consists of three simple steps:

1. The random system input vector given as a function of time $\mathbf{x}(t)$ is Fourier transformed.
2. The Fourier transform load vector $\widetilde{\mathbf{x}}(\omega)$ is multiplied by the FRF matrix as per Eq. (5.227).
3. The Fourier transform of the random response $\widetilde{\mathbf{y}}(\omega)$ is transformed back to time domain by the inverse Fourier transform to obtain the time domain response $\mathbf{y}(t)$.

The procedure is closely related to FFT filtering; see Chapter 8 on this issue. If the total number of data points is not too large to introduce memory allocation problems, then the FFT-based simulation can be performed as one big FFT; however, if the amount of data becomes a problem for the computer, the simulation can be performed using smaller data segment as described in Section 8.3.5.

References

[1] Newland, D.E.: *Mechanical vibration analysis and computation*. Dover Publications, Inc., 1989.
[2] Thompson, W.T.: *Theory of vibration with applications*. 2nd edition. George Allen & Unwin Ltd., 1981.
[3] Chopra, A.K.: *Dynamics of structures – theory and applications to earthquake engineering*, Prentice Hall, Boston, 4th edition, 2012.
[4] Bendat, J.S. and Piersol, A.G.: *Random data, analysis and measurement procedures*. 2nd edition. John Wiley & Sons, 1986.
[5] Caughey, T.H. and O'Kelly, M.E.J.: *Classical modes in damped linear systems. J. Appl. Mech ASME*, V. 32, p. 583–588, 1965.
[6] Hoen, C.: *An engineering interpretation of the complex eigensolution of linear dynamic systems*. In proceedings of the IMAC-XXIII, Jan 31–Feb 3, 2005.
[7] Hurty, W.C. and Rubinstein, M.F.: *Dynamics of structures*. Englewood Cliffs, New Jersey, Prentice Hall, 1964.
[8] Heylen, W., Lammens, S. and Sas, P.: *Modal analysis theory and testing*. Department of Mechanical Engineering, Katholieke Universiteir Leuven, 2007.
[9] Veletsos, A.S. and Ventura, C.E.: *Modal Analysis of Nonclassically Damped Linear Systems, J. Earthq. Eng. Struct. Dyn.*, V. 14, p. 217–243, 1986.
[10] Sestieri A., D'Ambrogio W.: *Structural dynamic modification. Encyclopedia of vibration*. London, Academic Press, 2001.
[11] Fox, R.L. and Kapoor, M.P.: *Rates of changes of eigenvalues and eigenvectors. AIAA J.*, V. 6, p. 2426–2429, 1968.
[12] Nelson, R.B.: *Simplified calculation of eigenvector derivatives. AIAA J.* 14(9), p. 1201–1205, 1976.
[13] Brincker, R. Skafte, A., López-Aenlle, M., Sestieri, A., D'Ambrogio, W. and Canteli, A.: *A local correspondence principle for mode shapes in structural dynamics. Mech. Syst. Signal Process.*, V. 45, No. 1, p. 91–104, 2014.
[14] Brincker, R. and López-Aenlle, M.: *Mode shape sensitivity of two closely spaced eigenvalues. J. Sound Vibr.*, V. 334, p. 377–387, 2015.
[15] O'Callahan, J., Avitabile, P. and Riemer, R.: *System equivalent reduction expansion process*. In 7th International Modal Analysis Conference (IMAC). Las Vegas, SEM, 1989.
[16] Kailath, T.: *Linear systems*. Prentice-Hall, Inc. 1980.

6

Random Vibrations

"We avoid the gravest difficulties when, giving up the attempt to frame hypotheses concerning the constitution of matter, we pursue statistical inquiries as a branch of rational mechanics."

– J.W. Gibbs

In this chapter, we shall consider how we can describe the random responses from a linear system that is loaded by environmental loads such as wind, traffic, or waves. We shall argue that in case of approximately white noise loading or in case of uncorrelated modal coordinates, the response of the system can be decoupled into SDOF degree-of-freedom systems, each of which is describing the random response of one mode.

In simple terms, this means that correlation function can be interpreted as free decays, and spectral densities can be interpreted as the corresponding frequency domain functions. This is the basic idea of all OMA.

6.1 General Inputs

6.1.1 Linear Systems

We are considering a linear system characterized by the impulse response function

$$h(t) \leftrightarrow H(s) \tag{6.1}$$

where $H(s)$ is the Laplace transform and the associated Fourier transform is $H(i\omega)$.[1] Instead of operating with the complex function $H(i\omega)$, it is sometimes normal to work with the real function $|H(i\omega)|^2$ representing the energy of the system as function of frequency. The corresponding time function

$$g(t) \leftrightarrow |H(i\omega)|^2 = H(i\omega) H^*(i\omega) \tag{6.2}$$

is called the deterministic correlation function of the system which can be considered as a measure of the system energy. From the transform theory using the time reversal and the convolution property we have

$$g(t) = h(-t) * h(t) \tag{6.3}$$

---

[1] In this section, we shall stay with the Laplace transform whenever we are considering the impulse response function, and in case we need the frequency response function we shall substitute the Laplace variable s with the variable $s = i\omega$ as we have done here.

Introduction to Operational Modal Analysis, First Edition. Rune Brincker and Carlos Ventura.
© 2015 John Wiley & Sons, Ltd. Published 2015 by John Wiley & Sons, Ltd.
Companion Website: www.wiley.com/go/brincker

In a single-input-single-output (SISO) case, the input $x(t)$ and the output $y(t)$ are related by

$$y(t) = h(t) * x(t) \tag{6.4}$$

In the following, we summarize some concepts of classical dynamics from Chapter 5. The reader is referred to Section 5.2.5 for more information about the following equations. In the general multiple-input-multiple-output (MIMO) case, the input-response relation is given by

$$\mathbf{y}(t) = \mathbf{H}(t) * \mathbf{x}(t) \tag{6.5}$$

where the matrix $\mathbf{H}(t)$ contains the impulse response functions of the system that are to be convoluted with the external loads contained in the force vector $\mathbf{x}(t)$. Taking the Laplace transform of both sides of Eq. (6.5) yields the well-known general input–output relation

$$\tilde{\mathbf{y}}(s) = \tilde{\mathbf{H}}(s)\tilde{\mathbf{x}}(s) \tag{6.6}$$

where $\tilde{\mathbf{H}}(s)$ is the transfer function matrix, which can be expressed as

$$\tilde{\mathbf{y}}(\omega) = \tilde{\mathbf{H}}(i\omega)\tilde{\mathbf{x}}(\omega) \tag{6.7}$$

in the frequency domain, where $\tilde{\mathbf{H}}(i\omega)$ is the FRF matrix and $\tilde{\mathbf{x}}(\omega)$, $\tilde{\mathbf{y}}(\omega)$ are the Fourier transforms of the input and output, respectively. As we have seen in Chapter 5, for a linear N-DOF system, the FRF matrix can always be written in the form, see Eq. (5.133)

$$\tilde{\mathbf{H}}(i\omega) = \sum_{n=1}^{N}\left(\frac{\mathbf{A}_n}{i\omega - \lambda_n} + \frac{\mathbf{A}_n^*}{i\omega - \lambda_n^*}\right) \tag{6.8}$$

and the corresponding impulse response function matrix is given by, see Eq. (5.136)

$$\mathbf{H}(t) = \sum_{n=1}^{N}\left(\mathbf{A}_n e^{\lambda_n t} + \mathbf{A}_n^* e^{\lambda_n^* t}\right) \tag{6.9}$$

where the residue matrices $\mathbf{A}_n$ are rank one matrices computed by the outer products of the mode shape vectors $\mathbf{b}_n$

$$\mathbf{A}_n = \frac{\mathbf{b}_n \mathbf{b}_n^T}{a_n} \tag{6.10}$$

where a_n are the generalized modal masses found by the inner products of the mode shape vectors over the state space matrix as given by Eq. (5.112). Instead of performing the summation over the N modes as done in Eqs. (6.8/6.9), we can perform the summation over the $2N$ modal parameters

$$\mathbf{H}(t) = \sum_{n=1}^{2N}\frac{\mathbf{b}_n \mathbf{b}_n^T}{a_n} e^{\lambda_n t} = \sum_{n=1}^{2N}\mathbf{A}_n e^{\lambda_n t} = \sum_{n=1}^{2N}\mathbf{b}_n \mathbf{b}_n^T h_n(t)$$

$$\mathbf{A}_n = \frac{\mathbf{b}_n \mathbf{b}_n^T}{a_n}; \quad h_n(t) = e^{\lambda_n t}/a_n \tag{6.11}$$

where mode shapes and other modal quantities all appear in complex conjugate pairs if all the modes are undercritically damped and $h_n(t)$ are the associated modal impulse response functions.

6.1.2   Spectral Density

The auto spectral density function for a time series $x(t)$ is defined as the Fourier transform of the correlation function $R_x(\tau)$

$$G_x(\omega) = \frac{1}{2\pi} \int_{-\infty}^{\infty} R_x(\tau) e^{-i\omega\tau} d\tau \qquad (6.12)$$

conversely the correlation function can be found from the inverse relation

$$R_x(\tau) = \int_{-\infty}^{\infty} G_x(\omega) e^{i\omega\tau} d\omega \qquad (6.13)$$

The most important property of the spectral density becomes apparent when we evaluate the initial value of the correlation function according to the expressions given by Eq. (6.13). Using the definition of the correlation function Eq. (2.27) we get

$$R_x(0) = E\left[x^2\right] = \int_{-\infty}^{\infty} G_x(\omega) d\omega \qquad (6.14)$$

This equation is a mathematical representation of the Parseval theorem. Since $\sigma_x^2 = E\left[x^2\right]$ for a zero mean signal, we see that the area under the spectral density equals the variance of the signal, see the top plot of Figure 6.1. If we consider the case where the time series $x(t)$ is bandpass filtered so that only

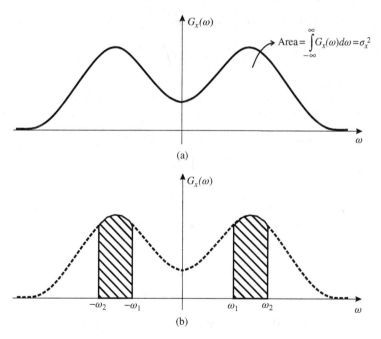

(a)

(b)

Figure 6.1   (a) Illustrates Parseval's theorem stating that the area under the auto spectral density equals the variance of the signal. (b) Illustrates that for a bandpass-filtered signal the area of the spectral density inside the band defines the variance of the filtered signal. The auto spectral density illustrates the distribution of energy as a function of frequency

the frequency content of the signal from frequency ω_1 to frequency ω_2 is kept (see Figure 1.6b) then the variance of the band filtered signal σ_{xb}^2 is simply

$$\sigma_{xb}^2 = 2 \int_{\omega_1}^{\omega_2} G_x(\omega) \, d\omega \tag{6.15}$$

The auto spectral density shows the distribution of energy as a function of frequency, and therefore the spectral density is often also called the power spectral density (PSD).

We should note that since we have made the choice to extend the Fourier coefficients (see the section about the complex Fourier series in Chapter 4) to negative frequencies, the spectral density is defined for both negative and positive frequencies. For some people, the interpretation of negative frequencies is problematic, so in some texts, this "problem" is removed by defining a one-sided spectral density that is twice the double-sided function. To keep things simple, we shall not use this definition of the one-sided function.

Similarly the cross spectral density function for two time series $x(t)$ and $y(t)$ is defined as the Fourier transform of the cross-correlation function $R_{xy}(\tau)$ – so we define

$$G_{xy}(\omega) = \frac{1}{2\pi} \int_{-\infty}^{\infty} R_{xy}(\tau) e^{-i\omega\tau} \, d\tau$$

$$R_{xy}(\tau) = \int_{-\infty}^{\infty} G_{xy}(\omega) e^{i\omega\tau} \, d\omega \tag{6.16}$$

In this case, Parseval's theorem gives us the relation

$$R_{xy}(0) = E[xy] = \int_{-\infty}^{\infty} G_{xy}(\omega) \, d\omega \tag{6.17}$$

For this case, the area under the cross spectral density equals the covariance $cov[x, y] = E[xy]$. For vector signals $\mathbf{x}(t)$ and $\mathbf{y}(t)$, the correlation function (CF) matrices are defined as given by Eq. (2.35)

$$\mathbf{R}_x(\tau) = E\left[\mathbf{x}(t)\mathbf{x}^T(t+\tau)\right]$$

$$\mathbf{R}_y(\tau) = E\left[\mathbf{y}(t)\mathbf{y}^T(t+\tau)\right] \tag{6.18}$$

$$\mathbf{R}_{xy}(\tau) = E\left[\mathbf{x}(t)\mathbf{y}^T(t+\tau)\right]$$

The corresponding spectral density (SD) matrices $\mathbf{G}_x(\omega), \mathbf{G}_y(\omega), \mathbf{G}_{xy}(\omega)$ are given by the Fourier transform, and similar to Eqs. (6.14/6.17) we have the corresponding implementation of Parseval's theorem

$$\mathbf{R}_x(0) = E\left[\mathbf{x}\mathbf{x}^T\right] = \mathbf{C}_x = \int_{-\infty}^{\infty} \mathbf{G}_x(\omega) \, d\omega$$

$$\mathbf{R}_y(0) = E\left[\mathbf{y}\mathbf{y}^T\right] = \mathbf{C}_y = \int_{-\infty}^{\infty} \mathbf{G}_y(\omega) \, d\omega \tag{6.19}$$

$$\mathbf{R}_{xy}(0) = E\left[\mathbf{x}\mathbf{y}^T\right] = \mathbf{C}_{xy} = \int_{-\infty}^{\infty} \mathbf{G}_{xy}(\omega) \, d\omega$$

where $\mathbf{C}_x, \mathbf{C}_y, \mathbf{C}_{xy}$ are the covariance matrices of the (zero mean) signals. Using the definition of the Fourier transform, Eq. (4.32), and the symmetry relation of the correlation function matrix, Eq. (2.35), it is easy to see that the spectral density matrix is always Hermitian, which means that any spectral matrix $\mathbf{G}(\omega)$ is always equal to its complex conjugate transpose

$$\mathbf{G}^H(\omega) = \mathbf{G}(\omega) \qquad (6.20)$$

Using the time reversal property of the Fourier transform, Eq. (4.34), the SD matrix has similar symmetry properties as the CF matrix, see Eq. (2.35)

$$\mathbf{G}^T(\omega) = \mathbf{G}(-\omega) \qquad (6.21)$$

Using the properties of the correlation function proved in Section 2.2.4 as well as the equations

$$\frac{d}{d\tau} R_{xy}(\tau) = \int_{-\infty}^{\infty} G_{xy}(\omega) i\omega e^{i\omega\tau} d\omega$$

$$\frac{d^2}{d\tau^2} R_{xy}(\tau) = \int_{-\infty}^{\infty} G_{xy}(\omega) \left(-\omega^2\right) e^{i\omega\tau} d\omega \qquad (6.22)$$

we have the following simple relations between the derivative of the correlation function and its spectral density

$$\dot{R}_{xy}(\tau) \leftrightarrow G_{xy}(\omega) i\omega$$

$$\ddot{R}_{xy}(\tau) = -R_{\dot{x}\dot{y}}(\tau) \leftrightarrow -\omega^2 G_{xy}(\omega) \qquad (6.23)$$

and it follows that

$$G_{\dot{x}\dot{y}}(\omega) = \omega^2 G_{xy}(\omega)$$

$$G_{\ddot{x}\ddot{y}}(\omega) = \omega^4 G_{xy}(\omega) \qquad (6.24)$$

Using the Parseval theorem (6.14) this gives the following expressions for the covariance of the velocity and acceleration signals

$$\sigma_{\dot{x}}^2 = \mathrm{E}\left[\dot{x}^2\right] = \int_{-\infty}^{\infty} \omega^2 G_x(\omega) d\omega$$

$$\sigma_{\ddot{x}}^2 = \mathrm{E}\left[\ddot{x}^2\right] = \int_{-\infty}^{\infty} \omega^4 G_x(\omega) d\omega \qquad (6.25)$$

Finally using the convolution property of the correlation function given by Eq. (2.48) and the convolution property and time reversal property of the Fourier transform, Eq. (4.34), we obtain the following important property of the spectral density under the assumption of periodic data (or very long data segments)

$$G_{xy}(\omega) = X^*(\omega) Y(\omega) \qquad (6.26)$$

6.1.3   SISO Fundamental Theorem

Now let us consider a single output $y(t)$ from a system loaded by a single input $x(t)$. The relation between the input and the output is given by Eq. (6.4), and according to Eq. (6.26) the output spectral density is given by

$$G_y(\omega) = Y^*(\omega) Y(\omega) \tag{6.27}$$

Taking the Fourier transform of the input–output relation given by Eq. (6.4) and using the convolution property of the transform we get

$$G_y(\omega) = Y^*(\omega) X(\omega) H(i\omega) = G_{yx}(\omega) H(i\omega)$$
$$G_{yx}(\omega) = Y^*(\omega) X(\omega) = G_x(\omega) H^*(i\omega) \tag{6.28}$$

And finally

$$G_y(\omega) = X^*(\omega) X(\omega) H^*(i\omega) H(i\omega) = G_x(\omega) |H(i\omega)|^2 \tag{6.29}$$

Together the three Eqs. (6.27–6.29) are called the fundamental equations, and Eq. (6.29) is called the fundamental theorem. Since these equations relate the input spectral densities with the output spectral densities in a very simple way, they are perhaps the most important equations in the whole theory of random vibration. However, since they are so important and in this first step have been derived under the simplified assumptions of periodic data, we will repeat the derivation under more general conditions.

Let us consider the response auto correlation function $R_y(\tau) = E\left[y(t) y(t+\tau)\right]$. Using Eq. (6.4) and the definition of the convolution, we have

$$R_y(\tau) = E\left[y(t) y(t+\tau)\right] = \int_{-\infty}^{\infty} E\left[y(t) x(t+\tau-\alpha)\right] h(\alpha)\, d\alpha \tag{6.30}$$

But since $E\left[y(t) x(t+\tau-\alpha)\right]$ is equal to the cross-correlation function $R_{yx}(\tau-\alpha)$ we have the convolution

$$R_y(\tau) = \int_{-\infty}^{\infty} R_{yx}(\tau-\alpha) h(\alpha)\, d\alpha = R_{yx}(\tau) * h(\tau) \tag{6.31}$$

Similarly for the cross correlation, we have

$$R_{yx}(\tau) = E\left[y(t) x(t+\tau)\right] = \int_{-\infty}^{\infty} E[x(t-\alpha) x(t+\tau)] h(\alpha)\, d\alpha \tag{6.32}$$

But since $E[x(t-\alpha) x(t+\tau)]$ is equal to the input auto correlation function $R_x(\tau+\alpha)$, by changing the integration variable to $\beta = -\alpha$, we have

$$R_{yx}(\tau) = \int_{-\infty}^{\infty} R_x(\tau+\alpha) h(\alpha)\, d\alpha = \int_{-\infty}^{\infty} R_x(\tau-\beta) h(-\beta)\, d\beta = R_x(\tau) * h(-\tau) \tag{6.33}$$

Or by combining Eq. (6.30) with Eq. (6.32) and Eq. (6.3) we get

$$R_y(\tau) = R_{yx}(\tau) * h(\tau) = R_x(\tau) * h(-\tau) * h(\tau) = R_x(\tau) * g(t) \tag{6.34}$$

Taking the Fourier transform of Eqs. (6.30) and (6.32), we have

$$G_y(\omega) = G_{yx}(\omega) H(i\omega)$$
$$G_{yx}(\omega) = G_x(\omega) H^*(i\omega) \tag{6.35}$$

If we combine these two equations, the following relation is obtained

$$G_y(\omega) = G_x(\omega) H^*(i\omega) H(i\omega) = G_x(\omega) |H(i\omega)|^2 \tag{6.36}$$

and the important fundamental equation in the general case of stationary signals is demonstrated.

6.1.4 MIMO Fundamental Theorem

The fundamental equations can be derived in a similar way for the general MIMO case. The correlation function matrix of the response is

$$\mathbf{R}_y(\tau) = E\left[\mathbf{y}(t)\,\mathbf{y}^T(t+\tau)\right] \tag{6.37}$$

Now using Eq. (6.5) and that the transpose of a matrix convolution follow the same rules as a simple matrix multiplication, that is

$$\mathbf{y}^T(t) = \mathbf{x}^T(t) * \mathbf{H}^T(t) \tag{6.38}$$

Using the definition of the convolution we obtain

$$\mathbf{R}_y(\tau) = \int_{-\infty}^{\infty} E\left[\mathbf{y}(t)\,\mathbf{x}^T(t+\tau-\alpha)\right] \mathbf{H}^T(\alpha)\,d\alpha \tag{6.39}$$

However, since $E\left[\mathbf{y}(t)\,\mathbf{x}^T(t+\tau-\alpha)\right]$ is the cross correlation matrix $\mathbf{R}_{yx}(\tau-\alpha)$ between the responses and the inputs

$$\mathbf{R}_y(\tau) = \int_{-\infty}^{\infty} \mathbf{R}_{yx}(\tau-\alpha)\mathbf{H}^T(\alpha)\,d\alpha = \mathbf{R}_{yx}(\tau) * \mathbf{H}^T(\tau) \tag{6.40}$$

Similarly, for the cross-correlation matrix, we find

$$\mathbf{R}_{yx}(\tau) = E\left[\mathbf{y}(t)\,\mathbf{x}^T(t+\tau)\right]$$
$$= \int_{-\infty}^{\infty} \mathbf{H}(\alpha) E\left[\mathbf{x}(t-\alpha)\,\mathbf{x}^T(t+\tau)\right] d\alpha \tag{6.41}$$

Since $E\left[\mathbf{x}(t-\alpha)\,\mathbf{x}^T(t+\tau)\right]$ is equal to the cross-correlation matrix $\mathbf{R}_x(\tau+\alpha)$ between the inputs, and by changing the integration variable to $\beta = -\alpha$ we can write

$$\mathbf{R}_{yx}(\tau) = \int_{-\infty}^{\infty} \mathbf{H}(-\beta)\mathbf{R}_x(\tau-\beta)d\beta = \mathbf{H}(-\tau) * \mathbf{R}_x(\tau) \tag{6.42}$$

By combining Eqs. (6.40) and (6.42) we have the general fundamental theorem

$$\mathbf{R}_y(\tau) = \mathbf{H}(-\tau) * \mathbf{R}_x(\tau) * \mathbf{H}^T(\tau) \tag{6.43}$$

Finally taking the Fourier transform of Eq. (6.43) and using the convolution property and the time reversal property of the transform, we arrive at the well-known fundamental theorem in the frequency domain

$$G_y(\omega) = \tilde{\mathbf{H}}^*(i\omega)\,G_x(\omega)\,\tilde{\mathbf{H}}^T(i\omega)$$
$$= \tilde{\mathbf{H}}(-i\omega)\,G_x(\omega)\,\tilde{\mathbf{H}}(i\omega) \tag{6.44}$$

The last equation follows from the identity $\widetilde{\mathbf{H}}^*(i\omega) = \widetilde{\mathbf{H}}(-i\omega)$ and from the fact that the transfer function is symmetric[2]. For the case where the FRF matrix is given numerically, the upper part of Eq. (6.44) is preferable because one has to take only the complex conjugate of all numbers in the matrix. However, for the case of an analytical problem where complex constants define the values of the matrix elements of the FRF matrix, it is not needed to take the complex conjugate of the involved complex constants. In such case, the lower part is preferable because one has to change only the sign on the frequency variable as indicated by the lower part of Eq. (6.44).

At this point it should also be noted that if we had defined the correlation function as $R_{xy}(\tau) = \mathrm{E}\left[x(t+\tau)y(t)\right]$ like it is done in some cases (see for instance Papoulis [1]) instead of the definition given in this book $R_{xy}(\tau) = \mathrm{E}\left[x(t)y(t+\tau)\right]$, then Eq. (6.26) becomes $G_{xy}(\omega) = X(\omega)Y^*(\omega)$ and Eq. (6.44) takes a somewhat simpler form $\mathbf{G}_y(\omega) = \widetilde{\mathbf{H}}(i\omega)\mathbf{G}_x(\omega)\widetilde{\mathbf{H}}^H(i\omega)$. However, in order to stay in line with common text books such as Bendat and Piersol [2] and Newland [3] concerning the definitions of the correlation functions we shall prefer the form given by Eq. (6.44).

6.2 White Noise Inputs

As explained in the introduction chapter, the concept of assuming a white noise input is essential in all OMA – even though we are not limited to this assumption as it was explained in the introduction. In this section, we shall consider how the important fundamental theorem derived in the aforementioned discussion is to be interpreted in terms of a white noise input.

6.2.1 Concept of White Noise

The basic idea of white noise is that if the scalar signal $x(t)$ is white noise, then at any time t there is no correlation with $x(t+\tau)$ except at $\tau = 0$. It is common to say that this means that the signal is zero mean and that the correlation function is a delta function, thus if $x(t)$ is white noise, then

$$R_x(\tau) = \mathrm{E}\left[x(t)x(t+\tau)\right] = 2\pi G_{x0}\delta(\tau) \tag{6.45}$$

where the constant G_{x0} is a scaling constant. If we then take the Fourier transform to get the spectral density and we use the fact that the Fourier transform of a delta function[3] is equal to $1/(2\pi)$, then we see that the corresponding spectral density is constant and equal to the value G_{x0}. Parseval's theorem shows us that the definition is useless in continuous time because the variance of the signal becomes infinite if the constant spectral density extends to infinity. A way to deal with this discrepancy is to imagine that the white noise is limited to a certain frequency band B going from DC up to the maximum frequency B. According to the Parseval theorem the variance is then determined as

$$\sigma_x^2 = 2G_{x0}B \tag{6.46}$$

The constant spectral density is given by

$$G_{x0} = \frac{\sigma_x^2}{2B} \tag{6.47}$$

---

[2] It follows from the fact that all impulse response function are real, or more precisely, it follows directly from $\mathbf{H}(-\tau) \leftrightarrow \widetilde{\mathbf{H}}(-i\omega)$ for a general complex function.

[3] The Laplace transform of the delta function is unity, but the Fourier transform has the well-known factor $1/(2\pi)$ on the transform; therefore, the Fourier transform of the delta function is $1/(2\pi)$.

and a white noise signal with variance σ_x^2 defined in the frequency band B is then defined by its correlation function given by

$$R_x(\tau) = 2\pi \frac{\sigma_x^2}{2B} \delta(\tau) = \pi \frac{\sigma_x^2}{B} \delta(\tau) \tag{6.48}$$

We must keep in mind that the delta function is a mathematical approximation to the fact that any signal with a limited area under the spectral density will have a correlation function with a finite initial value. But this is not the case when we model the correlation function as a delta spike.

Considering two white noise signals $x_1(t)$ and $x_2(t)$ we have $R_{12}(\tau) = E\left[x_1(t) x_2(t+\tau)\right] = 2\pi G_0 \delta(\tau)$ and since $E\left[x_1(t) x_2(t)\right]$ is the covariance between the two signals, then for a white noise vector signal we have the correlation function matrix

$$\mathbf{R}_x(\tau) = E\left[\mathbf{x}(t)\mathbf{x}^T(t+\tau)\right] = 2\pi \frac{\delta(\tau)}{2B} \mathbf{C} \tag{6.49}$$

where $\mathbf{C}$ is the covariance matrix for the signal. A white noise vector signal has a spectral matrix given by

$$\mathbf{G}_x(\omega) = \begin{cases} \dfrac{\mathbf{C}}{2B} & \text{inside the band} \\[2mm] 0 & \text{outside the band} \end{cases} \tag{6.50}$$

The spectral matrix of a white noise signal is always constant, real, positive definite, and symmetric inside the frequency band and a zero matrix outside of the band.

If we also assume that the components of the signal are independent, then the covariance matrix is diagonal. Furthermore, if all components also have the same variance σ_x^2 then we have the special case where

$$\mathbf{R}_x(\tau) = 2\pi \frac{\delta(\tau)}{2B} \sigma_x^2 \mathbf{I}$$
$$\mathbf{G}_x(\omega) = \frac{1}{2B} \sigma_x^2 \mathbf{I} \tag{6.51}$$

Going back to the simple case of a single-input-single output and assuming white noise input, we have according to Eqs. (6.36) and (6.45–6.47)

$$G_y(\omega) = G_x(\omega) H^*(i\omega) H(i\omega) = \frac{\sigma_x^2}{2B} H(-i\omega) H(i\omega) \tag{6.52}$$

From the transform theory using the time reversal and the convolution property, we then get the correlation function of the response

$$R_y(\tau) = 2\pi \frac{\sigma_x^2}{2B} h(-\tau) * h(\tau) \tag{6.53}$$

and we see that the deterministic correlation function given by Eq. (6.3) is proportional to the correlation of the response in case of white noise loading. This can also be derived directly from Eq. (6.34).

6.2.2 Decomposition in Time Domain

The decomposition in the time domain was described in the nineties by James et al. [4]. Their derivation was originally performed for normal modes and only a single external input. However, this derivation can easily be generalized to complex modes and general white noise input. Using the compact notation in Eq. (6.11) and the ideas of the derivation presented by James et al. [4] any response can be expressed by its modal decomposition as

$$\mathbf{y}(t) = \sum_{n=1}^{2N} \mathbf{b}_n q_n(t) = [\mathbf{b}_n] \mathbf{q}(t) \tag{6.54}$$

where $[\mathbf{b}_n]$ is the mode shape matrix and $\mathbf{q}(t)$ is a column vector holding the modal coordinates of the response. The modal coordinates are found from the convolution

$$q_n(t) = h_n(t) * p_n(t) = \int_{-\infty}^{t} h_n(t - \alpha) p_n(\alpha) d\alpha \tag{6.55}$$

where $p_n(t)$ is the modal load given by

$$p_n(t) = \mathbf{b}_n^T \mathbf{x}(t) \tag{6.56}$$

In Eq. (6.55), it has been assumed that all impulse response functions are vanishing for negative times, thus the contributions to the convolution integral is zero for $\alpha > t$. Then making use of the above equations and of Eq. (6.11) we obtain

$$\mathbf{y}(t) = \sum_{n=1}^{2N} \mathbf{b}_n \int_{-\infty}^{t} h_n(t - \alpha) p_n(\alpha) d\alpha$$

$$= \sum_{n=1}^{2N} \mathbf{b}_n \mathbf{b}_n^T \int_{-\infty}^{t} h_n(t - \alpha) \mathbf{x}(\alpha) d\alpha \tag{6.57}$$

$$= \sum_{n=1}^{2N} \mathbf{A}_n \int_{-\infty}^{t} e^{\lambda_n(t-\alpha)} \mathbf{x}(\alpha) d\alpha$$

Going back to the expression where the summation is performed over the set of modes, instead of over the set of modal parameters, and using the fact that poles and mode shapes appear in complex conjugate pairs we can write

$$\mathbf{y}(t) = \sum_{n=1}^{N} \left(\mathbf{A}_n \int_{-\infty}^{t} e^{\lambda_n(t-\alpha)} \mathbf{x}(\alpha) d\alpha + \mathbf{A}_n^* \int_{-\infty}^{t} e^{\lambda_n^*(t-\alpha)} \mathbf{x}(\alpha) d\alpha \right) \tag{6.58}$$

and the correlation function matrix is found from Eq. (6.18) taking into account that the residue matrices are symmetric

$$\mathbf{R}_y(\tau) = \mathrm{E} \left[\left(\sum_{n=1}^{N} \left(\mathbf{A}_n \int_{-\infty}^{t} e^{\lambda_n(t-\alpha)} \mathbf{x}(\alpha) d\alpha + \mathbf{A}_n^* \int_{-\infty}^{t} e^{\lambda_n^*(t-\alpha)} \mathbf{x}(\alpha) d\alpha \right) \right) \times \right.$$
$$\left. \times \left(\sum_{s=1}^{N} \left(\int_{-\infty}^{t} e^{\lambda_s(t+\tau-\beta)} \mathbf{x}^T(\beta) d\beta \mathbf{A}_s + \int_{-\infty}^{t} e^{\lambda_s^*(t+\tau-\beta)} \mathbf{x}^T(\beta) d\beta \mathbf{A}_s^* \right) \right) \right] \tag{6.59}$$

Since the expectation operator is only operating on the input vector the terms can be rearranged to obtain

$$
\mathbf{R}_y(\tau) = \sum_{n=1}^{N}\sum_{s=1}^{N}\left[\mathbf{A}_n \int_{-\infty}^{t}\int_{-\infty}^{t} e^{\lambda_n(t-\alpha)}e^{\lambda_s(t+\tau-\beta)}\mathrm{E}\left[\mathbf{x}(\alpha)\mathbf{x}^T(\beta)\right]d\beta d\alpha \mathbf{A}_s + \right.
$$

$$
+ \mathbf{A}_n \int_{-\infty}^{t}\int_{-\infty}^{t} e^{\lambda_n(t-\alpha)}e^{\lambda_s^*(t+\tau-\beta)}\mathrm{E}\left[\mathbf{x}(\alpha)\mathbf{x}^T(\beta)\right]d\beta d\alpha \mathbf{A}_s^* +
$$

$$
\tag{6.60}
$$

$$
+ \mathbf{A}_n^* \int_{-\infty}^{t}\int_{-\infty}^{t} e^{\lambda_n^*(t-\alpha)}e^{\lambda_s(t+\tau-\beta)}\mathrm{E}\left[\mathbf{x}(\alpha)\mathbf{x}^T(\beta)\right]d\beta d\alpha \mathbf{A}_s +
$$

$$
\left. + \mathbf{A}_n^* \int_{-\infty}^{t}\int_{-\infty}^{t} e^{\lambda_n^*(t-\alpha)}e^{\lambda_s^*(t+\tau-\beta)}\mathrm{E}\left[\mathbf{x}(\alpha)\mathbf{x}^T(\beta)\right]d\beta d\alpha \mathbf{A}_s^* \right)
$$

Using Eq. (6.45) we see that we can identify the expectation terms as proportional to the delta function

$$
\mathrm{E}\left[\mathbf{x}(\alpha)\mathbf{x}^T(\beta)\right] = 2\pi \mathbf{G}_x \delta(\beta - \alpha) \tag{6.61}
$$

where $\mathbf{G}_x$ is the constant input SD matrix. Using the properties of the delta function the inner integrals in Eq. (6.60) disappear and we have

$$
\mathbf{R}_y(\tau) = 2\pi \sum_{n=1}^{N}\sum_{s=1}^{N}\left(\mathbf{A}_n \mathbf{G}_x \mathbf{A}_s \int_{-\infty}^{t} e^{\lambda_n(t-\alpha)}e^{\lambda_s(t+\tau-\alpha)}d\alpha + \mathbf{A}_n \mathbf{G}_x \mathbf{A}_s^* \int_{-\infty}^{t} e^{\lambda_n(t-\alpha)}e^{\lambda_s^*(t+\tau-\alpha)}d\alpha \right.
$$

$$
\tag{6.62}
$$

$$
\left. + \mathbf{A}_n^* \mathbf{G}_x \mathbf{A}_s \int_{-\infty}^{t} e^{\lambda_n^*(t-\alpha)}e^{\lambda_s(t+\tau-\alpha)}d\alpha + \mathbf{A}_n^* \mathbf{G}_x \mathbf{A}_s^* \int_{-\infty}^{t} e^{\lambda_n^*(t-\alpha)}e^{\lambda_s^*(t+\tau-\alpha)}d\alpha \right)
$$

In order to evaluate the integrals in Eq. (6.62), it is needed to distinguish between the cases $\tau \geq 0$ and $\tau \leq 0$. Assuming $\tau \geq 0$, we can show that substituting $\gamma = \alpha - t$ and recognizing that the integrant vanishes for $\gamma \geq 0$

$$
\int_{-\infty}^{t} e^{\lambda_n(t-\alpha)}e^{\lambda_s(t+\tau-\alpha)}d\alpha = \int_{-\infty}^{0} e^{-\lambda_n\gamma}e^{\lambda_s(\tau-\gamma)}d\gamma = e^{\lambda_s\tau}\int_{-\infty}^{0} e^{(-\lambda_n-\lambda_s)\gamma}d\gamma
$$

$$
\tag{6.63}
$$

$$
= \frac{e^{\lambda_s\tau}}{-\lambda_n - \lambda_s}\left[e^{(-\lambda_n-\lambda_s)\gamma}\right]_{-\infty}^{0} = \frac{e^{\lambda_s\tau}}{-\lambda_n - \lambda_s}
$$

Equation (6.62) then reduces to

$$
\mathbf{R}_{y+}(\tau) = 2\pi \sum_{n=1}^{N}\sum_{s=1}^{N}\left(\frac{\mathbf{A}_n \mathbf{G}_x \mathbf{A}_s}{-\lambda_n - \lambda_s} + \frac{\mathbf{A}_n^* \mathbf{G}_x \mathbf{A}_s}{-\lambda_n^* - \lambda_s} \right)e^{\lambda_s\tau} + 2\pi \sum_{n=1}^{N}\sum_{s=1}^{N}\left(\frac{\mathbf{A}_n \mathbf{G}_x \mathbf{A}_s^*}{-\lambda_n - \lambda_s^*} + \frac{\mathbf{A}_n^* \mathbf{G}_x \mathbf{A}_s^*}{-\lambda_n^* - \lambda_s^*} \right)e^{\lambda_s^*\tau} \tag{6.64}
$$

Swapping the indices and defining the weighted sum of the residual matrices as

$$
\mathbf{B}_n = \sum_{s=1}^{N}\left(\frac{\mathbf{A}_s}{-\lambda_n - \lambda_s} + \frac{\mathbf{A}_s^*}{-\lambda_n - \lambda_s^*} \right) \tag{6.65}
$$

we have the modal decomposition of CF matrix for $\tau \geq 0$

$$\mathbf{R}_{y+}(\tau) = 2\pi \sum_{n=1}^{N} \left(\mathbf{B}_n \mathbf{G}_x \mathbf{A}_n e^{\lambda_n \tau} + \mathbf{B}_n^* \mathbf{G}_x \mathbf{A}_n^* e^{\lambda_n^* \tau} \right) \tag{6.66}$$

For the case $\tau \leq 0$, an equation similar to Eq. (6.63) can be developed recognizing that for $\tau \leq 0$ the integrant vanishes for $\gamma \geq \tau$. Hence,

$$\int_{-\infty}^{t} e^{\lambda_n(t-\alpha)} e^{\lambda_s(t+\tau-\alpha)} d\alpha = \int_{-\infty}^{\tau} e^{-\lambda_n \gamma} e^{\lambda_s(\tau-\gamma)} d\gamma = e^{\lambda_s \tau} \int_{-\infty}^{\tau} e^{(-\lambda_n - \lambda_s)\gamma} d\gamma$$

$$= \frac{e^{\lambda_s \tau}}{-\lambda_n - \lambda_s} \left[e^{(-\lambda_n - \lambda_s)\gamma} \right]_{-\infty}^{\tau} = \frac{e^{-\lambda_n \tau}}{-\lambda_n - \lambda_s} \tag{6.67}$$

Following the same approach as that for the case $\tau \geq 0$ we arrive at the modal decomposition of CF matrix for $\tau \leq 0$

$$\mathbf{R}_{y-}(\tau) = 2\pi \sum_{n=1}^{N} \left(\mathbf{A}_n \mathbf{G}_x \mathbf{B}_n e^{-\lambda_n \tau} + \mathbf{A}_n^* \mathbf{G}_x \mathbf{B}_n^* e^{-\lambda_n^* \tau} \right) \tag{6.68}$$

It follows from the definition given by Eq. (6.18) that the CF matrix satisfies the well-known symmetry identity $\mathbf{R}_y(-\tau) = \mathbf{R}_y^T(\tau)$. We must have $\mathbf{R}_{y-}(-\tau) = \mathbf{R}_{y+}^T(\tau)$, which we see from Eqs. (6.66/6.68). And this is actually the case because all residue matrices are symmetric and so are the weighted residue matrices given by Eq. (6.65).

6.2.3   Decomposition in Frequency Domain

The modal decomposition of the spectral density (SD) matrix can, of course, be found from the results just derived for the time domain because the SD matrix is the Fourier transform of the CF matrix. The SD matrix can be found as

$$\mathbf{G}_y(\omega) = \frac{1}{2\pi} \int_{-\infty}^{\infty} \mathbf{R}_y(\tau) e^{-i\omega\tau} d\tau$$

$$= \frac{1}{2\pi} \int_{-\infty}^{0} \mathbf{R}_{y-}(\tau) e^{-i\omega\tau} d\tau + \frac{1}{2\pi} \int_{0}^{\infty} \mathbf{R}_{y+}(\tau) e^{-i\omega\tau} d\tau \tag{6.69}$$

Using Eqs. (6.66/6.68) and the Fourier transform pairs[4]

$$e^{\lambda_n \tau} \leftrightarrow \frac{1}{2\pi} \frac{1}{i\omega - \lambda_n} \tag{6.70}$$

Eq. (6.69) can then be re-written as:

$$\mathbf{G}_y(\omega) = \sum_{n=1}^{N} \left(\frac{\mathbf{A}_n \mathbf{G}_x \mathbf{B}_n}{-i\omega - \lambda_n} + \frac{\mathbf{A}_n^* \mathbf{G}_x \mathbf{B}_n^*}{-i\omega - \lambda_n^*} + \frac{\mathbf{B}_n \mathbf{G}_x \mathbf{A}_n}{i\omega - \lambda_n} + \frac{\mathbf{B}_n^* \mathbf{G}_x \mathbf{A}_n^*}{i\omega - \lambda_n^*} \right) \tag{6.71}$$

where the first two terms correspond to the negative part of the CF matrix and the last two terms correspond to the positive part of the CF matrix.

---

[4] From Table 4.1 we have the Laplace transform pair $e^{\lambda \tau} \leftrightarrow 1/(s - \lambda)$; then we use that the Fourier transform can be calculated as the Laplace transform obtained for $s = i\omega$ and divided by 2π.

In the following sections, we will perform an independent derivation in the frequency domain to arrive at the same result. The decomposition in the frequency domain assuming a white noise input was first done by Brincker et al. [5] and [6] and later by Brincker and Zhang [7]. A similar derivation is given by Heylen et al. [8]. In this section, we shall explain the derivation in detail. To do this, we take origin in the fundamental theorem (see Eq. (6.44)). Using the lower part of Eq. (6.44) together with Eq. (6.8) gives

$$
\begin{aligned}
\mathbf{G}_y(\omega) &= \sum_{n=1}^{N}\left(\frac{\mathbf{A}_n}{-i\omega-\lambda_n}+\frac{\mathbf{A}_n^*}{-i\omega-\lambda_n^*}\right)\mathbf{G}_x\sum_{n=1}^{N}\left(\frac{\mathbf{A}_n}{i\omega-\lambda_n}+\frac{\mathbf{A}_n^*}{i\omega-\lambda_n^*}\right)\\
&= \sum_{n=1}^{N}\sum_{s=1}^{N}\left(\frac{\mathbf{A}_n}{-i\omega-\lambda_n}+\frac{\mathbf{A}_n^*}{-i\omega-\lambda_n^*}\right)\mathbf{G}_x\left(\frac{\mathbf{A}_s}{i\omega-\lambda_s}+\frac{\mathbf{A}_s^*}{i\omega-\lambda_s^*}\right)\\
&= \sum_{n=1}^{N}\sum_{s=1}^{N}\left(\frac{\mathbf{A}_n\mathbf{G}_x\mathbf{A}_s}{(-i\omega-\lambda_n)(i\omega-\lambda_s)}+\frac{\mathbf{A}_n\mathbf{G}_x\mathbf{A}_s^*}{(-i\omega-\lambda_n)(i\omega-\lambda_s^*)}+\frac{\mathbf{A}_n^*\mathbf{G}_x\mathbf{A}_s}{(-i\omega-\lambda_n^*)(i\omega-\lambda_s)}\right.\\
&\quad\left.+\frac{\mathbf{A}_n^*\mathbf{G}_x\mathbf{A}_s^*}{(-i\omega-\lambda_n^*)(i\omega-\lambda_s^*)}\right)
\end{aligned}
\tag{6.72}
$$

Using the Heaviside partial fraction expansion theorem, the fractions can be reformulated as

$$
\begin{aligned}
\frac{1}{(-i\omega-\lambda_n)(i\omega-\lambda_s)} &= \frac{1}{-\lambda_n-\lambda_s}\left(\frac{1}{-i\omega-\lambda_n}+\frac{1}{i\omega-\lambda_s}\right)\\
\frac{1}{(-i\omega-\lambda_n)(i\omega-\lambda_s^*)} &= \frac{1}{-\lambda_n-\lambda_s^*}\left(\frac{1}{-i\omega-\lambda_n}+\frac{1}{i\omega-\lambda_s^*}\right)\\
\frac{1}{(-i\omega-\lambda_n^*)(i\omega-\lambda_s)} &= \frac{1}{-\lambda_n^*-\lambda_s}\left(\frac{1}{-i\omega-\lambda_n^*}+\frac{1}{i\omega-\lambda_s}\right)\\
\frac{1}{(-i\omega-\lambda_n^*)(i\omega-\lambda_s^*)} &= \frac{1}{-\lambda_n^*-\lambda_s^*}\left(\frac{1}{-i\omega-\lambda_n^*}+\frac{1}{i\omega-\lambda_s^*}\right)
\end{aligned}
\tag{6.73}
$$

and we have

$$
\mathbf{G}_y(\omega)=\sum_{n=1}^{N}\sum_{s=1}^{N}\left(
\begin{aligned}
&\left(\frac{\mathbf{A}_n\mathbf{G}_x\mathbf{A}_s}{-\lambda_n-\lambda_s}+\frac{\mathbf{A}_n\mathbf{G}_x\mathbf{A}_s^*}{-\lambda_n-\lambda_s^*}\right)\frac{1}{(-i\omega-\lambda_n)}+\left(\frac{\mathbf{A}_n^*\mathbf{G}_x\mathbf{A}_s}{-\lambda_n^*-\lambda_s}+\frac{\mathbf{A}_n^*\mathbf{G}_x\mathbf{A}_s^*}{-\lambda_n^*-\lambda_s^*}\right)\frac{1}{(-i\omega-\lambda_n^*)}+\\
&+\left(\frac{\mathbf{A}_n\mathbf{G}_x\mathbf{A}_s}{-\lambda_n-\lambda_s}+\frac{\mathbf{A}_n^*\mathbf{G}_x\mathbf{A}_s}{-\lambda_n^*-\lambda_s}\right)\frac{1}{(i\omega-\lambda_s)}+\left(\frac{\mathbf{A}_n\mathbf{G}_x\mathbf{A}_s^*}{-\lambda_n-\lambda_s^*}+\frac{\mathbf{A}_n^*\mathbf{G}_x\mathbf{A}_s^*}{-\lambda_n^*-\lambda_s^*}\right)\frac{1}{(i\omega-\lambda_s^*)}
\end{aligned}
\right)
\tag{6.74}
$$

For the first term in the summation bracket, a matrix $\mathbf{B}_n$ can be defined such that

$$
\sum_{s=1}^{N}\left(\frac{\mathbf{A}_n\mathbf{G}_x\mathbf{A}_s}{-\lambda_n-\lambda_s}+\frac{\mathbf{A}_n\mathbf{G}_x\mathbf{A}_s^*}{-\lambda_n-\lambda_s^*}\right)=\mathbf{A}_n\mathbf{G}_x\mathbf{B}_n
\tag{6.75}
$$

where the matrix $\mathbf{B}_n$ is the same weighted sum of the residual matrices as given by Eq. (6.65) and can be used to simplify the first two terms of Eq. (6.74). For the last two terms in the summation bracket, we swap the summation indices and similarly we have for the third term in the bracket of Eq. (6.74)

$$
\sum_{s=1}^{N}\left(\frac{\mathbf{A}_s\mathbf{G}_x\mathbf{A}_n}{-\lambda_s-\lambda_n}+\frac{\mathbf{A}_s^*\mathbf{G}_x\mathbf{A}_n}{-\lambda_s^*-\lambda_n}\right)=\mathbf{B}_n\mathbf{G}_x\mathbf{A}_n
\tag{6.76}
$$

and the closed form solution for the modal decomposition of the spectral density matrix in the case of the white noise input finally takes the form

$$\mathbf{G}_y(\omega) = \sum_{n=1}^{N} \left(\frac{\mathbf{A}_n \mathbf{G}_x \mathbf{B}_n}{-i\omega - \lambda_n} + \frac{\mathbf{A}_n^* \mathbf{G}_x \mathbf{B}_n^*}{-i\omega - \lambda_n^*} + \frac{\mathbf{B}_n \mathbf{G}_x \mathbf{A}_n}{i\omega - \lambda_n} + \frac{\mathbf{B}_n^* \mathbf{G}_x \mathbf{A}_n^*}{i\omega - \lambda_n^*} \right) \tag{6.77}$$

We see that the result is in fact the same as the result found from the time domain, Eq. (6.71). Since the residual matrices are symmetric, we have

$$\mathbf{G}_y^H(\omega) = \sum_{n=1}^{N} \left(\frac{\mathbf{B}_n^* \mathbf{G}_x \mathbf{A}_n^*}{i\omega - \lambda_n^*} + \frac{\mathbf{B}_n \mathbf{G}_x \mathbf{A}_n}{i\omega - \lambda_n} + \frac{\mathbf{A}_n^* \mathbf{G}_x \mathbf{B}_n^*}{-i\omega - \lambda_n^*} + \frac{\mathbf{A}_n \mathbf{G}_x \mathbf{B}_n}{-i\omega - \lambda_n} \right) \tag{6.78}$$

and therefore comparing with Eq. (6.77) we see that each term in Eq. (6.78) correspond to a term in Eq. (6.77) and thus the spectral density is indeed Hermitian as it should be.

It should be noted that only the two middle terms in Eq. (6.77) are dominant for positive frequencies. The first of these two terms is dominant because the denominator $-i\omega - \lambda_n^*$ is equal to $-i\omega + i\omega_{dn} + \varsigma\omega_{0n}$ and converges to the small quantity $\varsigma\omega_{0n}$ when $\omega = \omega_{dn}$. Similarly the second of these two terms is dominant because the denominator $i\omega - \lambda_n$ is equal to $i\omega - i\omega_{dn} + \varsigma\omega_{0n}$ and converges to the small quantity $\varsigma\omega_{0n}$ when $\omega = \omega_{dn}$. The other two terms are dominant only for negative frequencies.

Example 6.1   Comparing the modal decomposition with the fundamental theorem

A possible input SD matrix must be positive definite. Such matrix can be formed by taking a the symmetric component $\mathbf{A}_s$ of any real matrix $\mathbf{A}$

$$\mathbf{A}_s = \left(\mathbf{A} + \mathbf{A}^T \right)/2 \tag{6.79}$$

The SVD of this matrix is

$$\mathbf{A}_s = \mathbf{USU}^T \tag{6.80}$$

we can force the eigenvalues to be nonnegative

$$\mathbf{G}_x = \mathbf{U}|\mathbf{S}|\mathbf{U}^T \tag{6.81}$$

Following this procedure, the input matrix for a 3×3 system is formed

$$\mathbf{G}_x = \begin{bmatrix} 1.3811 & 0.2944 & -0.7045 \\ 0.2944 & 2.9569 & 0.1392 \\ -0.7045 & 0.1392 & 1.6254 \end{bmatrix} \tag{6.82}$$

The modes of the system were chosen to be as shown in Table 6.1. The FRF matrix can then be calculated using Eq. (6.8) and the response SD matrix can then be calculated by using either the fundamental theorem according to Eq. (6.44) or by using the modal decomposition as given by Eq. (6.77). The two results for the SD matrix are shown in Figure 6.2 using the matrix entries row $r = 1$ and column $c = 2$ and as it appears, the results are identical.

Table 6.1   Modal parameters for Example 6.1

Mode	1	2	3
Mode shapes $\mathbf{b}_n$	$\{1\ 1\ 1\}^T$	$\{1\ -1\ 1\}^T$	$\{1\ 0\ -1\}^T$
Damping ratios ς_n (%)	0.20	0.10	0.10
Frequencies f_n (Hz)	29.5	50.0	71.0

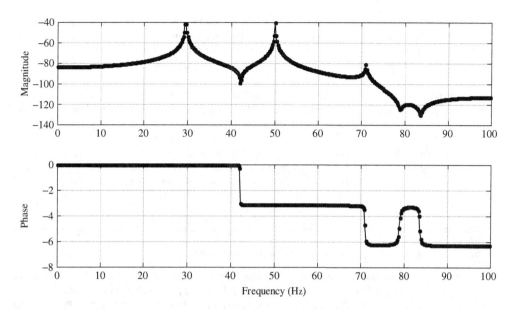

Figure 6.2   SD function for the considered system plotted for matrix entry row $r = 1$ and column $c = 2$. Results using the fundamental theorem according to Eq. (6.44) are plotted in with a solid line whereas the results of using the modal decomposition as given by Eq. (6.77) are plotted using dots

In this case, we see that the SD function has one visible zero between the first two modes and then two visible zeroes to the right of the highest mode. We see that the phase does not necessarily change around a modal peak, for the case shown in Figure 6.2 we see that no phase change takes place around the two first modal peaks whereas the phase changes 180° around all the visible zeroes and also around the highest mode.

6.2.4   Zeroes of the Spectral Density Matrix

By taking the contribution from the third term in Eq. (6.77) to the SD matrix we can express this as a rational fractional form

$$
\sum_{n=1}^{N} \frac{\mathbf{B}_n \mathbf{G}_x \mathbf{A}_n}{i\omega - \lambda_n} = \frac{\displaystyle\prod_{k=1}^{N} (i\omega - \lambda_k) \sum_{n=1}^{N} \frac{\mathbf{B}_n \mathbf{G}_x \mathbf{A}_n}{i\omega - \lambda_n}}{\displaystyle\prod_{n=1}^{N} (i\omega - \lambda_n)} = \frac{\displaystyle\sum_{n=1}^{N} \left(\mathbf{B}_n \mathbf{G}_x \mathbf{A}_n \prod_{k=1,k\neq n}^{N} (i\omega - \lambda_k) \right)}{\displaystyle\prod_{n=1}^{N} (i\omega - \lambda_n)}
\tag{6.83}
$$

where the large π-symbol stands for product. It is clear that the denominator polynomial is of order N and has the poles λ_n as its roots. Similarly, we can put all the four terms in Eq. (6.77) as a common denominator, and we see that this is a polynomial of order $4N$ and has the poles λ_n, λ_n^*, $-\lambda_n$, $-\lambda_n^*$ as its roots.

The numerator of the rational fraction in Eq. (6.83) has order $N - 1$, thus has $N - 1$ roots. Similarly if we put all the four terms in Eq. (6.77) as a common denominator, we will see that the numerator is a polynomial of maximum order $4N - 1$ and thus has maximum $4N - 1$ roots. The roots of the numerator polynomial are the zeroes of the spectral densities. The zeroes are placed symmetrically around DC and the number of visible zeroes seems to vary from zero approximately up to the number of modes. However, the number of visible zeroes can in fact be somewhat larger than the number of modes. Further, it seems like the zeroes are more visible in the cross-spectral density functions than in the auto spectral density functions.

What is important to note is that all the coefficients of the numerator polynomial depend on a matrix products of the form $\mathbf{B}_n\mathbf{G}_x\mathbf{A}_n$. The coefficients of the resulting numerator polynomial depend on the input spectral density matrix, and, therefore, the zeroes of the spectral densities will depend on the input spectral density matrix that is in turn proportional to the covariance matrix of the input.

Therefore, from the modal decomposition of the SD matrix, it is evident that the zeroes of the spectral density functions are not physical quantities like they are known to be for frequency response functions since the zeros for the SD matrix functions are dependent on the covariance matrix of the input.

Example 6.2   Illustrating variability of zeroes of spectral density

The same system is considered as in example 6.1 and the same SD functions are shown for 20 different randomly simulated input SD matrices in Figure 6.3. As discussed previously the location of the zeroes varies a lot according to the different input SD matrices. Further, it is clear that it is the rule more than the exception that the phase does not change around the spectral peaks as no phase changes takes place at any of the modal peaks for any of the 20 input SD matrices – for these cases, all phase changes take place around the zeroes.

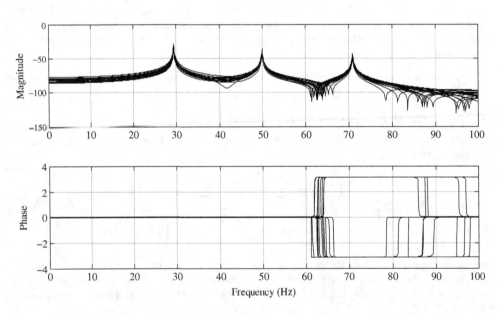

Figure 6.3   This plot shows the variability of the location of the zeroes for a 3 DOF system as the one considered in Example 6.1 for 20 different input SD matrices for matrix entry row $r = 1$ and column $c = 3$

6.2.5   Residue Form

Considering the third term of Eq. (6.77), and using Eq. (6.10) we get

$$\mathbf{B}_n \mathbf{G}_x \mathbf{A}_n = \mathbf{B}_n \mathbf{G}_x \frac{\mathbf{b}_n \mathbf{b}_n^T}{a_n} = \boldsymbol{\gamma}_n \mathbf{b}_n^T \tag{6.84}$$

where the modal participation vector $\boldsymbol{\gamma}_n$ is given by

$$\boldsymbol{\gamma}_n = \mathbf{B}_n \mathbf{G}_x \frac{\mathbf{b}_n}{a_n} \tag{6.85}$$

Each of the terms of the form $\mathbf{B}_n \mathbf{G}_x \mathbf{A}_n$ has rank one. Taking the transpose of Eq. (6.84), we have that

$$\mathbf{A}_n \mathbf{G}_x \mathbf{B}_n = \mathbf{b}_n \boldsymbol{\gamma}_n^T \tag{6.86}$$

Using Eq. (6.66), we see that

$$\mathbf{R}_{y+}(0) = \mathbf{C}_{yy} = 2\pi \sum_{n=1}^{N} \left(\mathbf{B}_n \mathbf{G}_x \mathbf{A}_n + \mathbf{B}_n^* \mathbf{G}_x \mathbf{A}_n^* \right) \tag{6.87}$$

So Eqs. (6.84/6.85) leads to the following expression for response covariance matrix

$$\mathbf{C}_y = 2\pi \sum_{n=1}^{N} \left(\boldsymbol{\gamma}_n \mathbf{b}_n^T + \boldsymbol{\gamma}_n^* \mathbf{b}_n^H \right) \tag{6.88}$$

Similarly, using Eq. (6.68) and Eq. (6.86) we obtain:

$$\mathbf{R}_{y-}(0) = \mathbf{C}_y = 2\pi \sum_{n=1}^{N} \left(\mathbf{A}_n \mathbf{G}_x \mathbf{B}_n + \mathbf{A}_n^* \mathbf{G}_x \mathbf{B}_n^* \right)$$
$$= 2\pi \sum_{n=1}^{N} \left(\mathbf{b}_n \boldsymbol{\gamma}_n^T + \mathbf{b}_n^* \boldsymbol{\gamma}_n^H \right) \tag{6.89}$$

which is in line with what we know about the covariance matrix: that is real and symmetric. Equations (6.88) and (6.89) do not mean that there is a modal contribution to the covariance matrix that is independent of the other modes, because the weighted sum of residue matrices given by Eq. (6.65) depends on all the modes. Further, the equations do not – as one would think – mean that the modal participation vector $\boldsymbol{\gamma}_n$ is proportional to the corresponding mode shape vector $\mathbf{b}_n$. Actually, for normal modes where the mode shapes are real, it is easy to verify that $\boldsymbol{\gamma}_n$ is in general complex and that even the real part is different from the corresponding mode shape. However, as it is shown in the following section, the modal participation vector is *approximately* proportional to the corresponding mode shape.

Using Eqs. (6.88/6.89), the CF matrix can be written as

$$\mathbf{R}_y(\tau) = \begin{cases} 2\pi \sum_{n=1}^{N} \left(\mathbf{b}_n \boldsymbol{\gamma}_n^T e^{-\lambda_n \tau} + \mathbf{b}_n^* \boldsymbol{\gamma}_n^H e^{-\lambda_n^* \tau} \right); & \tau \leq 0 \\ 2\pi \sum_{n=1}^{N} \left(\boldsymbol{\gamma}_n \mathbf{b}_n^T e^{\lambda_n \tau} + \boldsymbol{\gamma}_n^* \mathbf{b}_n^H e^{\lambda_n^* \tau} \right); & \tau \geq 0 \end{cases} \tag{6.90}$$

And the SD matrix is given by

$$\mathbf{G}_y(\omega) = \sum_{n=1}^{N} \left(\frac{\mathbf{b}_n \boldsymbol{\gamma}_n^T}{-i\omega - \lambda_n} + \frac{\mathbf{b}_n^* \boldsymbol{\gamma}_n^H}{-i\omega - \lambda_n^*} + \frac{\boldsymbol{\gamma}_n \mathbf{b}_n^T}{i\omega - \lambda_n} + \frac{\boldsymbol{\gamma}_n^* \mathbf{b}_n^H}{i\omega - \lambda_n^*} \right) \tag{6.91}$$

By taking the aforementioned observations into account, it matters if one takes the columns or the rows of the correlation matrix in order to identify the modes. If the positive part is used, then in Eq. (6.90) only the rows of the CF matrix are proportional to the mode shapes. Thus, using the positive part of the CF matrix, in order to pick columns out of the CF matrix that have the same form as a free decay[5] we have to use the transpose

$$\mathbf{R}_y^T(\tau) = 2\pi \sum_{n=1}^{N} \left(\mathbf{b}_n \boldsymbol{\gamma}_n^T e^{-\lambda_n \tau} + \mathbf{b}_n^* \boldsymbol{\gamma}_n^H e^{-\lambda_n^* \tau} \right); \quad \tau \geq 0 \tag{6.92}$$

In this case each column of the transpose CF matrix has the same form as a free response because each exponential term is proportional to the corresponding mode shape vector. If we, instead of the definition of the CF matrix given by Eq. (6.18) and by classical texts such as Bendat and Piersol [2] and by Newland [3], would follow for instance Papoulis [1] and define the CF matrix as

$$\mathbf{R}_x'(\tau) = E\left[\mathbf{x}(t+\tau)\mathbf{x}^T(t) \right]$$
$$\mathbf{R}_y'(\tau) = E\left[\mathbf{y}(t+\tau)\mathbf{y}^T(t) \right] \tag{6.93}$$

We see immediately that

$$\mathbf{R}_y'(\tau) = \mathbf{R}_y^T(\tau) \tag{6.94}$$

Therefore, using the positive part and the definition of the CF matrix as given by Eq. (6.93), each column of the CF matrix has the form of a free decay.

If we, for the form of the CF matrix given by Eq. (6.18), take the decays defined by the columns, then the columns do not have the same form as a free decay. If we assume that this is the case, then we have to accept that this is only an approximation as it is shown in the following section. Thus, a bias on the mode shape estimation is introduced.

Similarly, using the definition of the CF matrices as given by Eq. (6.18) leads to the closed-form solution for the modal decomposition of the SD matrix as given by Eq. (6.91). We see that the matrix columns of the first two terms of this solution – that are the terms that are related to the negative part of the CF matrix – are proportional to the mode shapes. The matrix columns of the last two terms of this solution – that are the terms that are related to the positive part of the CF matrix – are not proportional to the mode shapes. Therefore, for the SD matrix neither the columns nor the rows are proportional to the mode shapes. If instead we use the definition of the CF matrix as given by Eq. (6.93), then we arrive at the similar closed form solution for the SD matrix

$$\mathbf{G}_y'(\omega) = \sum_{n=1}^{N} \left(\frac{\mathbf{B}_n \mathbf{G}_x \mathbf{A}_n}{-i\omega - \lambda_n} + \frac{\mathbf{B}_n^* \mathbf{G}_x \mathbf{A}_n^*}{-i\omega - \lambda_n^*} + \frac{\mathbf{A}_n \mathbf{G}_x \mathbf{B}_n}{i\omega - \lambda_n} + \frac{\mathbf{A}_n^* \mathbf{G}_x \mathbf{B}_n^*}{i\omega - \lambda_n^*} \right) \tag{6.95}$$

and we see that the same two midterms are still dominant for positive frequencies. Also for this version of the SD matrix neither the columns nor the rows are proportional to the mode shapes. The expression in Eq. (6.95) is equal to the closed form solution given in Brincker and Zhang [7].

6.2.6 Approximate Residue Form

Considering the terms in the weighted sum of the residual matrices given by Eq. (6.65), we see that all the first terms are of the order of $1/2\omega$, where ω is a typical natural frequency for the system. The second terms have a varying weight depending on the frequency distance to the considered mode n due to the

---

[5] It follows from Eq. (5.133) and the whole concept of modal coordinates that a free decay is always proportional to the eigenvectors; this is also derived and explained in detail in Appendix C.

denominator $-\lambda_n - \lambda_s^*$. Considering two closely spaced modes with a frequency difference $\Delta\omega$ we can write

$$\lambda_n = -\varsigma\omega_{0n} + i\omega_{dn}$$
$$\lambda_s \cong -\varsigma\left(\omega_{0n} - \Delta\omega\right) + i\left(\omega_{dn} - \Delta\omega\right) \tag{6.96}$$

we can express the difference as

$$-\lambda_n - \lambda_s^* \cong 2\varsigma\omega_{0n} + i\Delta\omega \tag{6.97}$$

Therefore, the absolute value of the difference is

$$\left|-\lambda_n - \lambda_s^*\right| \cong \sqrt{4\varsigma^2\omega_{0n}^2 + \Delta\omega^2} \tag{6.98}$$

We can see that for $s = n$, we have the largest weighting term given by

$$\frac{1}{-\lambda_n - \lambda_n^*} = \frac{1}{2\varsigma\omega_{0n}} \tag{6.99}$$

If for mode n another mode is close so that the frequency difference $\Delta\omega$ is of the same order as $2\varsigma\omega_{0n}$ then the mode shape of the closely spaced mode will significantly contribute to the modal participation vector of mode n. On the other hand, if $\Delta\omega$ is of the same order as a typical natural frequency of the system, then the closely spaced mode is influencing the nth modal participation vector like all the other modes and therefore the influence on the participation vector from other modes is negligible. In the latter case, since the weighting term for mode n is approximately the $1/\varsigma$ larger than all other terms in the summation (6.65), as an approximation that is exact in the limit where the damping $\varsigma \to 0$ we have

$$\mathbf{B}_n \cong \frac{\mathbf{A}_n^*}{-\lambda_n - \lambda_n^*} = \frac{\mathbf{A}_n^*}{2\varsigma\omega_{0n}} \tag{6.100}$$

So that for instance the matrix factor in Eq. (6.86) is when we use the expression in Eq. (5.93) for the residue matrix,

$$\mathbf{A}_n\mathbf{G}_x\mathbf{B}_n = \frac{\mathbf{b}_n\mathbf{b}_n^T}{2i\omega_{dn}m_n}\mathbf{G}_x\frac{\mathbf{b}_n^*\mathbf{b}_n^H}{-2i\omega_{dn}m_n^*}\frac{1}{2\varsigma\omega_{0n}} = c_n^2\mathbf{b}_n\mathbf{b}_n^H \tag{6.101}$$

where the real and positive constant c_n^2 is given by

$$c_n^2 = \frac{\mathbf{b}_n^H\mathbf{G}_x\mathbf{b}_n}{8\omega_{dn}^2\left|m_n\right|^2\varsigma\omega_{0n}} \tag{6.102}$$

It should be noted that since the inner product $\mathbf{b}_n^H\mathbf{G}_x\mathbf{b}_n$ is always a real number (true also in the case of complex mode shapes) so is its complex conjugate, and therefore equals the similar inner product $\mathbf{b}_n^T\mathbf{G}_x\mathbf{b}_n^*$ that follows directly from Eq. (6.101).

For the case, where the damping is small, the modal participation vector is approximately proportional to the mode shape vector

$$\mathbf{\gamma}_n \cong c_n^2\mathbf{b}_n \tag{6.103}$$

Example 6.3   Illustrating mode shape bias

In this example, we illustrate mode shape bias that might occur in case we use a wrong form of the correlation function matrix as discussed in Section 6.2.5. We are again considering the same example as in Example 6.1 but now considering two different sets of modal parameters. One case (case I) refers to well-separated modes, and the other case (case II) is for closely spaced modes and mode shapes $\mathbf{b}_n$ as shown in Table 6.2.

Table 6.2 Parameters for the considered cases, cases I and II

Mode	1	2	3
Mode shapes $\mathbf{b}_n$	$\{1\ \ 1\ \ 1\}^T$	$\{1\ {-1}\ \ 1\}^T$	$\{1\ \ 0\ {-1}\}^T$
Damping ratio ς_n (%)	0.5	0.5	0.5
Frequencies f_n, case I (Hz)	39.5	50.0	61.0
Frequencies f_n, case II (Hz)	49.5	50.0	51.0

For cases I and II, the modal parameters including the identified mode shapes were first identified using the rows of the positive part of the CF matrix. The MIMO version of the Ibrahim time domain (ITD) technique was used for the identification, see Section 9.4 for information about this identification technique. In all cases, a small amount of white noise was added to the correlation functions equal to 10^{-6} times the maximum value of the CF matrix.

In all cases, the ID results are very close to the exact values when performing identification using the rows from the CF matrix as free decays. This is true for the mode shapes, the frequencies, and the damping. The differences are so small that the values are not important for this example.

A similar identification including the identification of the mode shapes $\mathbf{a}_n$ was then performed using the columns of the CF matrix as free decays, and as expected the bias on the mode shapes is large for case II with closely spaced modes, but moderate for case I with well separated modes. See the results in Tables 6.3 and 6.4 where only the real parts of the mode shapes are given. A clear tendency toward larger errors on the imaginary part of the mode shapes has been observed for case II.

Table 6.3 Identification results using the columns of the CF matrix as free decays, case I

Mode	1	2	3
Mode shapes $\mathbf{a}_n$	$\begin{Bmatrix} 0.5776 \\ 0.5766 \\ 0.5776 \end{Bmatrix}$	$\begin{Bmatrix} 0.5778 \\ -0.5695 \\ 0.5785 \end{Bmatrix}$	$\begin{Bmatrix} 0.7079 \\ -0.0022 \\ -0.7046 \end{Bmatrix}$
Damping ratios ς_n (%)	0.50	0.50	0.50
Frequencies f_n (Hz)	39.5000	50.0000	61.0000

Table 6.4 Identification results using the columns of the CF matrix as free decays, case II

Mode	1	2	3
Mode shapes $\mathbf{a}_n$	$\begin{Bmatrix} 0.6321 \\ 0.4462 \\ 0.6153 \end{Bmatrix}$	$\begin{Bmatrix} -0.6027 \\ 0.4743 \\ -0.6312 \end{Bmatrix}$	$\begin{Bmatrix} 0.6766 \\ 0.0482 \\ -0.7255 \end{Bmatrix}$
Damping ratios ς_n (%)	0.50	0.50	0.50
Frequencies f_n (Hz)	49.5000	50.0000	61.0000

6.3 Uncorrelated Modal Coordinates

In random vibrations, it is a commonly accepted approximation to assume that the modal coordinates are uncorrelated. This assumption can be used as a basis for modal decomposition of random signals that is approximately valid in cases where the input cannot be assumed to be white noise.

6.3.1 Concept of Uncorrelated Modal Coordinates

To support the assumption of uncorrelated modal coordinates, let us consider the case of a system excited by a large number of stochastically independent forces gathered in the excitation vector $\mathbf{x}(t)$. For simplicity, let us assume unit variance and a common scalar auto-correlation function $R_x(\tau)$, thus the correlation function matrix for the loads is proportional to the identity matrix

$$\mathbf{R}_x(\tau) = \mathrm{E}\left[\mathbf{x}(t)\,\mathbf{x}^T(t+\tau)\right] = \mathbf{I}R_x(\tau) \tag{6.104}$$

And the corresponding spectral density matrix is

$$\mathbf{G}_x(\omega) = \mathbf{I}G_x(\omega) \tag{6.105}$$

where $G_x(\omega)$ is the Fourier transform of $R_x(\tau)$. For simplicity, real modes are assumed, thus for each modal coordinate the modal force is found from Eq. (5.78)

$$p_n(t) = \mathbf{b}_n^T\mathbf{x}(t) \tag{6.106}$$

and the modal coordinate is found as the convolution between the modal impulse response function and the modal force

$$q_n(t) = h_n(t) * p_n(t) \tag{6.107}$$

By gathering the modal coordinates in the vector $\mathbf{q}(t)$ and the modal forces in the vector $\mathbf{p}(t)$ we have

$$\mathbf{p}(t) = \mathbf{B}^T\mathbf{x}(t)$$
$$\mathbf{q}(t) = \left[h_n(t)\right] * \mathbf{p}(t) \tag{6.108}$$

where we have gathered the mode shapes $\mathbf{b}_n$ in mode shape matrix $\mathbf{B}$ and where $\left[h_n(t)\right]$ is a diagonal matrix holding the modal impulse response functions. We can now consider the modal coordinates as the output of a system with the impulse response matrix $\left[h_n(t)\right]$. By using the fundamental theorem given by Eq. (6.44), we have that the correlation function matrix for the modal coordinates is given by

$$\mathbf{R}_q(\tau) = \left[h_n(-\tau)\right] * \mathbf{R}_p(\tau) * \left[h_n(\tau)\right] \tag{6.109}$$

Using Eq. (6.108) and Eq. (6.104), the correlation function matrix for the modal forces is then expressed as

$$\mathbf{R}_p(\tau) = \mathrm{E}\left[\mathbf{p}(t)\,\mathbf{p}^T(t+\tau)\right] = \mathbf{B}^T\mathrm{E}\left[\mathbf{x}(t)\,\mathbf{x}^T(t+\tau)\right]\mathbf{B}$$
$$= \mathbf{B}^T\mathbf{B}R_x(\tau) \tag{6.110}$$

If we assume unit length mode shape vectors and if the mode shape matrix is approximately geometrically orthogonal, that is $\mathbf{B}^T\mathbf{B} \cong \mathbf{I}$, then

$$\mathbf{R}_q(\tau) \cong \left[h_n(-\tau)\right] * R_x(\tau) * \left[h_n(\tau)\right] \tag{6.111}$$

Taking the Fourier transform of both sides of this equation, we obtain the spectral density matrix

$$\mathbf{G}_q(\tau) \cong \left[H_n(-i\omega)\right]G_x(\omega)\left[H_n(i\omega)\right] \tag{6.112}$$

That is also approximately diagonal. From the aforementioned discussion, the overall condition for this to be meaningful is that the correlation function matrix for the modal forces is approximately diagonal

$$\mathbf{R}_p(\tau) = \mathbf{B}^T \mathbf{R}_x(\tau) \mathbf{B} \cong \left[R_{x,n}(\tau) \right] \tag{6.113}$$

If this is the case, then the correlation function matrix and the spectral density matrix of the modal coordinates will also be approximately diagonal.

However, this is not the best argument for why the correlation function matrix of the modal coordinates is normally approximately diagonal. The best argument is related to our basic assumption that modal responses are narrow banded.

For simplicity, assuming real modes and taking the Fourier transform of Eq. (6.108), the modal coordinates are given by

$$\widetilde{\mathbf{q}}(\omega) = \left[H_n(i\omega) \right] \widetilde{\mathbf{p}}(\omega) = \left\{ H_n(i\omega) P_n(\omega) \right\} \tag{6.114}$$

where the components of the modal coordinate vector are given by, see Eq. (5.79)

$$Q_n(\omega) = H_n(i\omega) P_n(\omega) = \frac{P_n(\omega)}{m_n \left(i\omega - \lambda_n \right) \left(i\omega - \lambda_n^* \right)} \tag{6.115}$$

Using the property of the correlation function given by Eq. (2.48) and the definition of the CF matrix of the modal coordinates, we can express the CF matrix of the modal coordinates as a convolution

$$\mathbf{R}_q(\tau) = \mathrm{E} \left[\mathbf{q}(t) \mathbf{q}^T(t + \tau) \right] = \mathbf{q}(-t) * \mathbf{q}^T(t) \tag{6.116}$$

Taking the Fourier transform of this equation and using the time reversal property of the transform, we get

$$\mathbf{G}_q(\tau) = \left[G_{rc} \right] = \left[Q_r^*(\omega) Q_c(\omega) \right] \tag{6.117}$$

Inspecting Eq. (6.115), we see that for all off-diagonal elements in Eq. (6.117) the two narrow banded terms $Q_r^*(\omega)$ and $Q_c(\omega)$ have a common product that is vanishing for small damping because the two terms are peaking at different frequencies. One can say that this is true either for vanishing damping or for well-separated modes. Only the diagonal elements are large for this case because both terms in the product have a value different from about zero at some frequency.[6]

6.3.2 Decomposition in Time Domain

Under the aforementioned assumptions, we can now derive a decomposition for the random response. Assuming real mode shapes we have that any random response is given by

$$\mathbf{y}(t) = \sum_{n=1}^{N} \mathbf{b}_n q_n(t) = \mathbf{B}\mathbf{q}(t) \tag{6.118}$$

where the mode shape matrix $\mathbf{B}$ holds the N real mode shapes for the N modes and the vector $\mathbf{q}(t)$ holds the N modal coordinates. The correlation function matrix for the random response is then using Eq. (6.118)

$$\begin{aligned} \mathbf{R}_y(\tau) &= \mathrm{E} \left[\mathbf{y}(t) \mathbf{y}^T(t + \tau) \right] \\ &= \mathbf{B}\mathrm{E} \left[\mathbf{q}(t) \mathbf{q}^T(t + \tau) \right] \mathbf{B}^T \\ &= \mathbf{B}\mathbf{R}_q(\tau) \mathbf{B}^T \end{aligned} \tag{6.119}$$

---

[6] In this case the natural frequency of the considered modal coordinate.

If the CF matrix of the modal coordinates is diagonal as assumed in the preceding paragraph, then

$$\mathbf{R}_q(\tau) = \left[R_{q,n}(\tau) \right] \tag{6.120}$$

then the modal decomposition follows directly from Eq. (6.119)

$$\mathbf{R}_y(\tau) = \sum_{n=1}^{N} \mathbf{b}_n \mathbf{b}_n^T R_{q,n}(\tau) \tag{6.121}$$

Both rows and columns of the modal contributions to the correlation function matrix are proportional to the mode shapes.

In the general case of complex modes, the modal expansion can be found from Eq. (5.116)

$$\mathbf{y}(t) = \sum_{n=1}^{N} \left(\mathbf{b}_n q_n(t) + \mathbf{b}_n^* q_n^*(t) \right) = \mathbf{B}\mathbf{q}(t) + \mathbf{B}^*\mathbf{q}^*(t) \tag{6.122}$$

Using the property of the correlation function given by Eq. (2.48) and the definition of the CF function of the response

$$\begin{aligned}
\mathbf{R}_y(\tau) &= \mathrm{E}\left[\mathbf{y}(t)\mathbf{y}^T(t+\tau) \right] = \mathbf{y}(-t) * \mathbf{y}^T(t+\tau) \\
&= \left(\mathbf{B}\mathbf{q}(-t) + \mathbf{B}^*\mathbf{q}^*(-t) \right) * \left(\mathbf{q}^T(t)\mathbf{B}^T + \mathbf{q}^H(t)\mathbf{B}^H \right)
\end{aligned} \tag{6.123}$$

However, since only the diagonal terms are present according to the assumption of uncorrelated modal coordinates Eq. (6.123) reduces to

$$\begin{aligned}
\mathbf{R}_y(\tau) &= \mathbf{B}\left[q_n(-t) * q_n(t) \right] \mathbf{B}^T + \mathbf{B}\left[q_n(-t) * q_n^*(t) \right] \mathbf{B}^H \\
&+ \mathbf{B}^*\left[q_n^*(-t) * q_n(t) \right] \mathbf{B}^T + \mathbf{B}^*\left[q_n^*(-t) * q_n^*(t) \right] \mathbf{B}^H
\end{aligned} \tag{6.124}$$

that is a modal decomposition because each of the four terms can be modally decomposed as Eq. (6.121)

6.3.3   Decomposition in Frequency Domain

For normal modes, the decomposition in the frequency domain follows directly from the Fourier transform of Eqs. (6.119–6.121)

$$\mathbf{G}_y(\omega) = \mathbf{B}\mathbf{G}_q(\omega)\mathbf{B}^T \tag{6.125}$$

where the SD matrix for the modal coordinates $\mathbf{G}_q(\tau)$ is diagonal

$$\mathbf{G}_q(\omega) = \left[G_{q,n}(\omega) \right] \tag{6.126}$$

and the modal decomposition follows directly from Eq. (6.121)

$$\mathbf{G}_y(\omega) = \sum_{n=1}^{N} \mathbf{b}_n \mathbf{b}_n^T G_{q,n}(\omega) \tag{6.127}$$

In the case of general damping and complex mode shapes, decomposition in the frequency domain is easily done by taking the Fourier transform of Eq. (6.124) using the convolution property and the time reversal property of the transform

$$\begin{aligned}
\mathbf{G}_y(\omega) &= \mathbf{B}\left[Q_n(-\omega)Q_n(\omega) \right] \mathbf{B}^T + \mathbf{B}\left[Q_n(-\omega)Q_n^*(\omega) \right] \mathbf{B}^H \\
&+ \mathbf{B}^*\left[Q_n^*(-\omega)Q_n(\omega) \right] \mathbf{B}^T + \mathbf{B}^*\left[Q_n^*(-\omega)Q_n^*(\omega) \right] \mathbf{B}^H
\end{aligned} \tag{6.128}$$

Using the approximation that for any reasonably slowly varying function $g(\omega)$ the ratio of the function divided by a narrow band modal term, we can write

$$\frac{g(\omega)}{\left(i\omega - \lambda_n\right)} \cong \frac{g\left(\omega_{dn}\right)}{\left(i\omega - \lambda_n\right)} \tag{6.129}$$

and the diagonal terms of Eq. (6.128) can be approximated by

$$Q_n(-\omega)Q_n(\omega) = \frac{P_n(\omega)}{a_n\left(-i\omega - \lambda_n\right)}\frac{P_n(\omega)}{a_n\left(i\omega - \lambda_n\right)} \cong \frac{g_{1n}}{i\omega - \lambda_n}$$

$$Q_n(-\omega)Q_n^*(\omega) = \frac{P_n(\omega)}{a_n\left(-i\omega - \lambda_n\right)}\frac{P_n^*(\omega)}{a_n^*\left(-i\omega - \lambda_n^*\right)} \cong \frac{g_{2n}}{-i\omega - \lambda_n^*}$$

$$Q_n^*(-\omega)Q_n(\omega) = \frac{P_n^*(\omega)}{a_n^*\left(i\omega - \lambda_n^*\right)}\frac{P_n(\omega)}{a_n\left(i\omega - \lambda_n\right)} \cong \frac{g_{3n}}{i\omega - \lambda_n}$$

$$Q_n^*(-\omega)Q_n^*(\omega) = \frac{P_n^*(\omega)}{a_n^*\left(i\omega - \lambda_n^*\right)}\frac{P_n^*(\omega)}{a_n^*\left(-i\omega - \lambda_n^*\right)} \cong \frac{g_{4n}}{-i\omega - \lambda_n^*} \tag{6.130}$$

where the constant terms are

$$g_{1n} = \frac{P_n^2\left(\omega_{dn}\right)}{a_n^2\left(-i\omega_{dn} - \lambda_n\right)}$$

$$g_{2n} = \frac{P_n\left(\omega_{dn}\right)P_n^*\left(\omega_{dn}\right)}{a_n a_n^*\left(-i\omega_{dn} - \lambda_n\right)}$$

$$g_{3n} = \frac{P_n\left(\omega_{dn}\right)P_n^*\left(\omega_{dn}\right)}{a_n a_n^*\left(i\omega_{dn} - \lambda_n^*\right)} \tag{6.131}$$

$$g_{4n} = \frac{P_n^{*2}\left(\omega_{dn}\right)}{a_n^{*2}\left(i\omega_{dn} - \lambda_n^*\right)}$$

we can now write an expression for the approximate modal decomposition of the SD matrix from Eq. (6.128) as

$$\mathbf{G}_y(\omega) \cong \mathbf{B}\left[\frac{g_{1n}}{i\omega - \lambda_n}\right]\mathbf{B}^T + \mathbf{B}\left[\frac{g_{2n}}{-i\omega - \lambda_n^*}\right]\mathbf{B}^H + \mathbf{B}^*\left[\frac{g_{3n}}{i\omega - \lambda_n}\right]\mathbf{B}^T + \mathbf{B}^*\left[\frac{g_{4n}}{-i\omega - \lambda_n^*}\right]\mathbf{B}^H \tag{6.132}$$

It is seen that each of the matrix terms can be modally composed as in Eq. (6.127) and that Eq. (6.132) reduces to

$$\mathbf{G}_y(\omega) \cong \sum_{n=1}^{N}\left(\frac{\left(g_{1n}\mathbf{b}_n + g_{3n}\mathbf{b}_n^*\right)\mathbf{b}_n^T}{i\omega - \lambda_n} + \frac{\left(g_{2n}\mathbf{b}_n + g_{4n}\mathbf{b}_n^*\right)\mathbf{b}_n^H}{-i\omega - \lambda_n^*}\right) \tag{6.133}$$

We see that the result has a somewhat similar form as the dominant terms in the exact Eq. (6.77) for the modal decomposition of the SD matrix under the assumption of white noise input.

We see that the columns of the SD matrix are not proportional to the mode shapes, whereas the rows of the SD matrix are indeed proportional to the mode shapes. As for the white noise case, the columns are not exactly but approximately proportional to the mode shapes. For real modes and small damping, we have from Eq. (6.131) that all the terms $g_{1n}, g_{2n}, \ldots$ are equal for the same mode, that is, using the expression for the generalized modal mass as in Eq. (6.101) and assuming the modal mass and the damped natural frequency to be real we can write that

$$g_{1n} = g_{2n} = g_{3n} = g_{4n} = g_n = \frac{\left| P_n \left(\omega_{dn} \right) \right|^2}{\left| a_n \right|^2 \varsigma \omega_{0n}} = \frac{\left| P_n \left(\omega_{dn} \right) \right|^2}{4 m_n^2 \omega_{dm}^2 \varsigma \omega_{0n}} \tag{6.134}$$

and thus we have

$$\mathbf{G}_y (\omega) \cong \sum_{n=1}^{N} 2 g_n \left(\frac{\mathbf{b}_n \mathbf{b}_n^T}{i\omega - \lambda_n} + \frac{\mathbf{b}_n \mathbf{b}_n^T}{-i\omega - \lambda_n^*} \right) \tag{6.135}$$

which relates directly to the results of Section 6.2.6 (taking the two dominant middle terms of Eq. (6.91) and using the approximation given by Eqs. (6.102/6.103)).

References

[1] Papoulis A. and Pillai S.U.: *Probability, random variables and stochastic processes.* New York, McGraw-Hill, 2002.

[2] Bendat, J.S. and Piersol A.G.: *Random data: Analysis and measurement procedures.* New York, John Wiley & Sons Inc., 2010.

[3] Newland, D.E.: *An introduction to random vibrations, spectral & wavelet analysis.* Singapore, Longman Singapore Publishers Pte Ltd, 1997.

[4] James, G.H., Carne, T.G. and Lauffer, J.P.: *The natural excitation technique (NExT) for modal parameter extraction from operating structures. Int. J. Anal. Exp. Modal Anal.,* V. 10, No. 4, p. 260–277, 1995.

[5] Brincker, R., Zhang, L. and Andersen, P.: *Modal identification from ambient responses using frequency domain decomposition,* In Proceedings of the International Modal Analysis Conference – IMAC, p. 625–630, 2000.

[6] Brincker, R., Zhang, L. and Andersen, P.: *Modal identification of output-only systems using frequency domain decomposition, Smart Mater. Struct.,* V. 10, No. 3, p. 441–445, 2000.

[7] Brincker, R. and Zhang, L.: *Frequency domain decomposition revisited.* In Proceedings of the International Operational Modal Analysis Conference (IOMAC), p. 615–626, 2009.

[8] Heylen, W, Lammens, S. and Sass, P.: *Modal analysis theory and testing.* Kathoelieke Universiteir Leuven, Department of Mechanical Engineering, 1997.

7

Measurement Technology

"If anything can go wrong, then it will"

– Murphy

Measurements are one of the most important components of OMA. A successful OMA will greatly depend on the quality of the data obtained, and it is fair to say, that if this part of the total exercise is performed well then everything else that follows is relatively "easy and straightforward." The key steps to obtain good quality data are as follows:

- The testing session has been carefully planned and executed.
- Good quality signals have been captured for each test setup.

In this chapter, we will illustrate and explain how to accomplish these two important steps.

7.1 Test Planning

Some of the most important measurement activities should take place long before datasets are acquired. Planning for data acquisition and analysis should be initiated as early as possible, preferably during the proposal stage of a new test program. There are several issues that should be addressed during this planning, particularly the ultimate uses of the data, the selection of the number, and type of transducers and their frequency ranges, and whether the transducer is to measure more than one event or condition.

Detailed knowledge of the various instruments of the measurement system is required for their selection and utilization. In this section, we discuss how to properly plan OMA tests and prevent the major causes of errors when conducting vibration tests.

7.1.1 Test Objectives

Because of the complexity and cost of testing and also most likely due to time limitations to test a large structure, it is very important to establish clearly what the test objectives are. This should be done early in the test planning process. The questions that need to be answered at this stage are as follows:

What type of information is sought from the test?
What is the minimum amount of information that needs to be obtained?

Introduction to Operational Modal Analysis, First Edition. Rune Brincker and Carlos Ventura.
© 2015 John Wiley & Sons, Ltd. Published 2015 by John Wiley & Sons, Ltd.
Companion Website: www.wiley.com/go/brincker

Which additional information would be useful?
Are the resources adequate to achieve the above?

The test should be planned so that the absolutely necessary information is obtained early in the test. A SMART approach could be followed to achieve the desired goals of the test. That is, the test objectives should be as follows:

Specific to the type of structure being measured.
Measurable, so that it can be assessed whether or not the objectives were achieved.
Agreed upon by all the participants in the project.
Realistic.
Time limited, that is, can be achieved within a specified amount of time.

7.1.2   Field Visit and Site Inspection

During a field visit, many important details of the site or structure to test can be obtained. Although drawings and photos of the facility may provide useful insights into the structure, a site inspection can help prepare a good test plan and reduce the risk of unnecessary delays due to unforeseen circumstances. During a field visit, the logistics of sensor placing, cable arrangement if wired sensors are used, or locations for transmission of good quality wireless data if wireless sensors are used, should be investigated and all parts of the structure, which need to be instrumented, should be inspected. Particular attention should be given to the potential safety hazards that the test personnel may be exposed to. The same applies to OMA tests being conducted in a laboratory or testing facility.

7.1.3   Field Work Preparation

The preparation process for a field vibration test includes the following

- Testing strategy and data handling
- Administrative issues
- Equipment selection and condition

In the following sections, we briefly discuss these three important issues.

7.1.3.1   Testing Strategy and Data Handling

Based on the test objectives, a strategy for measuring a grid of points in the structure should be developed. Preliminary information about the dynamics characteristic obtained from analytical models of the structure aid in the selection of the optimal location of the equipment. After the locations for the reference sensors have been selected, the locations of roving sensor have to be planned. The setup plan should aim at the following:

- Obtain good mode shape definition early in the test.
- Minimize sensor movements between measurements.
- Prevent cable tangles when relocating sensors if cable-based system is being used.
- Ensure proper communication between devices if a wireless system is being used.

If two different measurement systems are being used for the reference sensors and the roving set of sensors, it must be considered how to take care of time synchronization between the two measurement systems, for instance, by using GPS.

A key component of the testing strategy is *flexibility* of the test plan. It is a fact that during testing of a large or complex structure some of the proposed sensor locations may have to be changed when conducting the actual field or laboratory work. Added flexibility to the testing plan is important and permits changes to be made to the setup plan at any time during the testing.

The test plan should also include a well-developed strategy on how to handle large amounts of collected data. It is advisable to prepare the computer files required for the on-site analysis ahead of the actual testing. An agreed-upon naming convention of data files should also be available at this time. When conducting field experimental work is always advisable to collect more data than what is needed, and introduce certain level of redundancy of the acquired data.

A test plan should include the following:

- A plan of the measurement locations (measurement grid);
- A sequence of tests to complete the grid;
- Identification of manpower needed to complete the tests;
- Safety and operational considerations;
- Data acquisition parameters, such as sampling frequency, number of data segments[1] per test, number of data points per segment, and so on.

7.1.3.2 Administrative Issues

Testing of an existing large structure requires careful coordination with the owner or operator of the facility. All authorities who have jurisdiction over the test structure must be informed about the planned tests. The local administrators and operators of the structure should be contacted to ensure that they are aware of the work to be conducted and that they will receive timely information. The project manager of the testing team should be aware of all the safety requirements and work procedures of the facility to be tested. Testing permits from relevant authorities should be obtained before beginning of the test work.

7.1.3.3 Equipment Selection and Condition

Since field testing generally involves the use of various types of equipment, it is best to use a detailed checklist in order to ensure that the appropriate equipment is selected for the test and that it is in good working condition. Electronic equipment should be tested prior to departure, preferably with the complete measurement chain. When the equipment is tested, make sure all tested equipment is put out ready to go (not back on the shelf). If it is necessary to rent or lease equipment, then these items should be organized well in advance and confirmed again shortly before the test date. All the personnel who will be participating in the field test should be familiar with the operation and use of the various pieces of equipment and should be aware of how to handle the electronic equipment, especially transducers and cables.

7.1.4 Field Work

The particular activities of any field test are specific to the type of structure and the test objectives. As an example, the following activities are typical of field ambient vibration testing.

1. Select and mark measurement locations (may be done during the site visit).
2. Install reference sensors.

---

[1] Data segments are discussed in detail in Chapter 8.

3. Start data acquisition of reference signals with initial data acquisition settings.
4. Install roving sensors for first setup.
5. Review data from reference sensors.
6. Adjust acquisition parameters if required.
7. Acquire signals from all channels.
8. Transfer new data from the data acquisition system to computer used for analysis.
9. Move sensors to new locations and simultaneously analyze the new data.
10. Start acquisition for the next setup and continue analysis of previous measurements.
11. Repeat steps 8–10 until all measurement points have been covered.
12. Review preliminary results before packing up the equipment.

During step 8, unacceptable data should be identified (for instance, performing a preliminary OMA) and the measurement repeated if necessary. It is also important to verify that the reference locations are indeed suitable for the test objectives. This may be done by comparing the spectra from all locations with those obtained at the reference locations. If the reference location PSDs contain all the peaks that are present in all the other spectra, then the reference location is considered suitable. If significant peaks are missing from all reference spectra then it may be necessary to relocate the reference stations and re-start testing. A complete backup of all the field data should be conducted before leaving the test site.

Field work can be considered as "taking the laboratory to the site," and therefore all good rules for how to organize laboratory work should be followed. A good guideline is the ISO/IEC 17025 standard [1] for testing and calibration laboratories.

7.2 Specifying Dynamic Measurements

In this section, we discuss about the specifications of dynamic measurements, and we provide more details of some of the important issues that should be considered when planning and conducting OMA tests.

7.2.1 General Considerations

Later in this chapter, we elaborate with detail on the issues of instrumentation features and proper application of instrumentation, but at this stage, we briefly introduce and discuss about the major aspects of these issues.

7.2.1.1 Sensors, Locations, and Directions

The initial task in dynamic measurement planning is the selection of the sensor locations and directions, which determines the total number of measurements. The choice can be based on experience or can also be based on computer simulations using finite element models of the structure to be tested or on predictions of the dynamic response of the structure based on simple beam, plate, or shell theories. For all types of measurements, each transducer should be as small and lightweight as possible in order to minimize the influence of the added mass from the sensors. And the sensors should be sensitive enough to pick up the expected operating signals.

If the test is costly or impossible to rerun and/or if the transducers need to be located in very noisy areas where the sources of vibration are not adequately known, and if the test results are highly dependent on accurate dynamic data, redundant adjacent transducers with different sensitivities may be needed to ensure that the desired test data are obtained even if some transducers fail. The important issue of having a channel count that is large enough to provide a suitable background for good identification is further treated in Section 7.2.2.

7.2.1.2 Frequency Ranges

After measurement numbers, locations, and directions have been selected, the next task is to establish the frequency range for each measurement, that is, the maximum and minimum frequencies to be recorded and analyzed. The selection of the maximum frequency is often the most challenging part of this task. It usually has the greatest impact on the total data or system bandwidth and on the various instruments of the measurement system. The maximum frequency is usually selected to equal the highest of the following:

- The maximum significant frequency of the operating response of the structure.
- A standard frequency for the data acquisition system, for instance, systems for testing of large structures are often having a standard Nyquist frequency of 50 or 100 Hz.

7.2.1.3 Dynamic Ranges

The dynamic range of a measurement system in dB units can be defined as

$$D = 20 \log \left(\frac{y_{\max}}{y_{\min}} \right) \tag{7.1}$$

where $y_{\max}$ is the maximum instantaneous value of the signal that can be transmitted by the system without distortion, and $y_{\min}$ is the minimum detectable value of the signal generated by the transducer at the output display.[2] The minimum detectable signal is dependent on

(a) the magnitude of undesired electrical noise in the measurement system;
(b) the quantization of the measurement by an AD converter (digital noise);
(c) low magnitude distortion of the signal; and/or
(d) interference from other measurements.

Factor (a) usually dominates to produce the background noise floor for the analog instruments of a measurement system, and (b) produces the theoretical noise floor of analog-to-digital (AD) converters. It is common practice to express signal-to-noise ratios for measurement systems in dB such as

$$SN = 20 \log \left(\frac{y_{\max}}{\sigma_n} \right) \tag{7.2}$$

where σ_n is the RMS value of the background noise. According to the American National Standard, ANSI S2.47, [2], the signal-to-noise ratio should not in general be less than 5 dB, and if it is between 10 and 5 dB a correction has to be applied to the data. In OMA, however, it is not recommended to use data with a smaller signal-to-noise ratio than 40 dB.

7.2.1.4 Measurement Duration

It is always desirable to measure the response of a structure during the complete duration of a specific excitation. For random signals, the choice of recording duration depends on the choice of permissible errors (bias and variance). For instance, according to ANSI S2.47, if the fundamental frequency of a building is f measured in Hertz, its modal damping factor is ζ, and if the bias error is to be 4% and variance error is to be 10%, then the recording duration T should be at least:

$$T = \frac{200}{\zeta f} \tag{7.3}$$

---

[2] The issue of defining the dynamic range is further discussed in Section 7.3.2.

This means that if one is planning to conduct tests of a mid-rise steel building with an estimated period of about 0.5 s (or 2 Hz frequency) and fundamental modal damping of about 1%, the duration of each test should be at least 10,000 s or 2 h and 47 min per test setup. For a taller building with a longer period, the measurement time would be correspondingly longer. For practical purposes, doing OMA tests, which require almost 3 h per test setup would be cost prohibitive and, unless really necessary to comply with this standard, or similar ones, shorter duration tests may be justifiable. The required duration of the tests can be also estimated as a function of the maximum correlation time in the response. The details of this alternative approach are given in Section 7.2.4.

7.2.1.5 Error Checklists

A checklist of errors should include a clear definition and tabulation of all potential errors in the data acquisition and analysis instrumentation, including the statistical estimation errors in the analysis of the signals. This checklist is valuable because

- It forces the planner to think about all possible sources of error before data are acquired and analyzed.
- It immediately identifies the primary sources of errors (the "weak links") in the system instrumentation.
- It allows the logical selection of the measurement durations. This will help ensure that the statistical sampling errors are consistent with the potential errors due to the data acquisition instruments.
- It permits an objective estimation of an overall system potential error in the final results of the experiment.

7.2.2 Number and Locations of Sensors

7.2.2.1 Minimum Number of Sensors (Rank of Problem)

How many sensors are needed to conduct an OMA test? The question is a simple one, but to answer it is a significant challenge. A further question is "why can we not just use one or two sensors?" The correct answer to this question is that usually we cannot do that; and if you try to use only one or two sensors for your measurement, you will fail in the most cases. In OMA, it is essential to use a reasonable number of sensors. The main point here is to understand why this is the case and to be able to estimate what a "reasonable number of sensors" really is.

The easiest way to understand how many sensors are needed for an OMA test is to consider the "rank of the problem" in connection to the frequency domain decomposition (FDD) theory, the details of which are presented in Chapter 10. Suffice to say here that in the FDD we estimate a spectral density (SD) matrix. The physical rank of this matrix is defined as the maximum number of modes that are contributing to the response in any frequency band. If the considered frequency band includes only two closely spaced modes, then the physical rank is two. If we have three closely spaced modes, then the physical rank is three, and so on. But we need to keep in mind that the calculation of the rank of the testing problem has to include also the influence of the noise that is always present. If, for example two noise sources are present, and we have only two closely spaced modes then the rank of the problem is $R_p = 2 + 2 = 4$.

It is important to understand that the measurement system should have the ability to clearly identify all important sources of noise and physical responses in such a way that the measurement system itself does not limit the rank of the problem. The rank of the SD matrix estimated from the recorded measurements is limited by the number of measurement points N since the SD matrix is $N \times N$, and the rank can never be larger than the smallest dimension of the matrix. However, if two measurement points are close to each other, these are representing the same information, and thus, the added channel does not in practice contribute to the matrix rank. Therefore, it is essential that the number of measurement channels is larger than the rank of the problem defined earlier, and secondly, that the measurement points are spread over the structure in such a way that each individual measurement point does not just mainly repeat any information in other channels. It is also important that no sensor is placed in a node point for any mode of interest because if that is the case, then this sensor cannot obtain any information about this mode.

One should never assume the rank of the problem to be smaller than four and therefore, if the sensors are to be placed at the correct locations, then four sensors is the absolute minimum number of sensors to be used in order to perform a successful OMA testing and analysis. However, there is always uncertainty on the correct selection of sensor locations away from any node points, so we would say that in practice a good rule of thumb is always to use at least five or six sensors.

7.2.2.2   Selection of the Best Location

Theoretical methods of finding the best location of sensors have never received much attention in practical test planning. A possible reason for this is that implementation in practice of these concepts is difficult and time consuming, and the results sometimes do not make sense or indicate that sensors should be placed at locations that are very difficult or impossible to reach. However, is it still of interest to discuss how these techniques work in general. More information about the subject can be found in Kirkegaard [3]. This reference provides a comprehensive treatment of the problem.

Finding "the best" location is an optimization problem, and as it is the case in most optimization problems the object function is flat around the optimum. Therefore, finding the optimum locations in practice is not that important, because if we are reasonably close to the optimum locations, trying to get closer to those locations will only result in marginally improvements in the identification procedure. In other words, the optimal solution is of limited practical interest. But what is of interest is to place the sensors in "a reasonable way," so that we can get "reasonable" information from the measurements.

A simple way to roughly get an idea of a set of reasonable sensor locations is the method presented by Ibanez et al. [4]. This method can be explained as follows. First, let us consider the modal decomposition of the dynamic response

$$\mathbf{y}(t) = \mathbf{A}\mathbf{q}(t) \tag{7.4}$$

Where the matrix $\mathbf{A}$ contains the true mode shapes of the structure and the vector $\mathbf{q}(t)$ contains the true modal coordinates. If we have an estimate $\widehat{\mathbf{A}}$ of the mode shape matrix, then in Eq. (7.4) we can take the pseudo inverse of $\widehat{\mathbf{A}}$ to obtain an estimate of the modal coordinates, that is,

$$\widehat{\mathbf{q}}(t) = \widehat{\mathbf{A}}^+ \mathbf{y}(t) \tag{7.5}$$

But this is equal to

$$\widehat{\mathbf{q}}(t) = \widehat{\mathbf{A}}^+ \mathbf{A}\mathbf{q}(t) \tag{7.6}$$

The matrix $\widehat{\mathbf{A}}^+ \mathbf{A}$ is a measure of how close the estimated modal coordinate vector is to its true value. The closer the product $\widehat{\mathbf{A}}^+ \mathbf{A}$ is to the identity matrix, the better is the sensor placement. We need a way to measure the deviation of the product from the identity matrix, such measure could, for instance, be $\det\left(\widehat{\mathbf{A}}^+ \mathbf{A}\right)$. Since $\det(\mathbf{I}) = 1$ and $\det\left(\widehat{\mathbf{A}}^+ \mathbf{A}\right)$ can be either smaller or larger than one, a positive measure of the deviation from identity can be obtained by calculating $\left|1 - \det\left(\widehat{\mathbf{A}}^+ \mathbf{A}\right)\right|$, and a measure of "the value" V of the sensor placement can be calculated as

$$V = 1 - \left|1 - \det\left(\widehat{\mathbf{A}}^+ \mathbf{A}\right)\right| \tag{7.7}$$

It is clear that using this approach, we must have more sensors than modes in the response given by Eq. (7.4). This approach forces us to think about how many modes we have to deal with and leads us to wonder what is the smallest frequency band that we can consider where the response is governed mainly by the modes in the frequency band of interest. A key question to answer is: how many modes do we have present in this band?

The estimate $\hat{\mathbf{A}}$ is not the same as the true mode shape matrix $\mathbf{A}$ because otherwise Eq. (7.7) would always give us a maximum (unit) value. We have to consider how accurately we can measure and estimate mode shapes and to do this we need a noise model. For the considered case, we can model each of the estimated mode shapes $\hat{\mathbf{a}}$ in $\hat{\mathbf{A}}$ as the true mode shape $\mathbf{a}$ plus some noise, that is,

$$\hat{\mathbf{a}} = \mathbf{a} + \{X_i\} \tag{7.8}$$

where $\{X_i\}$ is a random vector where each of the elements is a stochastic variable with standard deviation

$$\sigma = \varepsilon \max(a_i) \tag{7.9}$$

where a_i are the elements in the true mode shape $\mathbf{a}$ and ε is a relative uncertainty parameter. This simple approach will lead to a realistic picture of where sensors should be placed and, especially, where they should not be placed and how many sensors should be used. The following example will help the reader understand better how this technique can be used to determine suitable sensor locations.

Example 7.1   Sensor locations on a simply supported beam

Let us consider the need to identify, for instance, the first three bending modes of vibration of a simple simply supported beam and let us assume three different relative uncertainty parameters $\varepsilon = 0.01, 0.05$, and 0.20 describing different accuracies of mode shape estimation. We will investigate the use of either eight or four sensors for the identification. Let the dimensionless length of the beam be 100, and let a uniformly distribution of the eight sensors (Case A) be the locations 10, 20, 30, 40, 60, 70, 80, and 90 and let a second distribution of the four sensors (Case B) be the locations 20, 40, 60, and 80. Finally, let us consider a distribution of the eight sensors where the sensors are distributed more nonuniformly toward the ends of the beam (Case C) with the locations 10, 15, 20, 25, 65, 70, 85, and 90. Finally, let us consider a nonuniform distribution of the four sensors (Case D) with the locations 10, 15, 85, and 90. In order to analyze and compare these possibilities, we need an estimate of the mode shapes. In this case, we will use a simple harmonic wave form such as $a_n(x) = \sin(n\pi x/L)$ to approximate the mode shapes. In this expression, n denotes the mode number, x is the distance to the support, and L is the beam length. Using Eq. (7.7) and performing simulations of the estimated mode shapes according to Eq. (7.8) and implementing some averaging in order to remove uncertainty on the value parameter V, we obtain the results shown in Table 7.1.

Table 7.1   Measure of "the value" of the sensor placements estimated by Eq. (7.7)

Sensor distribution	$\varepsilon = 0.01$	$\varepsilon = 0.05$	$\varepsilon = 0.20$
Case A with locations 10, 15, 20, 25, 65, 70, 85, and 90	0.99	0.97	0.88
Case B with locations 20, 40, 60, and 80	0.99	0.96	0.85
Case C with locations 10, 15, 20, 25, 65, 70, 85, and 90	0.99	0.96	0.84
Case D with locations 10, 15, 85, and 90	0.94	0.64	0.09

From this analysis, it appears that it is reasonable to use only four sensors, but only if they are uniformly distributed along the beam. Further, we can see that identification methods with high spatial uncertainty (the case corresponding to $\varepsilon = 0.20$) should not be used at all, and that a nonuniform sensor distribution is acceptable in case eight sensors are used.

In general, however, a criterion for optimal sensor placement is more complicated than what has been discussed before. A commonly accepted approach is based on minimizing the expected uncertainty on the parameters to be estimated in the OMA test. In this approach, all modal parameters that are to be estimated are gathered in the vector $\boldsymbol{\theta} = \left\{\theta_1, \theta_2 \cdots \right\}^T$. This vector is a stochastic vector since all parameters are estimated with uncertainty, and thus it has a covariance matrix $\mathbf{C}_\theta$. This covariance matrix depends on the sensor location vector $\mathbf{z}$; so we can define $\mathbf{C}_\theta = \mathbf{C}_\theta (\mathbf{z})$. The next task is to find the value of the location vector that minimizes some scalar measure of the covariance matrix; for instance, the determinant.

Since it is complicated to establish the relation between the uncertainty on modal parameters and the location of the sensors, and also to apply a reasonable noise modeling (reflecting reality), this approach is difficult to implement in practice. A more practical approach would be to address the problem of where to locate sensors by simulations. The next section addresses this topic.

7.2.2.3   Simulations and Common Sense

One of the best ways to understand and investigate the challenges of selecting sensor locations is to perform simulations. Since we do not know the exact properties of the structure under consideration, we need to simulate a set of possible responses using different models and different placement of the sensors, and see which sensor locations provide the best estimates for all the models considered.

The key to having good results using this approach is to be guided by common engineering sense of what needs to be included in the models. Often some kind of FE model is available, but this model was probably developed by someone else (probably an analyst who was part of the design team) not involved with the testing of the structure. This means that the person who is planning the test might not necessarily know the reasons for the decisions made by the analyst when the FE model was developed. Although not knowing the rationale behind the FE model might be a challenge, this might also be an advantage to the test engineer. The test engineer will then be able to raise questions about the modeling assumptions, challenge the model, and make the necessary modifications to cover a reasonable set of possible responses.

It is instructive to have a good sense of what could be changed in an FE model in order to perform meaningful simulations. Often a rather small uncertainty is present due to mass density and Young's modulus, and a larger uncertainty is present due to incomplete information about connections and support conditions. In case a modal model is used for simulation, one might need to know the uncertainty on the modal parameters rather than the uncertainty related to the stiffness and mass distribution. In any case, our models are always uncertain, and, therefore, when performing simulations, different models should be used to account for the uncertainty in the model. Uncertainty values that have been used by the authors for some important parameters influencing simulation studies are given in Table 7.2.

When performing a simulation study based on an FE model, one would start out by choosing a sensor configuration and which parameters should be included in the random modeling of the FE model. The latter choice can be made based on the values from Table 7.2. For each simulated realization of the FE model, the true modal parameters are determined solving the eigenvalue problem, and a simulation is performed of the operational responses to the measured by the sensors. A reasonable measurement noise is to be added in order to model the accuracy of the sensors. Finally, OMA is performed on the simulated

Table 7.2 Uncertainty on the most important parameters of FE and modal models used for simulation studies

Physical quantity	Level of Uncertainty (%)
Young's modulus and mass density, steel, and other metals	1–5
Young's modulus and mass density, concrete, wood, fiber reinforced materials	5–20
Boundary conditions to the ground	10-infinity
Joints, bolted	10-infinity
Joints, welded	2–10
Total mass	1–5
Estimated Natural frequency	0.1–0.5
Measured response	0.2–2
Estimated mode shapes	2–5
Estimated damping	5–20
Mode shape scaling factor	5–30

operational responses, the identified modal parameters can be compared to the true values and finally it can be considered if the chosen sensor configuration can give us the needed modal parameters with the accuracy that is needed. In order to check that the sensor configuration is also meaningful for the real test, the simulations have to be carried out considering a reasonable variation of realizations of the FE model.

7.2.3 Sampling Rate

Sampling rate defines the upper limit of the frequency band that can be used for the analysis of the recorded signals. The sampling rate is the number of data samples acquired per unit of time and is commonly defined in terms of "samples-per-second" (sps). The sampling rate in "sps" is also denoted as the "sampling frequency." So if we say that data has been acquired at a sampling frequency of 200 Hz, it means that the data was acquired at a rate of 200 sps. The upper limit of the frequency band is given by the Nyquist frequency

$$f_v = f_s/2; \quad f_s = 1/\Delta t \tag{7.10}$$

where f_s is the sampling frequency, and Δt is the sampling time step. So, if the sampling frequency is 200 Hz, then the frequency band is limited to 100 Hz and any information about the structure beyond this frequency cannot be determined.[3] This means that the sampling rate must be chosen large enough so that all the modes of interest are properly detected from the measured signals. One should also consider that good quality data acquisition systems include antialiasing filters and that such filters have an effect for frequencies in the vicinity of the Nyquist frequency. In practice, it is considered that the effect of typical antialiasing filters results in an effective frequency band up to 0.8 times f_v, thus for practical purposes the recommended sampling frequency must be larger than approximately 1.2 times f_v. From Eq. (7.10), we obtain

$$f_s > 2.4 f_{max} \tag{7.11}$$

---

[3] The Nyquist frequency is a consequence of the principles of the discrete Fourier transform, see Chapter 4.

where f_{max} is the highest frequency of interest. If a sigma-delta technology is used in the analog-to-digital converter (ADC), then there is no reason to use a higher sampling rate than prescribed by Eq. (7.11). However, if this is not the case, a better signal-to-noise ratio is always achieved by using a higher sampling rate and then, as part of the signal conditioning process, decimating the signal down to the sampling rate given by Eq. (7.11). Using a higher sampling rate than what is needed according to Eq. (7.11) is called oversampling.

Using the Parseval theorem, it is easy to quantify the influence of oversampling. Noting that the Parseval theorem also applies to the noise $n = n(t)$, present in the signal, the noise spectrum can be assumed to be flat in the frequency band of interest, that is, $G_n(f) = G_{n0}$. Then from Eq. (6.14), we get

$$E\left[n^2\right] = \sigma_n^2 = 2G_{n0}f_v \tag{7.12}$$

If the uncertainty of each sample value obtained from the ADC, given by the variance σ_n^2, does not depend on the sampling rate as it is the case for simple[4] ADC's, then we can say that for two different Nyquist frequencies f_{v1} and f_{v2} and their corresponding noise floors G_{n01} and G_{n02}

$$2G_{n01}f_{v1} = 2G_{n02}f_{v2} = \sigma_n^2 \tag{7.13}$$

This means that the higher the sampling rate is selected, the lower the noise floor will be. Therefore, if one is using a simple ADC where the measurement uncertainty does not depend on the sampling rate, then it is more advantageous to over sample to acquire data and then decimate (down sample) before doing any further analysis of the data. This also simplifies and enhances the use of the antialiasing filters as these can be set to a fixed cutoff frequency corresponding to an adequately high and fixed sampling rate.

7.2.4 Length of Time Series

When identification problems arise in OMA these are often due to acquired time series that are too short. The length of the signal to be obtained can be estimated from the maximum correlation time in the response. If we assume that the longest correlation time is defined by the lowest natural frequency of the system,[5] $f_{min} = \omega_{min}/2\pi$, then, we can say that the auto correlation function of the corresponding modal coordinate is proportional to

$$R(\tau) = e^{-\varsigma\omega\tau}\cos\left(\omega_{min}t\right) \tag{7.14}$$

From this equation, we can see that the correlation function is reduced to about 4% at a time lag (the memory time of the system) T_{mem} corresponding to

$$T_{mem}\varsigma\omega_{min} = \pi \Rightarrow T_{mem} = 1/\left(2\varsigma f_{min}\right) \tag{7.15}$$

This provides some assurance that we will be able to estimate the damping with a minimal influence from leakage bias; see Chapter 8. To provide a reasonable estimate of the correlation function of the recorded motion, one could state that at least 20 data segments are needed (windowed data segments with 50% overlap and 40 frequency domain averages, see Section 8.4.2), thus a minimum time series length is estimated to about 20 times the memory of the lowest frequency of interest. The total time series length T_{tot} should, therefore, fulfill the requirement that

$$T_{tot} > \frac{20}{2\varsigma f_{min}} = \frac{10}{\varsigma f_{min}} \tag{7.16}$$

Using this equation, some examples of required minimum measurement times as a function of the expected damping of the system are presented in Table 7.3.

---

[4] ADC's that are not the sigma-delta type.

[5] Correlation functions have the same form as a free decay of the system, see Chapter 6.

Table 7.3 Minimum time series length's for different damping ratios

Damping (%)	Relative time series length $T_{tot} f_{min} = 10/\varsigma$
0.2	5000
0.5	2000
1.0	1000
2.0	500
5.0	200

For the example presented in Section 7.2.1, the use of Eq. (7.3) resulted in a record length of 10,000 s equal to about 2 $1/2$ h for a system with 1% damping and a period of 0.5 s (or $f_{min} = 2$ Hz). By contrast, Eq. (7.16) would result on a required record length of 500 s, or a little more than 8 min. Clearly, there is a trade-off in getting a record of shorter duration than that required by some established standards, but one has to keep in mind the practicality of conducting tests for OMA and the time and budget for the whole project.

7.2.5   Data Sets and References

The minimum number of sensors to be used for an OMA project can be determined by the rank of the problem as explained earlier. For most projects, this required number of sensors might not be very high, typically six to nine sensors.

Let us start the discussion of this subject considering a typical case of 3D displacement and let us discuss if good OMA testing can be performed using only two tri-axial (3D) sensors, one to be kept in the same point in all data sets, called the reference sensor, and the other one to be roved over the structure. The number of locations where the roving sensor will be placed will depend on the desired spatial resolution of the mode shapes to be identified.

If the reference is placed at a location where one of the modes has a nodal point, then this mode cannot be reliably estimated since the record of the reference sensor would not contain information about this mode (i.e., the spectral amplitude will be relatively small), and thus the mode shape coordinates of the roving sensor cannot be adequately scaled with respect to the reference sensor.

Good OMA studies can only be performed using only two 3D sensors if we can ensure that the reference sensor will not be placed at a node point for one of the modes of interest. Clearly, it is difficult to ensure that this is the case for complex structures, so to minimize the impact on the results of having a reference sensor at a nodal point, a better choice would be to use at least two 3D sensors as references and then one or more roving sensors. Similar consideration can be made for 2D and 1D cases. Guidelines for the number of reference sensors are summarized in Table 7.4.

Some examples of selection of reference and roving sensors for some typical 2D studies are shown in Figures 7.1 and 7.2. Figure 7.1 shows a typical testing plan for a bridge. A bridge is often a typical 2D study[6] because the main displacements are due to vertical and horizontal bending and to torsion, so three sensors will be needed at each cross section in order to estimate these three components of motion. As discussed earlier at least two different points should be used to place the reference sensors, thus two 2D

---

[6] This is true only in case the bridge behaves like a beam where the cross section does not deform, that is, only moves as a rigid body. If this is not the case and the bridge deck moves more like a plate, more sensors should be used over the cross section of the bridge.

Table 7.4   Recommended number of reference sensors

Type of study	Recommended number of reference sensors
3D	6 (two 3D sensors)
2D (for instance, beam problems)	4 (two 2D sensors)
1D (for instance, plate problems)	2 (two 1D sensors)

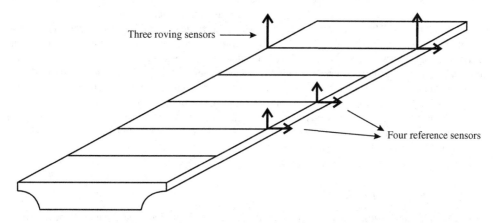

Figure 7.1   *Example of a 2D study:* Measurement plan for a straight bridge where the longitudinal displacements are neglected, thus only the transverse horizontal and vertical displacements and the torsions are to be estimated. Since the locations of the node points are uncertain, two 2D sensors are used in the reference set and then three sensors are used in the roving set. In this case, the minimum number of sensors needed to perform the measurements is seven and the number of data sets is equal to the number of cross sections where the mode shape is to be estimated

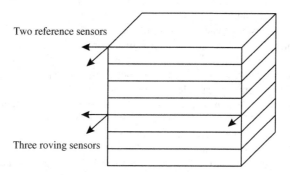

Figure 7.2   *Example of a 2D study:* Measurement plan for a simple rectangular "box type" building where the vertical displacements are neglected, thus only the two horizontal displacements and the torsion of the building is to be estimated. In this case, no modes have node points at the top corners, thus only one 2D reference sensor can be used and three sensors can be used in the roving set. In this case, the minimum number of sensors needed to perform the measurements is five and the number of data sets is equal to the number of floors where the mode shape is to be estimated

sensors could be used as the reference set. In this case, the minimum number of sensors is five: two 2D for the references, and a single 1D sensor for the roving set. However, using this minimum number of sensors will result in many data sets since the roving sensor will have to be installed at many locations in order to have a good spatial definition of the vertical, lateral and torsional modes. In order to reduce the number of data sets, a more practical way of planning the test would be to use more roving sensors. As suggested in Figure 7.1, a set of roving sensors are being used as a group of three for each transverse cross section of the bridge, the number of sensors are now seven instead of five and the total testing time would be reduced to one third. It should be noted that even though one would expect the vertical displacement over the supports to be small, this is not always the case for all the modes, and thus it is recommended to include also cross sections over and close to the supports in the measured data sets. This will also help in determining the actual level of flexibility of the supports of the structure being tested.

A simple "box type" building is also a 2D case where the vertical displacements are considered small and of minor importance (see Figure 7.2). The displacement of the structure can be considered analogous to a beam's behavior with horizontal displacements in two directions and torsional motion with respect to the vertical axis of the structure. The building behaves approximately like a vertical clamped beam, and such type of structures do not have a nodal point at the free end for any mode. In this case, we can use only one 2D sensor as a reference sensor if we place it at one of the corners of the top of the building. This approach assumes that the center of rotation is approximately in the middle of the building. For buildings of more complex geometry, where the center of rotation is more difficult to estimate, again two 2D sensors must be used for the reference set.

7.2.6 Expected Vibration Level

Before performing the actual test of a structure, it is important to have an idea about the expected levels of vibration. The selected sensors and measurement system must be adequate to measure the actual levels of vibration being present at the structure.

In civil engineering, the level of vibration must be estimated from the expected types of dynamic loads acting on the structure, such as wind, waves, and traffic. In many cases, the designer of the structure may be helpful in providing information about the types and amplitude of these loads.

For cases when there is very little or no information about the types and amplitude of the dynamic loads, such as the case of a bridge located in a remote location, or an empty building far away from any traffic in an area where the wind is often absent, the level of vibrations at the structure can be estimated as the response to micro seismic excitation (microtremors) from the ground supporting the structure. The minimum level of micro seismic excitation can be estimated from the Peterson low noise model (LNM), Peterson, [5], see Figure 7.3. The LNM values can be used in cases where it is essential to be able to perform OMA based on excitation caused by the microtremors. In order to get a better idea of the levels of motion from the LNM, some typical values are included in Table 7.5.

As many of the stations used in the Peterson investigation were in remote locations, and some in deep boreholes, the Peterson LNM can be considered as a conservative estimate for what would be expected in terms of microtremor excitation. Several factors influence the actual microtremor input; cultural noise (mainly vehicle traffic), site-amplification and site level (for instance, at Iceland microtremor level is significantly higher than average). In view of this, it is advisable to assume a somewhat higher level for the expected microtremor excitation if the LNM is used as a reference. Most of the station data show typical values of the order of $-140\,\mathrm{dB}$ (relative to $1\,\mathrm{m}/\left(s^2\sqrt{\mathrm{Hz}}\right)$. Thus, a reasonable estimate of the microtremor level might be obtained using this value.

The microtremor levels shown in Figure 7.3 are for the vertical response of the ground. However, since the Peterson investigation shows that minimum horizontal levels are similar or slightly higher than the vertical ones, the LNM model can be assumed to be representative of both vertical and horizontal minimum excitation levels.

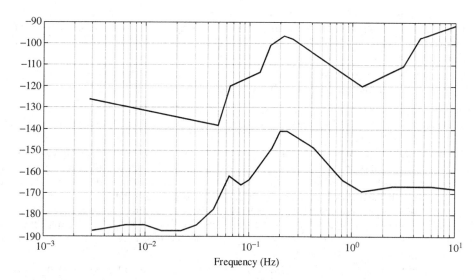

Figure 7.3 Minimum and maximum vertical seismic noise power spectral density (PSD) according to Peterson [5] in dB relative to $1\,\mathrm{m}/\left(\mathrm{s}^2\sqrt{\mathrm{Hz}}\right)$. The top plot is the Peterson high noise model (HNM) and bottom plot is the Peterson low noise model (LNM). The values were computed as 20 log of acceleration relative to $1\,\mathrm{m}/\left(\mathrm{s}^2\sqrt{\mathrm{Hz}}\right)$

Table 7.5   Minimum excitation levels

Vibration level	Frequency (Hz)		
	0.1	1	10
According to the Peterson high noise model (HNM)			
Acceleration [nm/nm/(s$^2\sqrt{\mathrm{Hz}}$)]	1,500	9,600	25,700
Velocity [nm/(s$\sqrt{\mathrm{Hz}}$)]	2,400	300	420
Displacement (nm/$\sqrt{\mathrm{Hz}}$)	3,900	48	6.7
According to the author's recommendation -140 dB [relative to $1\,\mathrm{m}/(\mathrm{s}^2\sqrt{\mathrm{Hz}})$] noise level			
Acceleration [nm/(s$^2\sqrt{\mathrm{Hz}}$)]	100	100	100
Velocity (nm/(s$\sqrt{\mathrm{Hz}}$))	160	16	1.6
Displacement (nm/$\sqrt{\mathrm{Hz}}$)	253	2.5	0.025
According to the Peterson low-noise model (LNM)			
Acceleration [nm/(s$^2\sqrt{\mathrm{Hz}}$)]	6.5	5	4.0
Velocity [nm/(s$\sqrt{\mathrm{Hz}}$)]	10	0.80	0.063
Displacement [nm/$\sqrt{\mathrm{Hz}}$]	16	0.13	0.0010

Using the microtremor acceleration level as minimum input provides a simple way of estimating the expected corresponding response. If we assume a system damping of 1%, then any structure will amplify the base acceleration about 35 dB around the natural frequencies of the structure. Since a signal-to-noise ratio of the order of 35 dB is just acceptable, the minimum input base excitation can be taken as an estimate of the sensor noise floor requirement. Therefore, for microtremor-based OMA, the acceleration sensors to be used should have a noise floor lower than -140 dB (relative to 1 m/ $\left(s^2 \sqrt{Hz} \right)$) corresponding to 10 ng/$\sqrt{Hz}$ over the frequency span of interest.[7] Minimum excitation levels according to the Peterson HNM and LNM and the recommended reference levels by the authors (the -140 dB rule) are summarized in Table 7.5.

For mechanical engineering applications, it is difficult to give similar guidelines, mainly because of the variety of structural systems to be tested and the conditions under which the structure is being tested. For instance, conducting OMA on an airplane in flight is totally different from conducting OMA on a satellite in space. In the first case, the loading is dependent on flight speed and airplane operation, while in the latter case the response might be associated to the small squeaks from temperature variation due to changes in sun radiation or it might be associated with impacts of micro particles. Using OMA for mechanical engineering applications requires good coordination between the designers of the structure and the engineers and technicians conducting the tests.

Since clipping should always be prevented, it is also necessary to consider the largest possible signals that can be expected. For civil structures, signals higher than 0.25 g are rarely seen under normal operating conditions; however, for mechanical structures a high-level signal depends a lot on the actual application and should be estimated case by case.

Whenever it is possible, the best way to estimate the expected level of vibration is to conduct pilot tests. The best option is, of course, to make several pilot tests covering a reasonable variety of loading situations so that a good understanding of what can be expected during the final OMA tests can be developed.

7.2.7   Loading Source Correlation and Artificial Excitation

The number of independent loading (input) sources is one of the factors limiting the rank of the problem. If we have only one independent loading source, then the rank of the problem is one. This means that, regardless of the system identification method being used, it will be difficult – if not impossible – to identify closely spaced modes or even to see if closely spaced modes are present. Therefore, a careful consideration of the number of independent input sources is "a must" in all OMA testing. A summary of typical identification problems in relation to number of independent inputs is given in Table 7.6.

To be precise, "the number of independent inputs" means the practical[8] rank of the correlation matrix of the inputs. Let us consider the input-output frequency domain relation, Eq. (5.227)

$$\tilde{\mathbf{y}}(f) = \tilde{\mathbf{H}}_{yx}(f)\tilde{\mathbf{x}}(f) \tag{7.17}$$

where $\tilde{\mathbf{H}}_{yx}(f)$ is the FRF matrix of the system. If the correlation matrix $\mathbf{C}_x$ of the corresponding time domain input $\mathbf{x}(t)$ has rank one or is close to rank one, then we would say that there is only one independent input. As discussed in Chapter 1, the input $\mathbf{x}(t)$ can be considered as the output of a linear filter describing the loading properties. Thus, the loading itself is considered as the response of the loading filter loaded by the independent white noise sources $\mathbf{e}(t)$

$$\tilde{\mathbf{x}}(f) = \tilde{\mathbf{H}}_{xe}(f)\tilde{\mathbf{e}}(f) \tag{7.18}$$

---

[7] It should be mentioned that sensors with a noise floor that low are difficult to obtain. This is one of reasons why it makes sense to consider sensors such as geophones, see Section 7.3.4.

[8] The rank can be estimated as the number of nonzero singular values. The practical rank means here to exclude the singular values related to the noise in this counting.

Table 7.6   Classification of typical problems in relation to number of independent inputs

Ideal cases	Medium level problematic cases	Highly problematic cases
Cases with many independent inputs such as • Moving stochastic loads • Large wind loaded structures • Structures loaded by traffic load or person load • Big machines with many moving parts	Cases where the number of independent inputs is close to one such as • Small wind loaded structures • Structures loaded only by ground motion • Structures loaded by sound pressure from a single source • Simple machines with a few moving parts	Cases where the loading is restricted to one input such as: • Hammer input in one point • Shaker input in one point

The loading filter modeled by the FRF matrix $\widetilde{\mathbf{H}}_{xe}(f)$ describes the coloring of the loading (deviation from white noise) and the geometrical distribution of the forces. However, if the vector of white noise sources $\widetilde{\mathbf{e}}(f)$ only contains a single white noise source, then the resulting vector $\widetilde{\mathbf{x}}(f)$ and its time domain equivalent $\mathbf{x}(t)$ are governed by this single source, and the covariance matrix is forced to be of rank one. From these considerations, we see that it is very possible that we might have many loads acting on a structure and still have a single independent input.

One can also see that a single source will limit the rank of the response spectral matrix to one. Using the fundamental theorem Eq. (6.44) on Eqs. (7.17) and (7.18), we get the following expression for the response spectral matrix

$$\mathbf{G}_y(f) = \widetilde{\mathbf{H}}_{yx}^*(f)\,\widetilde{\mathbf{H}}_{xe}^*(f)\,\mathbf{G}_e(f)\,\widetilde{\mathbf{H}}_{xe}^T(f)\,\widetilde{\mathbf{H}}_{yx}^T(f) \tag{7.19}$$

If there is only one source in the white noise vector $\widetilde{\mathbf{e}}(f)$, then all elements in the spectral matrix $\mathbf{G}_e(f)$ are zero except a single element in the diagonal, and the rank of the spectral matrix $\mathbf{G}_e(f)$ is one. However, the rank of a matrix product can never be larger than the minimum of the rank of the matrices in the product, so in this case the rank one input spectral matrix limits the response spectral matrix to rank one, as well.

The FDD technique, see Section 10.3, can be used to check the resulting rank of the response spectral matrix. This can be performed by taking a singular value decomposition of the spectral matrix, plot the singular values as a function of frequency and then check that a reasonable number of singular values are well separated from the noise floor. An example is shown in Figure 7.4 in which the importance of using an excitation with several independent inputs is illustrated. Note the various cases of closely spaced modes at 2 and 5.5 kHz that are easily seen in Figure 7.4a and not visible in Figure 7.4b.

The influence of the limited rank of the excitation source is not limited to the frequency domain. Time domain methods for OMA are also sensitive to this issue. A discussion on this is presented by Brincker et al. [6].

If the considered problem is close to an ideal case, where a large number of independent input sources are present, then the rank of the problem is not limited by the excitation of the structure, then there is no reason to look further into this problem, and OMA is a good tool to analyze such case.

In contrast, when the other extreme happens – where the testing technology forces the excitation to be a single independent input – OMA technology should not be used as the only tool to analyze data from such tests, because there is significant risk of missing important information.

Hammers and shakers should never be used in OMA testing. Using only one input, the excitation limits the rank of the problem to one – and we cannot see or identify closely spaced modes. However, this is not the only reason for not recommending using OMA techniques for these types of tests. One could try

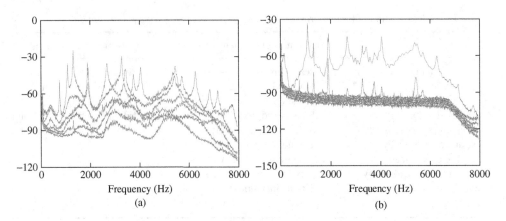

Figure 7.4 Singular value plots in dB of the response spectral matrix of a 30 cm-by-30 cm thin steel plate with 16 sensors in a 4×4 grid. (a) The result when a randomly moving load is used – in this case the back end of a pencil is being moved randomly over the entire surface of the plate. (b) The results when a stationary single input loading is being used instead. The distinct separation of the singular value lines in (a) indicates that the spectral matrix has rank of 5–6 over the whole frequency band, whereas in the (b) the rank is limited to one in nearly the whole frequency band. It is obvious that the case shown on the (b) has significantly less system information compared to the case shown on the (a) (Source: From Brincker et al. [6])

to overcome the rank one problem by using several hammers or several shakers at the same time, but there are still other problems associated with such an approach.

Hammer testing results in data with decreased quality because the hammer loading has a high crest factor, and therefore results in response data with a smaller signal-to-noise ratio than stationary random data. For medium size structures in the laboratory where a shaker might seem like an obvious choice, the shaker might add significant mass and damping to the structure, and as a result the computed modal parameters from this type of testing might be heavily biased.

There are a number of situations where a limited number of sources might seriously limit the rank of the problem. A typical example is when wind loading is considered in relation to model scaling. When a large structure, such as a high-rise building or a suspension bridge is being tested, the wind loading is assumed as consisting of many independent excitation sources. It is known that wind forces are correlated in space; however, since this correlation has a typical length scale – the correlation length – then, when the structure is large compared to the correlation length many independent excitation forces are acting on the structure, and the rank of the problem is not limited by the excitation forces. However, if a small scale model of the real structure is created and is subjected to the same wind field, then it would happen that the structure might be small compared to the correlation length, and we revert to the rank one problem because the number of independent excitation sources might be close to one.

Another example is a structure that is excited mainly by ground shaking. Ground motion is a multi-directional movement, and if these motions can be shown to be stochastically independent then OMA technique can be used effectively to identify the properties of the structure. If, for discussion purposes, we ignore the rotational components of the ground motion, then we may have three independent inputs, and we might think that we are then only limiting the rank of the problem to three. However, if all three components of the ground movement are mainly the result of the same incident wave, then the excitation is limited to a single independent source and the rank of the problem is again one. Factors that influence this include the type of earthquake mechanism, the proximity of the structure to the earthquake source, the directionality of the ground motion, the length of the rupture, the type of soil conditions underneath the building, and so on.

Similar arguments can be used for evaluating sound pressure loading created by a single loudspeaker, which is the typical case for many tests of small structures in the laboratory. In general, this kind of loading is not recommended for OMA. Again the reason is that a single source in principle reduces the excitation to one independent source, and limits the rank of the problem. In some cases where reflections from surrounding bodies might create complementary loading that might reduce the correlation, it is still not advisable to use OMA, unless it can be demonstrated that the rank of the problem has not been reduced so much that it significantly influences the quality of the results of the OMA.

In some cases, we might consider the possibility of applying additional artificial loading, this may be the case when

- the excitation of the structure is too small to create a signal of acceptable magnitude;
- the natural excitation is a single input or near to single input source;
- the experiments are being performed in the laboratory;
- identification techniques including both controlled and ambient input are to be used.

When natural response signals are too small, one should consider using artificial excitation. However, this should only be done when it has been verified that the natural excitation does not provide a response that has an acceptable signal-to-noise ratio.

When the natural excitation is limited to one single source, this source should be complemented with some artificial loading in order to increase the number of independent inputs to a reasonable level. In laboratory testing, artificial loading should be used for OMA.

The aforementioned discussion leads to the following question: "How should the artificial loading be created?" Since structures are of different sizes and shapes, it does seem reasonable to point to specific ways of creating the artificial loading. However, some guidelines can be given here. The main considerations are

- Induce a multiple input loading, that is, several random and independent inputs. A simple way to perform this kind of loading is to use a moving load (random in time and space).
- Use methods of loading the structure that do not significantly change the modal properties of the structure.

For small structures, a recommended approach is to use some kind of brushing or scraping device that is in contact with the structure all the time and is moved around randomly. As this device is moved around over the main parts of the structure, it causes a constant (stationary) excitation. Of course, the larger the structure, the larger the device should be. For instance, for an OMA test of a bridge, the simplest way to cause an artificial excitation is to drive a car up and down the bridge. For a building, the artificial loading might be created by having people moving around inside the various floors of the building. In any case, one must always keep in mind that if the loading method selected is likely to cause significant changes to the dynamic properties of the structure such as natural frequencies or damping ratios, then this method should not be used for the OMA tests. An example of such situation may be the use of heavy traffic to excite a bridge, as the additional mass of the vehicles may alter the mass of the structure.

In other cases, it might be worth considering performing the experiments with a combination of natural random loading and a controlled and known input. This type of testing technology and corresponding modal analysis techniques fall outside of the scope of this book. Readers interested in learning more about this type of testing are referred to Reynders et al. [7].

7.3   Sensors and Data Acquisition

7.3.1   Sensor Principles

The basic principle of the type of sensors normally used for dynamic response measurement can be roughly modeled as a spring-mass system as illustrated in Figure 7.5. We will assume that the frame around the spring-mass system is rigid and that it is attached to one point of the structure being measured.

The motion of the frame of the sensor with respect to a reference system is denoted as $x(t)$ (i.e., the base motion) and the relative motion of the mass[9] of the sensor with respect to its resting point within the frame is $y(t)$ (i.e., the relative motion).

In this SDOF model, the mass of the system, M is attached to a support frame through a spring with a stiffness constant k. If the relative displacement of the mass is y, then the force in the spring is $F = ky$. The natural frequency of this equivalent model is then given by:

$$f = \frac{1}{2\pi} \sqrt{\frac{k}{M}} \tag{7.20}$$

Many types of sensors are available for dynamic response measurements. The most common types are piezoelectric sensors, electromagnetic sensors (seismograph, geophone), and force-balance sensors. Typical characteristics of these types of sensors are summarized in Table 7.7.

In the piezoelectric type, the force of the spring is measured by the piezoelectric material that creates a charge proportional to the strain in the material. Normally, the sensor signal is amplified by a built-in charge amplifier. The power for this type of sensor is normally supplied through the signal cable and a high-pass filter is used to remove frequencies close to the DC level. When the support frame moves along the x-direction of motion (and we assume the relative movement to be vanishing), the inertial force induced by the acceleration $a = \ddot{x}(t)$ of the mass is $F = Ma$. Since the damping of this type of sensor is quite small, one can assume that the simplified model shown in Figure 7.5 is undamped, and by simple equilibrium of forces it can be shown that the deformation of the spring is proportional to the acceleration of the mass. This means that the strain of the piezoelectric material is proportional to the acceleration of the frame of the sensor, and since the frame of the sensor is attached to the structure at a certain point, then what is being measured by the change of strain of the piezoelectric material is the change of acceleration of the structure at the point of attachment of the sensor. The frequency band of this sensor for which this sensor is useful is generally well below its natural frequency. More information about this sensor type is given in Section 7.3.3.

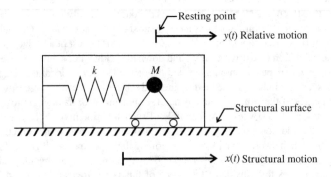

Figure 7.5   Simplified model of a sensor

---

[9] Often denoted the proof mass or the seismic mass of the sensor.

Table 7.7   Characteristics of commonly used sensors

Feature	Piezoelectric type	Electromagnetic type	Force balance
Type of measurement	Deformation (x)	Velocity ($\dot{x}$)	Force (F)
Built-in electronics	Charge amplifier	None	Displacement control
Lower frequency limit	Limited by ICP high-pass filter	Limited by correction procedure	No limit
Upper frequency limit	Limited by natural frequency	Limited by spurious spring modes	Limited by natural frequency
Cabling	Typical coaxial	Single twisted pair	Several twisted pairs

In the electromagnetic type, velocity is measured. The time derivative of the spring deformation is measured by a coil (normally also to be considered as the seismic mass) moving relative to a magnet attached to the supporting frame. Since the signal is produced solely by the coil moving relative to the magnet the sensor element does not need an external power supply to operate. For this type of sensor, the effective frequency band is mostly above the natural frequency of the instrument. The reason for this is that motion of the structure with frequency content above the natural frequency of the sensor causes insignificant movement of the coil and its relative motion with respect of the magnet attached to the supporting frame is essentially the same as that of the structure. However, information about the frequency content of signals below the natural frequency of the sensor might be recovered by inverse filtering. More details about this sensor type are presented in Section 7.3.4.

The force-balance accelerometer (FBA) has been designed on the principle that the net deformation of the spring must be zero. This is accomplished by adjusting the spring force continuously through the electronics of the device. The force needed to keep the spring deformation at zero is what is typically measured directly, and since the mass of the sensor is known, the acceleration can be inferred directly from the measured force. In theory, the sensor is able to provide a measure of the acceleration at zero-frequency.[10] The effective frequency band of the sensor is below its natural frequency. Since the natural frequency of the FBA is normally of the order of some hundredth of Hz and the damping is high, some phase distortion is present even at low frequencies. This problem has been partly solved by the use of MEMS[11] type force-balance sensors that have a natural frequency that is significantly higher than the traditional FBA. More information about this sensor is presented in Section 7.3.4.

Typical characteristics of the types of sensors described earlier are summarized in Table 7.7 and some of their commonly recognized advantages and disadvantages are summarized in Table 7.8.

7.3.2   Sensor Characteristics

The most important sensor characteristics that should be considered when planning an OMA test are the following: the sensitivity, noise floor, resolution, maximum input, and dynamic range.

Regardless of the type of sensor being used, the output is normally a voltage signal U approximately proportional to the measured physical quantity. For accelerometers, the sensitivity S is defined as

$$U = Sa \tag{7.21}$$

---

[10] At DC – which in practice means that this kind of sensor can be used to measure gravity.
[11] MEMS stands for micro electromechanical systems that leads to significantly smaller and less expensive sensors.

Table 7.8 Advantages and disadvantages of commonly used sensors

Feature	Piezoelectric type	Electromagnetic type	Force balance
Advantages	• Often small and easy to install • Good frequency response	• Low noise floor • Cheap and noise robust cabling	• Goes to DC (or close to it) • Low noise floor
Disadvantages	• Expensive and noise prone cabling • Fragile sensor element	• Bad frequency response • Low sensitivity in LF region	• Phase errors • Expensive heavy duty cabling • DC drift

and for velocity meters, the sensitivity is defined as

$$U = Sv \tag{7.22}$$

where v, a is the velocity and acceleration, respectively. The sensitivity is often also called the scaling factor or the calibration factor of the sensor.

The noise floor U_n of the sensor is normally given in terms of the square root of a flat spectral density $U_n = \sqrt{G_{n0}}$ (G_{n0} is the auto spectral density of the noise), thus U_n has units of $V/\sqrt{Hz}$. Alternatively, it can be given as the total noise δU_0 for a certain frequency band; for instance, in terms of V (RMS) from DC to a certain maximum frequency. According to the Parseval theorem for any flat spectral noise density and any frequency band B using Eq. (6.15) we have

$$\delta U_0 = U_n \sqrt{2B} \tag{7.23}$$

Since the quantity δU_0 is a measure of the smallest signal that we can measure in the frequency band B, it can be used as the resolution of the sensor in this frequency band.

The dynamic range D of a sensor is a measure of the ratio between the smallest signal δU_0 and the largest signal U_m that we can measure with a given sensor and is normally defined in terms of dB

$$D = 20\log\left(U_m/\delta U_0\right) = 20\log\left(\frac{U_m}{U_n\sqrt{2B}}\right) \tag{7.24}$$

Note that defining the dynamic range in this way, it is being defined for a certain frequency band. The smaller the frequency band, the smaller the noise and thus the larger the dynamics range. In contrast, the more generic expression presented in Eq. (7.1) is independent of the frequency band and provides less useful information than the aforementioned equation.

Other features of less importance must be mentioned. Some sensors with simple designs have an output that is highly dependent upon the power supply. For instance, some sensor signals are proportional to the voltage supply (as it is the case for a strain gauge). If the voltage supply is U_s, and if the output is proportional to acceleration, then the response is given by

$$U = CU_s a \tag{7.25}$$

where C is the calibration constant. In this case, the noise floor is heavily dependent upon the stability of the voltage supply.

Regardless of the sensor type, there is typically some dependence on the power supply. Instead of using Eq. (7.25) to characterize the output of the sensor, a more general formulation is used instead to better account for the variability of the voltage supply. The modified version of Eq. (7.25) is given as:

$$U = S_0 a + C \left(U_s - U_{s0} \right) a = \left(S_0 + C \left(U_s - U_{s0} \right) \right) a \tag{7.26}$$

where U_{s0} is the specified voltage supply (target value). If the actual power supplied is as specified by the manufacturer of the sensor, then the sensitivity is $S = S_0$, but if the power supplied is different than specified by the manufacturer, then the sensitivity is $S = S_0 + C \left(U_s - U_{s0} \right)$.

All DC sensors suffer of bias errors, which is a significant drawback for this kind of sensor. Bias error means that zero acceleration is not shown as zero output by the sensor. By modifying Eq. (7.26) to take this into account, the output signal is given by

$$U = \left(S_0 + C \left(U_s - U_{s0} \right) \right) \left(a - a_0 \right) \tag{7.27}$$

The response at $a = 0$ is the bias signal and is equal to $U_b = \left(S_0 + C \left(U_s - U_{s0} \right) \right) a_0$. From this expression, it appears that the bias signal contributes to the measured signal with a mean different from zero, and that by simply removing the mean value of the signal (DC value of the spectral density) its effect can be removed. However, since all quantities in Eq. (7.27), except the input acceleration a, normally must be considered to be temperature dependent, the bias signal introduces a significant noise that influences the whole frequency band. That part of the bias signal cannot be removed and the introduced noise in the lower frequency region can be significant. The temperature sensitivity can be investigated by differentiating Eq. (7.27) with respect to the temperature T. Assuming $(d/dT) \left(U_s - U_{s0} \right)$ to be negligible, and by setting $U_s - U_{s0} = 0$, we obtain

$$\frac{dU}{dT} \approx \frac{dS_0}{dT} \left(a - a_0 \right) - S_0 \frac{da_0}{dT} \tag{7.28}$$

The scaling factor temperature sensitivity is then defined for $a_0 = 0$ and is often given in terms of the relative value

$$\frac{1}{U} \frac{dU}{dT} = \frac{1}{S_0} \frac{dS_0}{dT} \tag{7.29}$$

The bias temperature sensitivity is defined for $a - a_0 = 0$ and is normally given in terms of the acceleration bias sensitivity

$$\frac{da_0}{dT} = -\frac{1}{S_0} \frac{dU}{dT} \tag{7.30}$$

An example of scaling factor and bias temperature sensitivity can be found in Gannon et al. [8]. The paper by Gannon et al. describes the development of a $\pm 2\,g$ MEMS (Micro Electro Mechanical Systems) analog force balance accelerometer (FBA) with a noise floor at $100\,\mathrm{ng}/\sqrt{\mathrm{Hz}}$ and a dynamic range better than 115 dB. The scaling factor temperature sensitivity of the sensor was reported to $200\,\mathrm{ppm}/^\circ\mathrm{C}$. This mean that if the temperature changes $50\,^\circ\mathrm{C}$, then this introduces an error on the sensitivity of the sensor equal to 1%. This does not represent any serious problem for practical use except that it illustrates the necessity to calibrate the sensor under the conditions where it is going to be used. They also reported the bias temperature sensitivity to $200\,\mu\mathrm{g}/^\circ\mathrm{C}$. This temperature sensitivity represents a much more serious problem, because if the temperature is just fluctuating $0.05\,^\circ\mathrm{C}$ in a certain frequency band with a band width of 1 Hz, this increases the noise floor in that band to $10\,\mu\mathrm{g}$, which is a value that is 100 times larger than what is measured under constant temperature conditions (the reported 100 ng in a 1 Hz band).

For DC sensors, the bias temperature sensitivity should always be carefully considered and measures to reduce its influence, such as isolation from direct sun heating and so on, should be implemented before performing the tests. The noise value is often relatively high even for good quality sensors; typical values are of the order of several hundred $\mu\mathrm{g}/^\circ\mathrm{C}$ for the MEMS sensor discussed here.

Nearly all sensors are affected by the so-called $1/f$ effect, which introduces undesirable noise in the lower frequency region. Typically, sensors have a frequency band where the noise has white noise characteristics (flat noise spectrum) and a low frequency region where the noise is governed by the $1/f$ effect or pink noise; see for instance, Figures 7.18 and 7.19.

For a sensor designed such that there is insignificant influence of the power supply, temperature, and any other ambient condition on the data, and that it has been designed as a perfect linear system, the calibration constant given by Eq. (7.21) or (7.22) is still just an approximation, even if its optimal value has been achieved. We need to recognize that an accelerometer (or velocity sensor) is a dynamic system. The linear input-output relationship can be expressed in the frequency domain as

$$\tilde{U}(f) = \tilde{S}(f)\tilde{a}(f) \tag{7.31}$$

where the symbols $\tilde{U}, \tilde{S}, \tilde{a}$ represent the Fourier transform of the corresponding quantities. In this more general approach, the sensitivity is interpreted to be a frequency response function (FRF) $H(f)$ of the form:

$$H(f) = \tilde{S}(f) = \tilde{a}(f)/\tilde{U}(f) \tag{7.32}$$

The FRF is a complex quantity and it is common practice to represent it in graphical form as shown in Figure 7.6 where the magnitude value and the phase of the complex FRF are plotted as a function of frequency. As shown in Figure 7.6, the frequency response of a typical sensor might be limited by a cutoff filter close to DC and by the natural frequency of the sensor in the upper frequency region. In between these two "disturbances" in the frequency response, a useful frequency band is defined. For the case shown in figure 7.6, a useful frequency band can be considered as, for instance, from 10% to 50% of the dimensionless frequency band shown in the figure.

Other factors that may have important influence on the quality of the data measured by a sensor include ambient factors such as humidity and air pressure. For instance, air pressure might influence highly sensitive sensors measuring vertical movements due to buoyant forces (see Section 7.3.4.1). Another

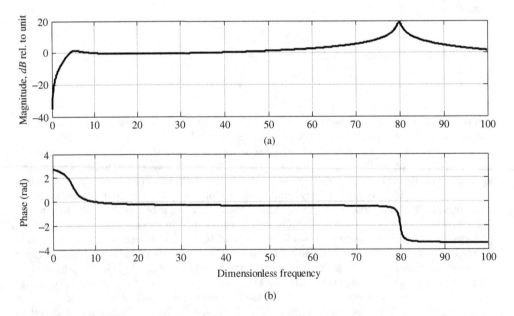

Figure 7.6 Typical form of the FRF of a sensor. Since the FRF is a complex quantity, it is common practice to present it in two companion plots: one plot of the magnitude of the FRF (a) and one plot of the phase between the imaginary and real components of the FRF (b). As illustrated the frequency response of a typical sensor might be disturbed by a cutoff filter close to DC and by the natural frequency of the sensor in the upper frequency region

consideration might be the influence on the data quality of the electromagnetic fields or internal modes. For instance, the geophone has spurious modes due to the leaf spring support of the coils that influence the response of this sensor in the higher frequency region. One also needs to keep in mind that maximum loads, such as maximum temperature range and shock resistance, can have a significant effect on the quality of the signals. Shock resistance might be essential if there is any significant probability that the sensor might be dropped accidentally during testing.

7.3.3   The Piezoelectric Accelerometer

This kind of accelerometer is relatively small, has a good frequency response over a large frequency band, and is reasonably priced and easy to use. It is widely used in both mechanical and civil engineering testing.

The word "piezo" comes from the Greek verbum $\pi\iota\acute{e}\zeta\omega$ that means "a press" or a stress that is applied. Therefore, piezoelectric refers to changes in the electric properties of the material due to an induced stress.

It is well known that some materials change resistance when strained, and this is what is used as the basic principle in strain gauges.[12] In the common piezoelectric accelerometer, however, a material is used that creates a charge when a stress is induced on the material. A large variety of materials have piezoelectric properties, but since the materials normally used in sensors are crystals of different kinds,[13] we discuss the piezoelectric material as "the crystal" in the following sections.

In principle, the induced stress can be bending, shear, or simple uniformly distributed normal stress. The latter one will be considered here. We will explain how a charge is built up when a normal stress is induced on the crystal, and how the charge is converted into a voltage due to the dielectric properties of the same material. It should be noted that the explanations presented here are rather simplified since crystals are non-isotropic materials where both strain and charge density depend on the three-dimensional state of both the stress and the electric field. More information about the complex behavior of crystals used in piezoelectric sensors can be found in Gautschi [10].

In order to illustrate the operational principles, let us consider a crystal bar with length L and cross section area A loaded by the axial force F as shown in Figure 7.7(a). When the force is applied the crystal elongates by ΔL to the deformed length $L + \Delta L$. The uniform axial strain of the crystal is then

$$\varepsilon_x = \Delta L / L \tag{7.33}$$

The uniaxial normal stress σ_x induced on the bar is $\sigma_x = F/A$, and if we assume that the material is linear elastic, the application of Hooke's law gives us the well-known relation between the normal stress and the stain, that is

$$\sigma_x = E\varepsilon_x \tag{7.34}$$

where E is the Young's modulus. As we indicated earlier, we are considering a material that creates a charge when subjected to a load, and the charge will built up at the ends of the crystal bar. It is a basic assumption in all continuum mechanics that any material response only depends on strain and temperature history. Therefore, it is natural to assume that if the temperature is constant, then the total charge Q on the surfaces ($-Q$ on one surface and $+Q$ on the other surface) is linear dependent on the current strain and on the area of the crystal

$$Q = CA\varepsilon_x \tag{7.35}$$

---

[12] Such materials are often called piezo-resistant.
[13] Quartz, topaz, and others.

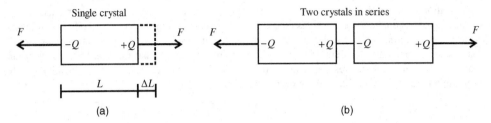

Figure 7.7 When a bar of piezoelectric material is loaded by an axial force, a negative charge is built up at one end and a positive charge at the other end. (a) A single crystal is deformed. (b) When two crystals are connected in series the charges from the two crystals are canceled out at the joint

where C is a material constant. The proportionality with respect to the area is easily demonstrated by placing two similar crystals side by side, and then realizing that each of them would carry the same charge and the combination of the crystals will at the same time double the area and the total charge.

One might think that a similar argument could be used to assume that the charge could be proportional to the length of the crystal, but this is not the case. Considering two crystals mechanically in series as shown in Figure 7.7(b), we see that when connected like this, both force and charge will cancel out at the common cross section, and thus, doubling the length will have no effect on the charge at the ends of the resulting crystal.[14] Combining Eqs. (7.34) and (7.35) and using the relation $F = A\sigma_x$, we have that the total charge is then given by

$$Q = CA\frac{\sigma_x}{E} = \frac{C}{E}F = d_xF \qquad (7.36)$$

where the constant $d_x = C/E$ is denoted as the uniaxial piezoelectric coefficient. We have now explained how the charge is created; now let us consider how the charge is turned into a voltage by the same crystal. It is well known from capacitor physics, that when a charge is present at two surfaces of a dielectric material,[15] then the charge density q (charge per area of the capacitor) is given by

$$q = e\frac{V}{L} \qquad (7.37)$$

where e is the permittivity (with unit $C/(Vm)$) of the material and V is the electric potential between the two surfaces. Since the charge is distributed uniformly over the section of the crystal with the area A we have that the total charge is given by

$$Q = e\frac{V}{L}A \qquad (7.38)$$

The voltage over the crystal is then given by

$$V = \frac{QL}{eA} \qquad (7.39)$$

or using Eq. (7.36)

$$V = \frac{L}{eA}d_xF = \frac{Ld_x}{e}\sigma_x \qquad (7.40)$$

---

[14] It is worth noting, that if the same two crystals are joined with an insulation disc in between the two crystals at the joint so that the negative charge at one crystal cannot cancel out with the positive charge at the other crystal, then the total charge can be doubled if the two crystals are electrically connected in parallel.

[15] Insulator that can be polarized by an electric field.

Finally, we see that we obtain a voltage over the two end surfaces of the crystal that is proportional to the stress acting on the material. Comparing with Eq. (7.35), we see that when the charge is proportional to strain and area and not dependent on the length of the crystal. It is kind of the opposite with the voltage over the crystal that is proportional to stress and the length of the crystal and does depend on the area.

The spring constant k to be used in order to find the natural frequency of the sensor according to Eq. (7.20) is easily determined as $k = EA/L$.

If the excitation frequency is well below the natural frequency of the sensor, then the SDOF sensor system behaves like a static system, and thus the force is proportional to the acceleration of the sensor. Piezoelectric sensors generally have a high natural frequency, so they provide a direct measure of acceleration over a broad frequency band. Since the mechanical system of this type of accelerometer has often a relatively low damping, the phase distortion between the input and the output is concentrated in a narrow frequency band around the natural frequency of the sensor, and it has little effect in the frequency band of application.

As the signal path from the crystal voltage to the amplifier (often a charge amplifier) is noise prone, the amplifier is normally built into the sensor. In order to provide the built-in sensor amplifier with power, the sensor is feed with a DC current through the signal cable, and thus, the sensor signal has to be filtered by a high-pass filter in order to prevent mixing the sensor signal with the DC power signal to the amplifier. This design is often denoted ICP.[16]

A typical state-of-the-art accelerometer that is commonly used for OMA on smaller specimens and for cases where the response is not too low, is the 4508 accelerometer from Brüel & Kjær. This piezoelectric accelerometer has a combination of high sensitivity, low mass and small physical dimensions that make it suitable for modal measurements of middle to small size structures. The sensor has titanium housing with integrated connector for a coaxial cable. It has a sensitivity of the crystal equal to 5 pC/g and a built-in ICP compatible charge amplifier that gives more than 100 dB dynamic range. The amplifier has low output impedance so that relatively long cables can be used. The accelerometer can be chosen with sensitivities from 10 mV/g to 1 V/g and can be supplied with TEDS.[17] The natural frequency of the sensor is at 25 kHz and the applicable frequency range is 0.3–8000 Hz. The FRF of a typical 4508 sensor is shown is Figure 7.8.

7.3.4   Sensors Used in Civil Engineering Testing

Typical sensors for civil engineering are in general the same types of sensors used for geophysical and seismic testing, thus information about these sensors can be found in publications on these subjects, for instance, in Wieland [11, 12] and Havskov and Alguacil [13].

7.3.4.1   The Geophone

The geophone consists of a coil suspended around a permanent magnet. According to Faraday's law, the output from the sensor is proportional to the relative velocity between the coil and the magnet. The geophone will often have two coils such that the signal from the geophone is created as the difference signal between the two coils, see Figure 7.9. The signal is created solely by the coil moving relative to the magnet so no power is needed in order to create the sensor output.

---

[16] "ICP" is an abbreviation for "integrated circuit piezoelectric." Since it a registered trademark of the vendor, PCB, the acronym "IEPE," which stand for "integrated electronics piezoelectric" is often used in instead. IEPE also refers to the standard for piezoelectric transducers.

[17] TEDS means "transducer electronic data sheet." This means that the information about the sensor, for instance, the calibration factor of the sensor, is stored in the sensor, so that the measurement system can read the calibration factor and measurements can be given in the right physical units.

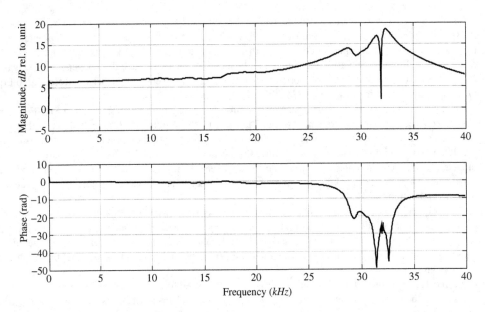

Figure 7.8 Typical FRF of the 4508 accelerometer from Brüel and Kjær (Source: Data provided by courtesy of Brüel and Kjær)

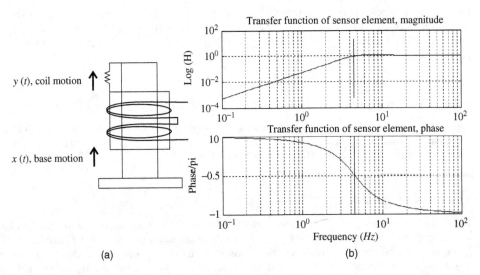

Figure 7.9 (a) Definition of base and coil displacements. (b) Theoretical transfer function for the SM6 geophone sensor element, from Brincker et al. [14]

The main limitation of the geophone sensor is that its linear frequency range is limited to frequencies above the natural frequency of the sensor, which is typically around 4–12 Hz. On the other hand, the fact that the sensor measures velocity is advantageous since displacement, or acceleration, can be obtained more readily by differentiating or integrating the signal only once.

Describing the suspended coil as a SDOF system and making use of Faraday's law, the FRF between velocity of the base and sensor output can be determined as (see for instance Brincker et al. [14])

$$H(\omega) = \frac{\bar{\ddot{y}}(\omega) - \bar{\ddot{x}}(\omega)}{G\bar{\ddot{y}}(\omega)} = \frac{1}{G} \frac{\omega^2}{\omega_0^2 + 2i\omega_0\omega\varsigma - \omega^2} \qquad (7.41)$$

where ω is the cyclic frequency of excitation, ω_0 is the corresponding natural frequency $\omega_0 = 2\pi f_0$ and G is the generator constant with units of Vs/m. The damping ratio ς of the suspended system is given by

$$\varsigma = \frac{c}{2\sqrt{kM}} \qquad (7.42)$$

where c is the equivalent viscous damping of the sensor system that is mainly determined by the electrical properties of the coil, and as before M and k are the mass and the stiffness, respectively, of the SDOF sensor system. It is easy to show that the following simple results for the phase ϕ can be obtained at the natural frequency

$$\phi = \pi/2 \qquad\qquad \frac{d\phi}{df} = \frac{1}{\varsigma f_0} \qquad (7.43)$$

These relations can be used for a simple identification of the natural frequency and damping of the sensor element.

One of the popular geophones for vibration measurements, the SM6 from Input/Output, Inc., USA, has the following properties, $G = 28.8$ Vs/m, $f_0 = 4.5$ Hz, $\varsigma = 0.56$. The noise properties of this simple sensor are excellent. For the SM6 sensor, the inherent noise floor can be estimated from the resistance of the coil $R = 375\ \Omega$. Using Eq. (7.59) and dividing by the generator constant G, the noise floor is found to be $\delta v_{in} = 0.085$ nm/s$\sqrt{Hz}$, that corresponds to an electrical signal of $V_{noise} = 2.5$ nV/$\sqrt{Hz}$, which is also the value stated by the vendor. Part (b) of Figure 7.9 shows the theoretical transfer function for the SM6 geophone sensor.

It is important to note that if the geophone signal is corrected by inverse filtering, the inherent flat noise floor in the frequency region below the natural frequency is amplified accordingly. It should further be noted that the inherent noise floor due to the Brownian motion given by Eq. (7.60) might be dominant in the lower frequency region. Special measures could be considered to reduce this influence by providing a vacuum around the sensor element. For vertical sensors, the influence of buoyant forces should be reduced by proper sealing of the sensor casing, Wieland and Streckeisen [15]. Digital sensors based on the SM6 geophone are described in Brincker et al. [16, 17].

7.3.4.2 The Force Balance Accelerometer

In the FBA, the spring in Figure 7.5 is replaced by an active electronic circuit that restraints the relative movement of the seismic mass, that is, trying to satisfy that $y = 0$.

In general, all force balance sensors consist of the following three main components: a displacement sensor that measures the displacement of the seismic mass, an actuator that creates the force to keep the seismic mass in position, and finally a servo amplifier that creates the control signal to the actuator based on the measured displacement signal.

The displacement sensor can be based on different principles[18] the discussion of which is outside the scope of this book. But what is worth mentioning is that friction forces are minimized, that the sensitivity is adequately high, that noise is low and that the frequency response is good over the frequency band where the sensor is to be used. The actuator is often an electromagnetic assembly where a coil attached to the seismic mass is forced by a stationary permanent magnet as shown in Figure 7.10 – in some cases the permanent magnet can actually be the seismic mass and then the coil is the stationary component.

---

[18] Often capacitive displacement sensors are preferred because they can be made very small and without any friction forces in between the two ends of the sensor.

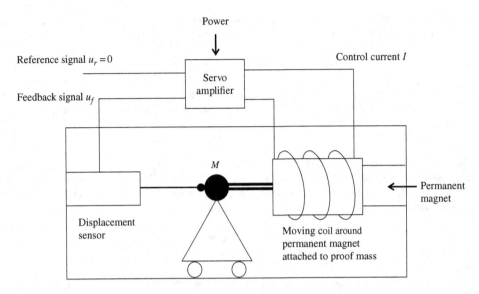

Figure 7.10 Measurement principle for a FBA where the spring in Figure 7.5 is replaced by an active servo system consisting of a displacement sensor, an actuator often realized by a coil around a permanent magnet and finally a servo amplifier creating the control current to the actuator as proportional to the measured displacement of the seismic mass

Let us say that the seismic mass has a displacement y different from zero, see Figure 7.5. We will then measure a displacement signal that is proportional to the displacement

$$u_f = C_u y \tag{7.44}$$

where we assume that the feedback signal u_f is measured in Volts (V), and the calibration constant C_u has the units V/m. The force F that is created by the actuator is proportional to the supplied current I to the coil

$$F = C_f I \tag{7.45}$$

where the units of the actuator sensitivity[19] C_f are Newton per ampere (N/A). The servo amplifier creates the control current I based on the feedback signal u_f from the displacement sensor. In general, when using a proportional servo system, a reference signal u_r is specified for the system to follow, and the control current is then proportional to the error $u_f - u_r$. A proportional servo control current would, therefore, in general have the form

$$I = C_I \left(u_f - u_r \right) \tag{7.46}$$

As we try to keep the seismic mass at $x = 0$, the reference signal is $u_r = 0$, and the control equation simplifies to

$$I = C_I u_f \tag{7.47}$$

where the proportional control constant C_I has the units $A/V = \Omega$. Using Eqs. (7.44), (7.45), and (7.47), we can obtain the expression for the force on the seismic mass as

$$F = C_f C_I C_u x \tag{7.48}$$

---

[19] Also sometimes denoted the force sensitivity.

At this point, we realize that the control system can be represented by the spring in Figure 7.5, and that the spring constant is given by

$$k = C_f C_I C_u \tag{7.49}$$

The sensor natural frequency is then given by Eq. (7.20). As the output from the sensor it is possible to use either the signal u_f from the displacement transducer or the current I driving the coil. If the sensor is used to measure signals well below the natural frequency of the sensor, the SDOF sensor system acts approximately as a static system, and both signals will be proportional to the force acting on the seismic mass, and also proportional to the acceleration of the sensor. In practice, however, one does not want to deal with the feedback signal (good practice in all servo applications), so it is normal practice to use the control current or a voltage proportional to the control current as the output from the FBA.

Because the FBA uses a servo principle instead of a mechanical spring, it is also often called a servo accelerometer. The actual force acting on the seismic mass is a very good measure of the acceleration of the seismic mass at low frequencies – even down to DC – which makes this type of accelerometer useful for low-frequency measurements.

One of the advantages of using the servo principle instead of a spring with the same stiffness and then measure the extension of the spring with the displacement sensor is that a mechanical spring would have much larger temperature sensitivity, and, therefore, the noise floor at low frequencies would be significantly higher.

Normally, FBA sensors have a natural frequency around a few hundred Hz, although natural frequencies of accelerometers of the MEMS type might be a factor 10 higher. Another important fact about the FBA is that it normally has a highly damped natural frequency. This causes a significant phase distortion in the frequency band below the natural frequency. Therefore, the user should consider phase errors in the measured signals when working with these accelerometers.

An important factor to consider is the noise properties of a FBA. This depends mainly on the noise properties of the electrical circuits of the sensor and the servo amplifier, but also on the cabling. The cabling often consists of two twisted pairs, one carrying the power supply to the servo amplifier, the other carrying the signal from the sensor to the measurement system.

A sensor often used for measuring low frequent signals of large structures is the Kinemetrics EPI sensor that is a good example of a state-of-the-art FBA. In this servo accelerometer the relative displacement of the seismic mass with respect to the sensor case is measured by a capacitive displacement transducer. The natural frequency of the sensor is 224 Hz, the damping ratio at the natural frequency is approximately 70%, and the −3 dB frequency of the sensor is at 200 Hz. An empirical model of the sensor system uses two pairs of conjugate poles to represent the transfer function of the instrument. Using this model of the sensor, the FRF of the sensor can be graphically represented as shown in Figure 7.11. As it appears from this figure, the amplitude errors are quite small for frequencies less than 50 Hz, but a relatively large phase distortion is present even at low frequencies. At low frequencies, the phase error is approximately a linear function of frequency with a phase error at about 90° at 50 Hz. Deviations of the phase distortion predicted by the model are reported to be smaller than 2.5° in the 0–100 Hz band. The noise properties of the sensor are quite good with a dynamic range reported to be better than 155 dB. The sensor has a user selectable maximum range, and if the range is selected to 0.25 g (the smallest possible value for the sensor) the sensor meets the $-140\,\text{dB}$ (rel $1\,\text{m}/(\text{s}^2 \sqrt{\text{Hz}})$ requirement for low excitation defined in Section 7.2.6. For the EPI sensor, the bias temperature sensitivity is reported to be better than $500\,\mu\text{g}/^\circ\text{C}$.

7.3.5   Data Acquisition

Before OMA analysis techniques can be used to analyze the data, the analog signals produced by the sensors must be converted into numbers. This is done by an analog-to-digital converter (ADC).

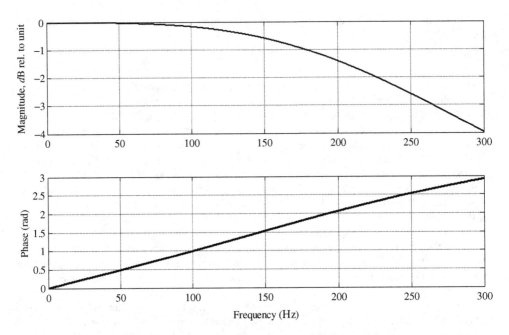

Figure 7.11 FRF of the Kinemetrics EPI sensor. The natural frequency of the sensor is at 224 Hz, but due to the high damping of the sensor, no peak is visible in the magnitude and a large deviation of the phase is present. Information about the sensor provided by courtesy of Kinemetrics

The capability of the ADC to perform this conversion is generally defined in terms of the number of bits available in the internal processor.

Let the ADC have N_{bit} number of bits, and let us assume that the input voltage is symmetric $\pm U_{mADC}$, (for instance ± 5 V or ± 10 V), then the ADC has $N_{val} = 2^{N_{bit}}$ numbers available to digitize the signal, and the digital resolution ΔU_{ADC} of the ADC becomes

$$\Delta U_{ADC} = 2U_{mADC}/ \left(N_{val} - 1\right) \qquad (7.50)$$

Denoting r_0 and r_1 as the minimum and maximum output digital values from the ADC respectively and r as the reading from the ADC, then the actual value of N_{val} is given by $N_{val} = r_1 - r_0 + 1$, and the corresponding measured analog voltage is found as

$$U = \Delta U_{ADC} r - U_0 = 2U_{mADC} \frac{r}{r_1 - r_0} - U_0 \qquad (7.51)$$

where U_0 corresponds to the reading of the converter for a zero input voltage. Note that sometimes the output r from an ADC is given as an unsigned integer, so in this case the readings will start at $r_0 = 0$. In other cases, the readings are a signed integer, where zero input voltage corresponds to the reading zero, and in this case $U_0 = 0$.

Because of noise in the measurement circuits the measurement uncertainty δU_{0ADC} (RMS value for a certain frequency band, see Eq. (7.23)) for the ADC is somewhat higher than the resolution ΔU_{ADC}. The measurement uncertainty defines the number of the least significant bit (LSB) according to the equation

$$2^{LSB} = \delta U_{0ADC}/\Delta U_{ADC} \qquad (7.52)$$

where the LSB is counted from the right and gives an estimate of the number of bits influenced by the measurement uncertainty (noisy bits). The number of effective bits N_{eff} is then defined as

$$N_{eff} = N_{bit} - LSB \tag{7.53}$$

and the dynamic range for the ADC is given by

$$D_{ADC} = 20 \log \left(U_{mADC} / U_{0ADC} \right) = 20 \log \left(2^{N_{eff}-1} \right) = \left(N_{eff} - 1 \right) 20 \log 2$$
$$\approx 6 \left(N_{eff} - 1 \right) dB \tag{7.54}$$

For instance, if the ADC has 21 effective bits, then the dynamic range is approximately 120 dB.

Note that the dynamic range is again defined for a certain frequency band. Thus, if the noise floor is constant as it is for a sigma-delta ADC, the smaller the frequency band, the smaller the RMS noise in the considered band, and thus the higher the dynamic range. This is not the case for a simple ADC where it is normally assumed that the RMS noise is constant and thus independent of the frequency band. The properties of a simple ADC and the sigma-delta-based ADC are further illustrated in the following two examples.

Example 7.2 Simple 24 bit ADC with 17 effective bits

We assume that the input range is ±5 V, the actual input is a constant zero volt signal, the sampling rate is $f_s = 10.000$ Hz, and the measurement time is $T = 1$ s. In this case, the number of readings is $N = f_s T$ = 10.000. The digital resolution is calculated according to Eq. (7.50) to $\Delta U_{ADC} = 0.596 \mu V$ and the measurement uncertainty given by Eqs. (7.52) and (7.53) to $\delta U_{0ADC} = 2^{(24-17)} \Delta U_{ADC} = 76.3 \mu V$. The dynamic range of the ADC can be calculated using the first part of Eq. (7.54) to 96.33 dB or using the last part (approximation) of the same equation to 96 dB. The noise properties of the ADC can be modeled by a zero mean Gaussian sequence with N data points and with a standard deviation equal to the measurement uncertainty. Taking the mean of the 10,000 readings will give us an answer around zero with a typical deviation of ±1 μV. This is as expected since the standard deviation of the mean can be approximated as $76.3 \mu V / \sqrt{N} = 0.763 \mu V$.

Example 7.3 Sigma-delta 24 bit ADC with 50 kHz sampling

We will assume that the basic noise property of the ADC is the same as in Example 7.2, that is, the ADC has an uncertainty on all readings $\delta U_{0ADC} = 76.3 \mu V$. The sigma-delta technology implies sampling at the maximum rate $f_{s0} = 50$ kHz followed by subsequent decimation to the user defined sampling frequency f_s. Setting the measurement uncertainty equal to the standard deviation σ_n of the noise $\sigma_n = \delta U_{0ADC}$ and using the Parseval theorem from Eq. (7.12), we obtain the noise floor $G_{n0} = \sigma_n / \sqrt{f_{s0}} = 0.341 \mu V / \sqrt{Hz}$. The measurement uncertainty δU_s for the user defined sampling frequency f_s can then be found using the Parseval theorem once more and we get that $\delta U_s = G_{n0} f_s$. The corresponding LSB, the effective number of bits and finally the dynamic range D_s can be obtained from Eqs. (7.52) to (7.54). The results for different user defined sampling frequencies are given in Table 7.9.

As expected, the maximum sampling rate corresponds to the assumed number of effective bits equal to 17 (as assumed in Example 7.2), and in this case, the dynamic range is also 96 dB as found in Example 7.2. From Table 7.9, we see that if the sampling rate decreases the number of effective bits increases as well as the dynamic range. Actually we see that the effective number of bits can be larger than the available 24 bits in the ADC. This might seem strange but is known to be a fact for many commercially

Table 7.9 Dynamic range D_s, LSB and effective number of bits for different user defined sampling frequencies for the ADC considered in Example 7.3

f_s (Hz)	50,000	5000	500	50	5.0	0.5
D_s (dB)	96	106	116	126	136	146
LSB	7.0	5.4	3.7	2.0	0.4	−1.3
N_{eff}	17.0	18.7	20.3	22.0	23.6	25.3

available sigma-delta ADC's. It is due to the high degree of averaging that is involved when performing decimation of the data, which results in a significant reduction of the uncertainty on the averaged data. We can see that we have more than 24 effective bits when the sampling frequency is lower than about 5 Hz.

7.3.6 Antialiasing

The time step Δt between data points defines the sampling frequency f_s and the Nyquist frequency f_v according to the following relationship,

$$f_s = 2f_v = 1/\Delta t \qquad (7.55)$$

Sampling with a time step of Δt only allows information to be included in the frequency band from DC to the Nyquist frequency f_v. Parseval's theorem,

$$E\left[x^2\right] = \sigma_x^2 = 2\int_0^{f_v} G_x(f)\,df \qquad (7.56)$$

clearly shows, that if a signal $x(t)$ has frequency components outside of the Nyquist band, then the energy of these frequencies is artificially forced into the Nyquist band. This is known as "aliasing" of the signal outside of the frequency band and leads to erroneous interpretation of the data. In order to remove this measurement error, antialiasing filters are used to remove all signal components above the Nyquist frequency.

If the antialiasing filter is not a built-in facility in the measurement system, then the user of the system should incorporate an analog filter between the sensor and the ADC. A graphical representation of this concept is shown in Figure 7.12.

The antialiasing filter is a low-pass filter that is designed in a way that the energy of the filtered signal is negligible for frequencies beyond the Nyquist frequency. If the user needs flexibility on the selection of the sampling frequency for the OMA test, but the antialiasing filter has a fixed cutoff frequency, the filter should have a fixed cutoff frequency higher than the highest sampling frequency of interest. The signal can then be sampled using a sampling frequency high enough to include the antialiasing filter, and the sampled signal can then always be decimated to a target sampling frequency. Depending on the quality of the ADC one could consider to over sample the signal as much as possible for the reasons given in Section 7.2.3.

7.3.7 System Measurement Range

The complete measurement system determines the quality of the measured data, not the individual elements by themselves. It does not help much if high quality sensors are used but the quality of the data

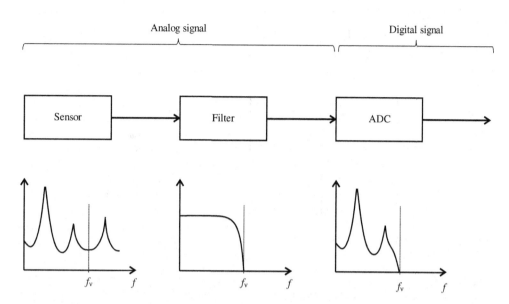

Figure 7.12   An analog antialiasing filter must always be applied between the sensor and the ADC in order to effectively remove the energy of the signal outside of the Nyquist band. The left plot shows the signal from the sensor that might include signal components outside of the Nyquist band. Applying the low-pass filter shown in the middle will cut of these components, and finally the signal that reaches that ADC will have only components in the Nyquist band

acquisition system and the antialiasing filter is low. As a practical rule one can say that the lowest quality component in the measurement chain defines the quality of the signals being measured.

For instance, if we consider a measurement chain consisting of a sensor with noise floor δU_0 and an ADC with noise floor $\delta U_{0\mathrm{ADC}}$, the two sources can be modeled as two independent stochastic variables and the noise floor in the measured data is then the sum of the variances as $\sqrt{\delta U_0^2 + \delta U_{0\mathrm{ADC}}^2}$. If one of the noise sources are dominating, then the results is approximately the highest noise floor of the two; that is, the system noise floor $\delta U_{0\mathrm{sys}}$ is approximately given by

$$\delta U_{0\mathrm{sys}} = \max\left(\delta U_0, \delta U_{0\mathrm{ADC}}\right) \qquad (7.57)$$

Similarly the maximum voltage that can be measured is the minimum of the maximum inputs for the two components so the corresponding dynamic range of the total system is approximately

$$D_{\mathrm{sys}} = 20\log \frac{\min\left(U_m, U_{m\mathrm{ADC}}\right)}{\max\left(\delta U_0, \delta U_{0\mathrm{ADC}}\right)} \qquad (7.58)$$

From this expression, we see that it is extremely important that sensors and data acquisition system have matching maximum input and noise floor characteristics. The matching problem is illustrated in Figure 7.13.

7.3.8   Noise Sources

In OMA, it is important to consider electrical noise sources because these noise sources define the smallest signals that can be measured – we call this the noise floor of our equipment. If the noise floor of

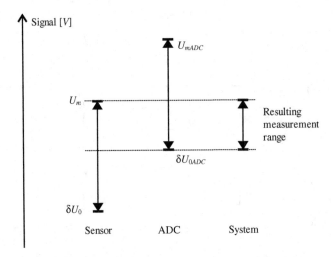

Figure 7.13 Matching problem of sensors and ADC. A nonmatching maximum input and noise floor might significantly reduce the measurement range of the combined system consisting of the sensor and ADC. The ideal case is where the two measurement elements match reasonably so that the system noise floor and system dynamic range are approximately equal to the same properties of the individual elements

the equipment is comparable to our sometimes very weak operational signals this might result in a bad signal-to-noise ratio and to serious difficulties during the analysis of the data.

When choosing electronic components with a low noise floor, several main issues are to be considered. First, it is desirable to minimize the noise to which the measurement system is exposed. Second, it is important to use electronic components that are constructed to be as insensitive to incoming noise as possible. Finally, it is essential that the applied components create a minimum amount of self-noise.

When minimizing the noise exposure two possibilities exist. The electronic components can be shielded or the sensor element can be placed in a non-noisy environment. Only the first option is feasible, since in practice, electrical noise is often present, as is the case of buildings with many electrical installations. When shielding is used, ideally (in fixed installations where it is a natural solution) the entire signal path should preferably be placed within a solid shield (such as a tube around the cable) due to its superior shielding efficiency compared to braided shielding [18, 19]; however, such shielding easily becomes expensive when including both the sensor, cabling to the data recorder and the recorder front end. An example where this kind of shielding is used is on offshore structures where it must be used in any case for safety reasons.[20]

The second way of minimizing the noise in the system is to minimize the number of components, which are sensitive to incoming electromagnetic noise. Often amplifiers and filters are used for signal conditioning, and one should be aware that these components can be excellent noise collectors and therefore the noise properties of these components should be carefully considered. Inductors and wire-wound resisters can work as loop antennas, collecting HF noise into the signal path. In general, it is desirable to minimize the number of components within the signal path to an absolute minimum. More information about electrical noise radiation and suppression can be found in Ott [18] and Williams [19].

In OMA equipment, the self-noise represents important noise sources that have to be taken into account. For example a standard resistor, which is used in most electronic equipment, the thermal noise voltage

---

[20] In such applications, it would be mandatory to use the so-called intrinsically safe instrumentation, which means that measures are taken to limit the possibilities for ignition that might cause explosions.

(RMS open-circuit) generated from such a resistor due to the Brownian motions of the electrons can be obtained using classical statistical mechanics and the Maxwell–Boltzmann distribution law to give

$$\delta U_0 = \sqrt{4kTBR} \tag{7.59}$$

where k is the Boltzmann constant, T is the absolute temperature, R is the resistance and B is the considered frequency bandwidth; see Ott [18]. The Boltzmann constant has the dimension energy divided by temperature, thus the term kTB has dimension of Watts, which is the same as electrical potential times electrical current. Since resistance is potential over current we end up with the right unit of dimension – potential squared – inside the parenthesis. This noise source – also denoted Johnson noise – is omnipresent and only depends on the absolute temperature. Working around room temperature and assuming that temperature changes are small, it can be considered to be constant. Further, since the noise is proportional to $\sqrt{B}$ Eq. (7.59) implies that the spectral density is assumed to be flat, and thus, the noise can also be specified as $V/\sqrt{Hz}$ by leaving out the frequency band B in Eq. (7.59).

$1/f$ noise is a common description of a broad range of noise sources, which have in common that the power spectral density is approximately proportional to $1/f$. This kind of noise is often referred to as "pink noise" because it lies in between the white noise (proportional to $1/f^0$) and red noise (proportional to $1/f^2$), thus $1/f$ noise means all noise that has a power spectral density approximately proportional to $1/f^\alpha$ where $0 < \alpha < 2$. This kind of noise is present in nearly all electronic devices. The principal sources of $1/f$ noise in electronic devices are typically due to slow fluctuations of properties of the materials of the devices, such as those due to temperature variations.

The cabling and the grounding of the measurement system often represent problems of noise pickup especially 50 or 60 Hz noise from the power supply. To minimize this kind of noise, cabling with twisted pairs are often used in order to cancel out noise from electromagnetic radiation; see Figure 7.14.

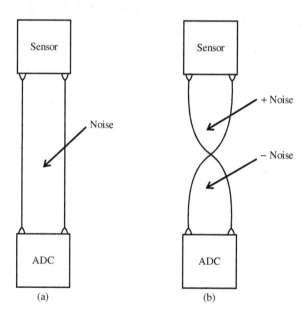

Figure 7.14   Noise cancelation using twisted pairs. (a) Shows a wire without twisting in which the full influence of the electromagnetic radiation is picked up by the wire acting as a loop antenna. (b) Shows a pair with a single twist. If the noise is the same in the two loops, then the twist creates two noise signals of opposite polarity in the signal path and the noise is canceled out

To minimize noise due to the grounding of the measurement system, ground loops should be prevented. Ground loops might occur whenever more than a single grounding is used in a measurement system, and a typical case is when several sensors are connected to the same steel structure, see Figure 7.15. In this figure, it is shown how two sensors are connected through the ground and how the wires to the sensors together with the ground connection constitutes an antenna for a strong noise pickup. The problem can be prevented by isolating (electrically) the sensors from the structure or by disconnecting the shielding of the cable to the sensor casing (the two connections marked "x" in Figure 7.15).

Even if electrical noise is reduced and the influence from this kind of noise is prevented, influence from mechanical noise can often be a problem. Normally, a sensor element includes an active mass. This mass is influenced by impacts due to the Brownian motion of the surrounding air molecules, and thus an inherent acceleration background noise is introduced. From classical statistical mechanics and the Maxwell–Boltzmann distribution law, the acceleration background noise δa_{in} can be obtained according to Usher [20] as

$$\delta a_{in} = \frac{\sqrt{4kTBc}}{M} \tag{7.60}$$

where M is the seismic mass, c is the damping constant caused by interaction with the air, and k, T, and B is the Boltzmann constant, the absolute temperature, and the frequency band respectively (as in Eq. (7.59)). The constant kTB has the dimension of Watts equal to Nm/s and since the damping is force per velocity the dimension of the square root is Newton and finally using the second law of Newton the dimension of the noise comes out right as acceleration.

If the excitation spectrum from the Brownian motion of the air molecules is considered to be flat, and if we are considering a frequency band well below the natural frequency of the sensor, the response spectrum will also be approximately flat. This is in agreement with Eq. (7.60) since the background noise is proportional to the square root of the frequency band. Therefore, the noise can also be specified as acceleration per square root Hz by omitting the frequency band B in the equation.

From Eq. (7.60), it can be inferred that the larger the active mass the lower the noise floor due to Brownian motion of the air molecules. Therefore, one way to reduce this noise source is to use an active mass of a proper size. It is worth noticing, however, that the air damping c must be proportional to the

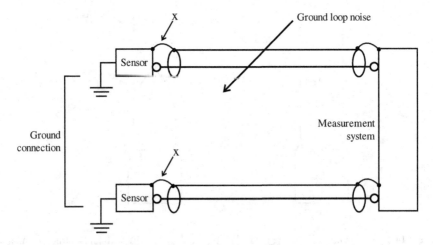

Figure 7.15 Ground loops occur when several groundings are active in the same measurement setup. The problem can be solved by disconnecting the shielding to the sensor casing (the two connections marked "x") or by isolating the sensor casing from the ground

surface area of the seismic mass, thus if the seismic mass has the linear size L, the air damping can be expressed as $c = \alpha L^2$ where α is a shape factor for the seismic mass. Similarly the mass can be expressed as $M = \beta \rho L^3$ where ρ is the mass density of the seismic mass material and β is another shape factor. Therefore, we have that $\sqrt{c}/M = L^{-2}\sqrt{\alpha}/(\rho\beta)$, and the Brownian noise floor is proportional to L^{-2} and not to L^{-3} as one might immediately think considering Eq. (7.60).

Another way to reduce the noise is to reduce the number of impacts from Brownian motion of the air molecules. This can be done by suspending the seismic mass in vacuum resulting in a smaller damping ratio c caused by the air.

When the air pressure is fluctuating, the buoyant force on the seismic mass is changing, and this influences the sensor element in vertical measurement applications. Wieland and Streckeisen [15] state that if the active mass is not protected against air pressure fluctuations, buoyant forces from air pressure fluctuations will at least be three orders of magnitude larger than seismic background noise. Therefore, for vertical sensors, it is essential to shield the sensor element from air fluctuations. This can be done by using an airtight instrument casing. Noise introduced by buoyant forces is concentrated in the low frequency region.

7.3.9 Cabled or Wireless Sensors?

Because of the many complicated noise sources the primary goal in noise reduction is to limit the number of places where noise can enter into the measurement system; for instance, through the analog cabling system. Just for this reason, whenever wireless and digital solutions can be implemented they should be applied. However, it is also important to reduce cabling for two other reasons: cost and simplicity.

Wireless sensing technology has made significant advances in recent years and one may wonder why so many people involved in vibration testing are still using wired sensors. There are some good reasons why cables are still being used. These reasons are as follows:

- Time synchronization of the distributed ADC's.
- Power supply problems with wireless units.
- Reliability of wireless systems not comparable to that of wired systems.

When wireless systems are being used, normally digital information is transferred wirelessly. This means that the A/D conversion should be performed at the sensor. In vibration testing, we need to sample the signals simultaneously, and this causes the problem of time synchronization between the different ADC's located at the sensors. If the maximum frequency of interest is f_{max}, and if the maximum phase error is set to be, for instance, 1 deg, then the maximum allowed synchronization error δt can be quantified as

$$\delta t = 1/\left(360 f_{max}\right) \qquad (7.61)$$

This demand is easily met by GPS synchronization (that is normally within micro second uncertainty). For cases where a GPS signal can be picked up, then wireless can be used without concern for this problem. In cases where a GPS signal cannot be picked up, for instance, inside heavy structures such as mines and tunnels, it might be too difficult to synchronize the signals (long wires to the GPS antenna) and thus, a wired system is to be preferred. Alternatively, if the modes to be identified are expected to be real or nearly real, the modal responses of the first modes can be used to synchronize the signals, see Brincker and Brandt [21].

The power supply is also a potential source of problems, because a measurement system must in many cases be able to function over extended periods of time without any maintenance. Bridges in remote places and offshore structures are good examples of structures where it is not practical to have a system that often needs to be serviced due to power supply failure. So, if power is available at the site, a wired system might be preferred to have a reliable power supply to each sensor. If power is not available at the site, the problem of power supplying the units must be solved in any case, and having individual sensors, each with its own power supply might be a possible solution. Possible sources of power for individual sensor are solar panels, small wind turbines, or each node might have a built-in power unit harvesting the energy from the ambient vibrations, Priya and Inman [22].

There are many reasons for the reduced reliability of wireless systems compared to the wires ones. Simple reasons are loss of connection between the sensors and the data acquisition system due to changes of the signal path such as people or cars moving in the signal path and loss of GPS connection so that data cannot be synchronized.

If a cabled system is used, the type of the cabling should be carefully investigated. In principle, three main types of analog cables exist:

- Coaxial cables typically used for piezoelectric sensors.
- Shielded twisted pairs typically used for simple sensors such as geophones.
- Multiple twisted pairs typically used for force-balanced sensors.

Coaxial cables should only be used in short cable installations with low electrical noise exposure. The other types of cables can be used as long cable applications and under more noisy conditions. Since there is always a noise pickup in analog cables, only digital signals should be used in cases of long cable runs in environments with high electrical noise exposure.

When the ADC is located in the sensor itself and digital signals are transmitted by the cables, the sensor is often called a sensor node. The advantages of using digital signals in the cables are as follows:

- No cable noise.
- Cheap cabling.
- More intelligent sensors (systems check, self-calibration).
- Several sensors can be present on the same cable.

Systems with digital cables and good time synchronization are available; see, for instance, Brincker et al. [16] and [17].

7.3.10   Calibration

Sensors are normally calibrated by the manufacturer or supplier, or can be sent to the supplier for calibration. However, anybody conducting OMA measurements should consider performing in-house calibration of the sensors and cables – or at least perform a check of proper sensor operation before conducting a test.

A classical calibration where the input is controlled and known can be done as shown in Figure 7.16, where the sensor to be calibrated is subjected to a broad band input from an exciter.

Figure 7.16a shows the displacement of the sensor being measured by a laser system. In this case, the traceability can be assured by calibrating the laser system using a dial micrometer. The velocity of the sensor is found by numerical differentiation of the laser signal. In this case, care should be taken to consider effects of the time delay in the laser as well as noise generated by the numerical differentiation.

Another and somewhat easier way of calibration is shown in Figure 7.16b where the input is measured by a reference accelerometer. In this case, the traceability can be assured through a calibration of the reference accelerometer at a recognized calibration laboratory.

Assuming that the considered sensor is an accelerometer, the broad band input acceleration $x(t)$ and the sensor response $y(t)$ is measured, and traditional modal analysis methods are used to estimate the sensor FRF $H(f)$. The classical way to do this is to obtain two different FRF estimates denoted $\hat{H}_1(f)$

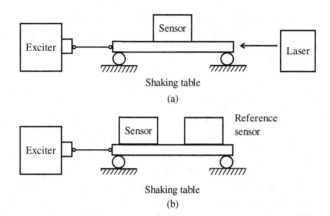

Figure 7.16 A simple way to perform a traceable calibration of a vibration sensor is to use a broad input from a shaker and measure the response with a displacement laser (a) or a reference sensor (b)

and $\hat{H}_2\,(f)$ from the auto- and cross-power spectral densities as

$$\hat{H}_1\,(f) = \frac{\hat{G}_{xy}\,(f)}{\hat{G}_{xx}\,(f)}$$

$$\hat{H}_2\,(\omega) = \frac{\hat{G}_{yy}\,(\omega)}{\hat{G}_{yx}\,(\omega)}$$

(7.62)

The corresponding coherence between the two signals is then calculated as

$$\gamma^2_{xy}\,(\omega) = \frac{\hat{H}_1\,(\omega)}{\hat{H}_2\,(\omega)} = \frac{\hat{G}_{xy}\,(\omega)\,\hat{G}_{yx}\,(\omega)}{\hat{G}_{xx}\,(\omega)\,\hat{G}_{yy}\,(\omega)}$$

(7.63)

We will only use the result of the calibration where the coherence is close to unity, Bendat and Piersol [23]. If the coherence is not acceptable, then the influence of noise in the calibration test has to be reduced. For instance, a higher excitation of the shaking table might be applied to achieve a better signal-to-noise ratio of the signals.

Even if a sensor has been properly calibrated, it is advisable always to ensure that the sensor is operating as expected when performing a test. A simple way of checking that the instrument is working as assumed should be considered. Taking into account that such check should be easy to perform on-site one could consider

- Simple response check.
- Relative calibration.

A simple response check can, for instance, be a turning a DC accelerometer upside down to check that the total DC shift during the turn is equal to the change of gravity. A geophone can be checked by providing a current over the coil and then recording the free response, see Wieland [11]. A general way to perform a simple check of any sensor is measuring its response to a standardized pulse.

In OMA, the absolute calibration is of lesser importance than a relative calibration. A relative calibration between sensors can be carried out by simply taking operational response measurements placing all sensors in the same location on the structure (a collocation test). The procedure is as follows:

1. Place all sensors in "the same" location (as close as possible).
2. Take a simultaneous time record of all sensors (according to Eq. (7.16)).
3. Decimate the signal to include only the lowest modes of the structure.
4. Calculate the RMS value from each signal.
5. Use the ratio between the RMS values as an indicator for relative calibration.

In a collocation test, we cannot place the sensors in exactly the same point and therefore the sensors might experience slightly different signals. The influence can be minimized by decimating the signal to include only the modes with mode shapes that have a small change in value over the set of DOF's where the sensors are placed in the collocation test.

In cases where it is difficult to conclude where the sensors should be placed in order to minimize the influence of different signals, it is recommended that the collocation test is repeated at several locations and then checking that the calibrations from the different locations provide consistent results.

Assuming that a single mode is dominating the measured motion, the influence of measurement errors can easily be analyzed based on the simple frequency domain identification method, see Section 10.2. Let us consider just two measurement points with the response signals

$$s_1(t) = a_1 q(t)$$

$$s_2(t) = a_2 q(t) \tag{7.64}$$

where $a = \{a_1, a_2\}^T$ is the mode shape (considered to be real) and $q(t)$ is the modal coordinate. Assuming that the sensors have the FRF's $\tilde{h}_1(\omega)$ and $\tilde{h}_2(\omega)$ we will then measure the frequency domain signals

$$\tilde{y}_1(\omega) = a_1 \tilde{q}(\omega) \tilde{h}_1(\omega)$$

$$\tilde{y}_2(\omega) = a_2 \tilde{q}(\omega) \tilde{h}_2(\omega) \tag{7.65}$$

The corresponding auto- and cross-spectral densities are given by (see Eq. (8.44))

$$G_{11}(\omega) = \tilde{y}_1^*(\omega) \tilde{y}_1(\omega) = a_1^2 \tilde{q}^*(\omega) \tilde{q}(\omega) \tilde{h}_1^*(\omega) \tilde{h}_1(\omega) = a_1^2 G_q(\omega) \left| \tilde{h}_1(\omega) \right|^2$$

$$G_{12}(\omega) = \tilde{y}_1^*(\omega) \tilde{y}_2(\omega) = a_1 a_2 \tilde{q}^*(\omega) \tilde{q}(\omega) \tilde{h}_1^*(\omega) \tilde{h}_2(\omega) = a_1 a_2 G_q(\omega) \left| \tilde{h}_1(\omega) \right|^2 \frac{\tilde{h}_2(\omega)}{\tilde{h}_1(\omega)} \tag{7.66}$$

As it appears, the cross-spectral density is biased by the factor $\tilde{h}_2(\omega)/\tilde{h}_1(\omega)$. Since the natural frequency and the damping are to be estimated from the properties around the peak of the response spectral density, and if we assume that the bias factor is reasonably flat around the peak, then the bias will not significantly influence the frequency and damping estimates.

However, since the mode shape is estimated from any column (or row) in the spectral density matrix, the mode shape estimate normalizing the first coordinate to unity is obtained as

$$\hat{a}_1 = 1$$

$$\hat{a}_2 = \frac{G_{21}(\omega)}{G_{11}(\omega)} = \frac{\tilde{h}_2(\omega)}{\tilde{h}_1(\omega)} \frac{a_2}{a_1} \tag{7.67}$$

and as it appears, the bias factor has full influence on the mode shape coordinates. Using relative calibration means that all sensors do not necessarily have the right FRF, that is, $\tilde{h}(\omega) \equiv 1$, but all have approximately the same FRF. In this case, the influence of the bias factor disappears on the mode shape estimate as it is seen from Eq. (7.67).

7.3.11   Noise Floor Estimation

Noise can be identified by measuring the same signal with two or more sensors. First, let us consider the simple case where only one noise source is to be identified. In this case, we can follow the ideas in Bendat and Piersol [24] and Brincker et al. [9] and assume that two measured signals y_1, y_2 are given by

$$y_1(t) = s(t) + n_1(t)$$
$$y_2(t) = s(t) + n_2(t)$$
(7.68)

where $s(t)$ is the signal and $n_1(t)$, $n_2(t)$ are uncorrelated noise sources with the same properties. Since the noise sources are uncorrelated and also are assumed to be uncorrelated with the signal, the noise spectrum can be found as

$$G_n(f) = G_{y_1}(f) - G_s(f)$$
(7.69)

We could also use the other measured signal as basis for estimating the auto spectrum for the measured output, and since the signal spectrum can be estimated as the absolute value of the cross spectrum between the two outputs, we can define

$$G_n(f) = \sqrt{G_{y_1}(f)\, G_{y_2}(f)} - \left| G_{y_1 y_2}(f) \right|$$
(7.70)

giving

$$G_n(f) = \left(1 - \gamma_{y_1 y_2}(f)\right) G_y(f)$$
(7.71)

where

$$G_y(f) = \sqrt{G_{y_1}(f)\, G_{y_2}(f)}$$
(7.72)

and where $\gamma_{y_1 y_2}(f)$ is the coherence between the two measured channels. High coherence means that the two measured signals are well correlated and thus has a low influence of noise. Low coherence indicates that the signals are strongly influenced by noise.

Figure 7.17 shows an example of a spectral density of a single piezoelectric accelerometer influenced by noise (a) and the coherence (b) with a similar sensor measuring the same signal. In this example, the coherence shows clearly that the signals measured by the sensors are highly influenced by noise in the entire frequency band. The signal spectral density and the coherence indicate that smaller frequency bands around 20–40 Hz and 640–670 Hz might include some reliable physical information about the actual input to the sensors.

The signal spectrum can be found from Eqs. (7.71) and (7.72)

$$G_{ss}(f) = \gamma_{y_1 y_2}(f)\, G_{yy}(f)$$
(7.73)

Figure 7.18 shows the calculated noise spectrum using Eq. (7.71) and the signal spectrum using Eq. (7.73) for the same data as shown in Figure 7.17. This approach gives a well-defined estimate of the sensor self-noise spectrum. Concerning estimation of the signal spectrum leaves some uncertainties. In the mid-frequency span, it looks like we might have estimated something meaningful; however, this might not be the case, since the spectral values in this frequency band could be controlled by the noise floor of the measurement amplifier. This is easier to see if we use more measurement channels and perform a singular value decomposition of the response spectral density matrix.

Measuring several channels of data simultaneously and then calculating the spectral density matrix, the spectral density matrix can be decomposed by FDD, see Section 10.3.

$$\mathbf{G}_{yy}(f) = \mathbf{U}(f)\left[s_n^2\right]\mathbf{U}^H(f)$$
(7.74)

When we use the FDD for OMA, we interpret the singular vectors to be estimates of the mode shapes of the measured structure, and the singular values to be auto spectral densities of the modal coordinates.

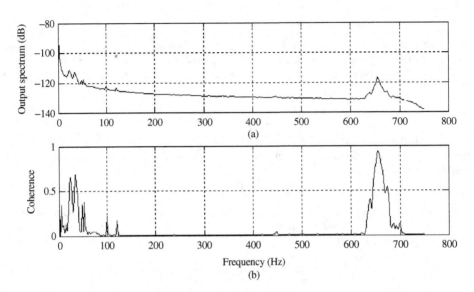

Figure 7.17 (a) Spectral density function of a signal from a piezoelectric accelerometer that is influenced by noise, in dB (rel. to $V/\sqrt{Hz}$). (b) Coherence between two simultaneous measurements using the same type of piezoelectric accelerometer in the same location (Source: From Brincker and Larsen [9])

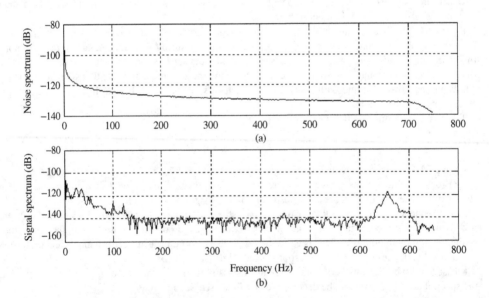

Figure 7.18 (a) Calculated sensor self-noise spectrum using Eq. (7.71). (b) Calculated signal spectrum using Eq. (7.73), both in dB (rel. to $V/\sqrt{Hz}$). From Brincker and Larsen [9].

However, if there are more measurement channels present than the physical rank of the data (the number of physical signals present at a certain frequency), then the different noise sources will be modeled.

When we place all sensors in the same location, we know that only one physical signal is present. In the earlier considered data, we had two different components, the physical signal and the sensor noise. If we like to determine the noise floor of the measurement amplifier, then we have to add at least one more measurement channel. In the present example, this was done by adding two more sensors (geophones), with a considerable lower noise floor. The computation of the singular values according to Eq. (7.74) results in the plot shown in Figure 7.19.

In this plot, one can clearly see the noise floor of the two piezoelectric accelerometers (compare with Figure 7.18a), the ambient vibration signal as well as the noise floor of the measurement amplifier (lowest singular value). As it appears, the noise floor of the measurement amplifier actually lies below the ambient response spectrum in the mid frequency band. Using this approach, it is possible to check the noise floor of a group of sensors and the measurement amplifier onsite (collocation test). See Section 7.3.10 for more on information on the collocation test.

In this example, there is a clear evidence of $1/f$ noise; under 100 Hz for the measurement amplifier and under 200 Hz for the piezoelectric accelerometers. The measurement amplifier noise spectrum shows significant noise components from the 50 Hz power supply (spikes at 50, 150 Hz etc.). The measurement amplifier shows in this case a noise floor (well away from DC) of about -150 dB (rel. $1 \text{V} / \sqrt{\text{Hz}}$), or about $30 \text{nV} / \sqrt{\text{Hz}}$. The noise floors of the two added geophones are not visible in Figure 7.19 and the reason for this is that the noise floor of these sensors[21] is well below the noise floor of the measurement amplifier.

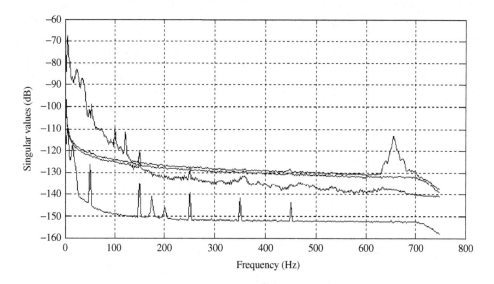

Figure 7.19 Plot of the singular values of the spectral matrix in dB (rel. to $\text{V} / \sqrt{\text{Hz}}$) of four data channels measuring the same signal. Two sensors with high noise level (piezoelectric accelerometers) are visible in the data, the noise floor of the two other sensors (geophones) is not visible since the noise level of these sensors falls below the noise floor of the measurement amplifier, which is indicated by the lowest singular value (Source: From Brincker and Larsen [9])

---

[21] The added sensors are geophones with an inherent noise floor of approximately 2.5 nV/$\sqrt{\text{Hz}}$.

7.3.12   Very Low Frequencies and Influence of Tilt

In civil engineering, we deal with large structures with very low frequencies. For instance, long span bridges or large offshore structure can easily have natural periods of 10 s or longer corresponding to frequencies lower than 0.1 Hz.

Using accelerometers at such low frequencies introduces two major problems; one is the increasing sensor noise in the low frequency region, the other one is the influence of the tilt of the structure introducing gravity effects on the acceleration measurements. The first problem has already been discussed in Section 7.3.8, thus, the present section is devoted to the errors that might be introduced by structural tilt.

To better understand the effect of tilt, let us consider the example of a classical pendulum. The purpose of this simple exercise is to illustrate a problem that can be observed when trying to measure the motion of a pendulum like structure with an accelerometer attached to the mass of the structure. It can be observed from experiments that no meaningful signal related to the acceleration of the mass is being measured, although the mass is in motion.[22]

The equation of motion for a classical pendulum as shown in the left part of Figure 7.20 is given by

$$M l^2 \ddot{\theta} + Mgl \sin(\theta) = 0 \tag{7.75}$$

where M is the mass of the pendulum, l is the length of the pendulum, g is the gravity, θ is the angle of rotation of the pendulum. If the angle is small enough, one can justify the approximation $\sin\theta \cong \theta$ and the equation of equilibrium can be conveniently re-written as $l\ddot{\theta} + g\theta = 0$.

Let us assume that a sensor with a seismic mass m and stiffness k as described in Figure 7.5 is attached to the mass of the pendulum with the intention to measure its tangential acceleration. The right part of Figure 7.20 illustrates the idea. The mass of the sensor is very small as compared to the mass of the pendulum ($m \ll M$) and for the purpose of this discussion we will neglect the damping of the system and of the sensor unit.

The measurement signal in the sensor element can be found by formulating the equation of motion of the sensor when subjected to an acceleration corresponding to the tangential acceleration of the mass of the pendulum, and by taking into account the effect of gravity at any instant of time during the motion

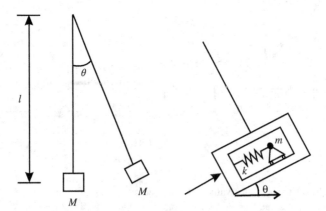

Figure 7.20   Classical pendulum with a sensor attached to the mass

---

[22] It is obvious, that when a pendulum performs a swing, the mass is accelerating, and one would expect that putting an accelerometer on the swinging mass, the acceleration can be measured. However, due to gravity influence from the tilt, this is not the case.

of the pendulum. From equilibrium forces acting along the axis of the sensor, we get:

$$m\left(a_T + a\right) + kx - mg \sin(\theta) = 0 \qquad (7.76)$$

in which a is the acceleration of the mass of the sensor (movement relative to the casing) and a_T is the tangential acceleration of the pendulum mass given by the well-known expression

$$a_T = g \sin(\theta) \qquad (7.77)$$

Upon substituting Eq. (7.77) into Eq. (7.76) and cancelling common terms, we obtain

$$ma + kx = 0 \qquad (7.78)$$

This equation indicates that the effect of gravity cancels out the force on the mass of the sensor by the tangential acceleration and that the sensor will not measure any vibrations unless the mass of the sensor is subjected to an initial displacement, an initial velocity, or a combination of both (i.e., free vibration motion). No matter what is the intensity of the acceleration of the pendulum, or the angle of rotation, the sensor will not be able to measure any motion of the mass of the pendulum.

We can then conclude that placing a sensor on a classical pendulum with the intention to measure the acceleration of the pendulum will not result in anything meaningful if the influence of the gravity is not removed. This example also illustrates that the influence of tilt of the structure being tested can be rather large on the measurements using sensors that work on the concept illustrated in Figure 7.5.

In the aforementioned example with the classical pendulum, the influence of the tilt reduces the measured response to zero or close to zero. However, the inverted pendulum is generally used to represent simple structures, such as a tower or a building, and the effects of tilt on the measured motions can also be explained by reference to a simple model such as that shown in Figure 7.21. If we take the pendulum shown in Figure 7.20 and invert it, one can readily show that the gravity effect on Eq.(7.76) is additive and that the tangential acceleration of the pendulum will be biased by the effect of the acceleration of gravity, and this effect is being larger as the natural frequency of the structure decreases.

This can be illustrated by a simple example. If we have a $l = 100$ m high tower that acts as a vertical clamped beam where the lowest mode is at $f = 0.1$ Hz, and the amplitude at the top is $\delta = 1$ m (see Figure 7.21), and let us assume that the deformation of this tower is due to its first mode of vibration only. The actual acceleration a_1 due to the modal movement (the acceleration that should be measured by the sensor) can be determined to be $a_1 = (2\pi f)^2 \delta = 0.395$ m/s^2. For a case like this, the tilting angle θ is approximately $\theta = 2\delta/l$, thus the influence from gravity on the measured acceleration is $a_2 = g\theta = 0.196$ m/s^2. Therefore, the actual acceleration that the sensor will measure is $a = a_1 + a_2$. This means

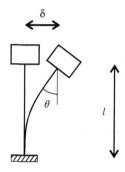

Figure 7.21 A tower primarily acting as a clamped beam with a sensor at the top. In this case, the tilt will significantly increase the measured acceleration if modes are present in the low frequency band

that what has been measured corresponds to a signal that is 50% larger than the acceleration of the structure (note that the error is independent of the size of the deformation). Since the tilting angle is proportional to deformation but the acceleration is proportional to $\omega^2 \times$ deformation, the influence of tilt increases as the natural frequency of the structure decreases, and this is why this is an important issue for low frequency systems. Extensive work has been conducted on the effect of tilt on measured motions due to earthquake shaking. See, for instance, the work by Boroschek and Legrand [25] and Graizer and Kalkan [26] on this subject matter.

This problem can be treated in several ways. The best way is to perform an independent measurement of the tilting angle of the sensor using a tilt meter and then correct the measured acceleration for the bias error introduced by the tilt. However, separate measurement of tilt may significantly increase the cost of the measurement system. One could also consider using a finite element model of the system to correlate the measured acceleration with tilt and then correct the measurements accordingly. Since there is only a unique relationship between tilt and acceleration for a single mode, the signal must be dominated by a single mode, or the modes must be well separated so that they can be isolated by filtering. If none of these simplified approaches can be used, the signal must be modally decomposed by using modal filtering as discussed in Section 12.5.

7.4 Data Quality Assessment

No matter how well the testing has been planned, how good the sensors are and how well we have checked the total measurement system, there is always a risk that something unforeseen may happen during the actual test. Sometimes, this may have a catastrophic influence on the quality of the acquired data. Further, for a typical OMA tests, we have to go to the structure, which may be difficult to access or may require special arrangements to access. Therefore, it is essential that the quality of the data is checked before leaving the test site.

Typical problems that are commonly experienced and that should be checked before leaving the test site are as follows:

- Clipping
- Excessive noise
- Outliers

The first two are normally due to improper setting of the sensitivity of the data acquisition system, whereas the last one is usually due to physical damages or malfunction of the data acquisition system.

It should be noted that quality check preferably should be carried out on the raw data. Preferred raw data is the data (raw numbers) directly obtained from the ADC.

7.4.1 Data Acquisition Settings

Clipping means that the signal saturates the ADC. This check is easy to conduct as we just have to check if any of the acquired data fulfill the clipping conditions

$$r = r_0 \text{ or } r = r_1 \tag{7.79}$$

where r_0 and r_1 is the minimum and maximum reading from the ADC. Clipping is prevented by setting the maximum input voltage value U_{mADC} high enough so that clipping is prevented.

However, if the maximum input voltage U_{mADC} is set too high then the measured signal might be too close to the minimum input level and thus it might be influenced by either the noise floor of the ADC or by the resolutions of the ADC. Therefore, setting the maximum input voltage level is always a trade-off between the risk of clipping and excessive noise.

It is recommended to conduct pilot tests to estimate a reasonable value of the maximum input voltage before the final OMA test is performed.

7.4.2 Excessive Noise from External Equipment

Many times tests are performed under circumstances where new noise sources might show up that we have not seen before. That might be, for instance, machines starting/stopping or other significant electric equipment that influence the measurement through the power supply, or by electromagnetic radiation.

This kind of noise might lead to spikes in the signal, or slowly varying noise due to changes in the power supply characteristics. The first kind could be treated like outliers, or since an outlier will result in higher noise floor, it could be checked that the resulting noise floor is as it should be expected.

Slowly varying influences due to changes of the power supply caused by big electric machines are difficult to detect and difficult to prevent once the influence is present. Therefore, this kind of excessive noise should be prevented by using a stabilized or a battery-based power supply for all measurement equipment.

7.4.3 Checking the Signal-to-Noise Ratio

The signal-to-noise ratio should always be evaluated before leaving the test site. A simple way of performing this evaluation is to perform a FDD of the measured data and make a plot of the singular values of the spectral density matrix as a function of the frequency, see Section 10.3 and Figures 7.4 and 7.19.

The case shown in the left plot of Figure 7.4 can be used as an example. In this case, 16 sensors where used for the measurements, but only six columns in the spectral matrix were used in the singular value decomposition; thus the plot shows six singular values. The top 3–4 singular values give us information about the physical system that we are testing (the modal contributions) whereas the bottom 1–2 singular values are related to the noise. The distance between the maximum values of the modal peaks and the minimum singular value is a measure of the signal-to-noise ratio in each frequency band. For instance, in the lower frequency band from DC to 2000 Hz, the signal-to-noise ratio is of the order of 70–80 dB, which is a good signal-to-noise ratio for an OMA test.

A signal-to-noise ratio of 30–40 dB should be considered as minimum for an acceptable OMA test.

7.4.4 Outliers

In statistics, an outlier is an observation that is numerically distant from the rest of the data. In OMA data, typical outliers are dropouts and spikes – both kinds difficult to detect by simple means. Dropouts are large deviations toward the mean, and spikes are large deviations toward higher values.

Typical reasons for outliers in dynamic testing are problems with plugs and cables (bad connections) and spikes caused by stopping and starting electric machines. Only spikes can be safely detected by statistical outlier testing because only the spike will tend to give too high values. Dropouts might not be detected by standard outlier testing.

In any case, even though statistical measures are being used in order to check for outliers, the simple way to take a visual look at the data should always be used. When visually inspecting the data, it is helpful to look at several channels of data at the same time (synchronous recordings), but limit the time span in each window to a reasonable size. That size could be, for instance, the system memory time T_{mem} as given by Eq. (7.15).

Another important visual inspection of the data is to perform an operating deflection shape (ODS) animation in the time domain. In the ODS time domain animation, all the movements of the structure are shown in a way (normally blown up and shown in slow motion) that can be followed by the naked eye.

Any non-consistencies in the data will be seen clearly as an "impossible" movement of the structure. A similar check can be performed making a preliminary OMA an animate the mode shapes to see if the estimated mode shapes make good sense.

If significant outliers are present, then the test should either be repeated to obtain better data, or the data should be cleaned for outliers by applying the same time window to all channels similarly, see Section 8.3.7. The latter solution can remove the influence of outliers but will at the same time reduce the information in the signals, thus this solution should only be used if plenty of data is available.

7.5   Chapter Summary – Good Testing Practice

At this point, we will conclude on the most important issues to consider in order to perform a good and reliable OMA testing.

Overall we can say that planning is needed to ensure that the test is well organized and that the equipment is properly selected and calibrated. The goal is to obtain measurements that contain information that allows us to extract the needed modal parameters of the structure.

The essence of good testing practice in OMA can be reduced to the following main points:

- Before the test, develop a finite element model of the structure and compute its dynamic characteristics. Then make variations of the model to investigate the influence of possible uncertainties of the model elements.
- Perform simulations in order to confirm that the measurement plan includes a good choice of measurement locations and that it can provide the needed accuracy of the modal estimates.
- If a model cannot be made available ensure at least that there is a good understanding of the dynamic characteristics of the structure to be tested.
- Estimate the expected vibration level or perform a pilot test in order to ensure that the correct type of sensors is being used and that the testing system can measure the expected signals accurately.
- Consider possible problems related to very small responses and if your sensors are able to give you an acceptable signal-to-noise ratio under these conditions.
- Evaluate the number of natural independent input sources and consider the option of using artificial loading to ensure a more suitable number of independent input sources.
- Check all equipment, and know noise floors and calibration factors – and even so, whenever possible perform a collocation test on the test site for relative calibration.
- Consider possible problems related to very low frequencies such as excessive noise and tilt problems.
- Evaluate if phase distortion from sensors and other components in the measurement chain might influence your results (mode shapes).
- Document the measurement plan so that the project tasks can be properly organized, that is, plans and tasks can be handed out to the people who will perform the test.
- Make sure that all personnel are well trained, and that each person has a clear understanding of who is responsible for what.
- Make use of the SMART approach to plan the tests – and be prepared for the possibilities of changes in the test plan.
- Make use of field notes and photography to document the locations where the measurements are being taken and the measurement directions of all sensors. (As measured – not as planned).
- Be sure to document the whole measurement session in all details of importance.
- Collect more data than needed so that you are able later to discard some data and still be able to perform a good OMA on the remaining data.
- Check the data quality before leaving the test site.
- Take care of the safety of personnel and equipment.
- Follow regulations and standards and advice from owners and local authorities.

References

[1] *General requirements for the competence of testing and calibration laboratories.* ISO/IEC 17025 standard, 2nd edition, 2005.

[2] *Guidelines for the measurement of vibrations and evaluation of their effects on buildings.* American National Standard, Vibration of Buildings – September 1990 (R2001).

[3] Kirkegaard, P.H.: *Optimal design of experiments for parametric identification of civil engineering structures.* Ph.D. thesis, Department of Building Technology and Structural Engineering, Aalborg University, Denmark, 1991.

[4] Ibanez, P., Vasudevan, R. and Smith, C.B.: *Experimental and theoretical analysis of buildings.* ASCE-EMD Specialty Conference, UCLA Extension, p. 412–430, 1976.

[5] Peterson, J.: *Observation and modeling of seismic background noise levels.* U.S. Geological Survey Technical Report. 93-322, p. 1–95, 1993.

[6] Brincker, R, Ventura, C., Andersen, P.: *Why output only modal analysis is a desirable tool for a wide range of practical applications.* In: Proceedings of IMAC-21: Orlando, Kissimmee, Florida, Society for Experimental Mechanics. p. 265–272, 2003.

[7] Reynders, E., Teughels, A., DeRoeck, G.: *Finite element model updating and structural damage identification using OMAX data.* Mech. Sys. Signal Process., V. 24, p. 1306–1323, 2010.

[8] Gannon J., Pham, H. and Speller, K.: *A robust low noise MEMS servo accelerometer,* In Proceedings of the ISA Emerging Technologies Conference, Houston, TX, September 10–12, 2001.

[9] Brincker, R., Larsen, J.A.: *Obtaining and estimating low noise floors in vibration sensors.* In Conference Proceedings of IMAC-XXIV, Bethel, CT, USA, 2007.

[10] Gautschi, G.: *Piezoelectric Sensorics.* Springer, 2002.

[11] Wielandt E.: *Seismic sensors and their calibration,* presented in "Manual of Observatory Practice" edited by Peter Bormann and Erik Bergmann, available on the internet.

[12] Wielandt E, *Seismometry.* In *International handbook of earthquake and engineering seismology,* Vol. 1, edited by William Hung Kan Lee, International Association of Seismology. 3rd edition, 2000.

[13] Havskov J., Alguacil, G.: *Instrumentation in earthquake seismology,* Springer, 2004.

[14] Brincker, R., Lagö, T., Andersen, P., Ventura, C.: *Improving the classical geophone sensor element by digital correction.* In Proceedings of the International Modal Analysis Conference. Orlando, FL, USA, January 31–February 3, 2005.

[15] Wieland, E. and Streckeisen, G.: *The leaf-spring seismometer: design and performance,* Bull. Seismo. Soc. Am., V. 72, No. 6, p. 2352, 1982.

[16] Brincker, R, Larsen, J.A, Ventura, C.: *A general purpose digital system for field vibration testing.* In Proceedings of the International Modal Analysis Conference. Orlando, FL, USA, February 19–22, 2007.

[17] Brincker, R, Bolton B, Brandt A.: *Calibration and processing of geophone signals for structural vibration measurements,* In Conference Proceedings of IMAC 2010, February 1–5, Jacksonville, FL, 2010.

[18] Ott, H.W., *Noise reduction techniques in electronic systems,* 2nd edition, John Wiley & Sons, 1988.

[19] Williams, T.: *EMC for product designers,* 3rd edition, Newnes, 2001.

[20] Usher M.J.: *Developments in seismometry.* J. Phys. E: Sci. Instrum., V. 6, p. 501–507, 1973.

[21] Brincker, R. and Brandt, A.: *Time synchronization by modal correlation.* In Proceedings of the International Operational Modal Analysis Conference (IOMAC), Istanbul, May 9–11, 2011.

[22] Priya, S., Inman, D.: *Energy harvesting technologies,* Springer-Verlag, 2008.

[23] Bendat, J.S. and Piersol, A.G.: *Random data – analysis and measurement procedures,* 2nd edition, John Wiley & Sons, 1986.

[24] Bendat, J.S. and Piersol, A.G.: *Engineering applications of correlation and spectral analysis,* 2nd edition, John Wiley & Sons, Inc, 1993.

[25] Boroschek, R.L. and Legrand, D.: *Tilt motion effects on the double-time integration of linear accelerometers: An experimental approach,* Bull. Seismol. Soc. Am., V. 96, No. 6, p. 2072–2089, 2006.

[26] Graizer, V. and Kalkan, E.: *Response of pendulums to complex input ground motion.* Soil Dyn. Earthq. Eng., V. 28, No. 8, pp. 621–631, 2008.

8

Signal Processing

"All we know of the truth is that the absolute truth, such as it is, is beyond our reach."

– Nicholas of Cusa

In general, one can say that the purpose of signal processing is to amplify some elements of the measured data creating a more clear picture of the physical problem that we are dealing with. This is extremely important in OMA, because here all the modal information (or "physics" of the problem) information normally is buried in the chaos created by the randomness of the measured signals. Without signal processing, nothing of importance is clearly visible, but the good news is that with a reasonable choice of signal processing techniques enough becomes visible so that we can extract the most important information.

As explained in the introduction and as it appears from the chapter on Random vibration, what we in general want to do in an OMA identification process is to estimate either correlation functions or spectral densities and then from these functions extract the physical information about the system, that is, the modal parameters for each mode (the mode shape, the natural frequency, and the damping).

As we have argued in the chapter on probability, normally random responses are Gaussian distributed, and therefore since we do not need the mean value of the measured signals, all information in the signals are concentrated in the second order properties, which are described completely by correlation functions or spectral densities. Both functions carry the same information because they form a Fourier transform pair. Correlation function are time domain functions and spectral densities are frequency domain functions.

The main aim of this chapter is to explain how we can estimate the correlation functions and the spectral density functions and what we should be aware of in this process. Also, it is important to known what issues we should go through before we start estimating these functions, because in some cases we cannot start the estimation process right away. We need to take some actions first.

This chapter is mainly based on the approach followed by commonly accepted text books such as Bendat and Piersol [1], Oppenheim and Schafer [2] and Proakis and Manolakis [3]. Recent publications such as Brandt [4] have been useful for gathering this material.

8.1 Basic Preprocessing

The first step in the signal processing task is to check the data quality and to convert it to a standard form. In the following sections we will go through the most important steps involved in this process.

Introduction to Operational Modal Analysis, First Edition. Rune Brincker and Carlos Ventura.
© 2015 John Wiley & Sons, Ltd. Published 2015 by John Wiley & Sons, Ltd.
Companion Website: www.wiley.com/go/brincker

8.1.1   Data Quality

The most important step – the step that comes before anything else – is to check the data quality. Ideally this check should be done "on-the-spot" when the data has been taken. It is often a very resource-demanding task to carry out OMA tests, and thus it would not be good if after the data has been collected and the test is over, and the data is being analyzed at the office, that it becomes clear that the data does not have the quality needed for the subsequent OMA analysis.

The most basic quality checks should be performed on the raw data as it comes out of the ADC. This step should include checks for clipping, dropouts and spikes. Clipping means that the signal saturates the ADC. This check can be made easily according to Eq. (7.79). Dropouts and spikes are more difficult to detect by simple means. They share the same quality of being large deviations from the true signal; drops-outs are large deviations towards zero, and spikes are large deviations toward higher values. As argued in Section 7.4.4 to check for these kinds of errors, the user should always make a visual inspection of the data and should consider to perform a time domain ODS.

Another simple way of checking the data for drop-outs and spikes is to calculate the moving average $\mu(\tau)$ and moving standard deviation $\sigma(\tau)$

$$\mu(\tau) = \frac{1}{T} \int_{\tau-T/2}^{\tau+T/2} y(t)dt$$

$$\sigma(\tau)^2 = \frac{1}{T} \int_{\tau-T/2}^{\tau+T/2} (y(t) - \mu(t))^2 dt$$

(8.1)

and then watch out for large deviations from the typical values of these quantities. Here $y(t)$ is the measured signal and T is the size of the averaging window.[1]

It should be mentioned that kurtosis is often used for outlier analysis because this measure is more sensitive to outliers than the mean value, and that using predictor error methods[2] the variance of the prediction error can also be used for outlier detection.

The final test that the data are good enough for OMA is of course that an OMA analysis of reasonable quality can be performed. A good way of testing the data is to perform a simple OMA to see if all modes that are expected to be present can be accurately estimated from the test data. A simple and fast OMA for such purpose could be the simple peak picking FDD.[3]

8.1.2   Calibration

The raw data that comes out of the ADC is just a digital number without any physical unit. As it is explained in the chapter on Measurement technology, the digital number can be converted to the voltage that has been measured in the ADC.

In principle, for OMA it is enough to perform a relative calibration of the instruments as explained in Section 7.3.10, because that is what is needed to ensure error free mode shape estimation.

However, in order to be able to make spectral density plots using the right units, individual calibration factors should be available for each sensor[4] and the physical signals must be estimated accordingly.

---

[1] See Section 8.3.2 for examples of moving average filters.
[2] See the section on ARMA models in Chapter 9.
[3] The reader will find more information on frequency domain decomposition (FDD) in Section 10.3.
[4] In case of complex sensors, such as for instance, the geophone, then the FRF of the sensor must be available and the signal must be corrected accordingly.

Although in principle it does not matter for the OMA results, OMA preferably should be carried out using physical signals such that spectral densities can be presented using the right units of measure.

The reader is referred to Section 7.3.10 in the measurement chapter on how to perform a calibration of a sensor and how to estimate the physical signal from the measured signal.

8.1.3   Detrending and Segmenting

Detrending means to remove DC offsets[5] in the signals so we force the signals to have a zero mean. The reason why we do this is that normally we cannot trust the DC component in the signal from sensors due to excessive noise in the low frequency region, as explained in the chapter on measurement technology. For instance, we can think of the DC drift of the signal as due to temperature effect in the electronics. Another reason is that in OMA we are dealing with the dynamics of the system and thus, we are simply not interested in the response significantly below the lowest frequency of the structure.

Detrending is in general done by digital high-pass filtering that must be used in all cases of large DC-related errors. Digital filters can be used for this task as explained later in this chapter, but since IIR[6] filters tend to become unstable when the cut-off frequency is close to DC, FIR filters are often chosen for this kind of filtering. However, both kinds of filters might disturb the signal up in the Nyquist band to a degree that we might not be willing to accept, and thus other ways of dealing with this task might have to be considered.

An alternative way to perform detrending in cases of small DC errors is to segment the signal in overlapping data segments of duration T and simply remove the mean value of each data segment. In order to reduce leakage in subsequent Fourier transforms and to minimize discontinuities between data segments, each data segment is tapered by applying a window $w(\tau)$ so that the windowed data segment $y_k(\tau)$ at time t_k is defined by, see Figure 8.1

$$y_k(\tau) = y(t_k + \tau)w(\tau) - \mu_k \qquad (8.2)$$

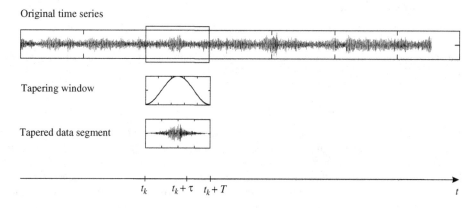

Figure 8.1   For detrending, but also for subsequent signal processing it might be practical to segment the data as shown in this figure. Each data segment is captured by overlapping and window tapering such that the sum of all data segments is equal to the original signal

---

[5] A fixed value plus a slowly varying component. For instance, in the case of a straight line the slope of the line can be considered as the trend.

[6] IIR and FIR filters are introduced in Section 8.3.1.

where μ_k is the mean value of the windowed data segment $y(t_k + \tau)w(\tau)$, τ is the local time and the absolute time is $t = t_k + \tau$. Removing the mean of the windowed data segment is equivalent to removing the DC value of its Fourier transform.[7] Thus, removing the mean is a simple way of performing a sharp high-pass filtering.

It is practical to perform the windowing and overlapping between segments in such a way that the windowed data segments $Y_k(t)$ without detrending, but defined in terms of the absolute time

$$Y_k(t) = \begin{cases} y(t)\,w(t - t_k); & t \in [t_k;\ t_k + T] \\ 0 & ;\quad t \notin [t_k;\ t_k + T] \end{cases} \tag{8.3}$$

can be assembled to form the original time series

$$y(t) = \sum_k Y_k(t) \tag{8.4}$$

The size T of the data segment is taken as the memory estimate T_{mem}.[8] It should be noted that in case of segmented data, it is also easy to handle cases where the sensor has a complex and frequency dependent calibration constant (transfer function). In this case the calibration is most easily dealt with in the frequency domain where each data segment can be adjusted dividing by the FRF of the sensor and then taken back to time domain.

If follows from Eqs. (8.3/8.4) that the overlapped window most add to unity, except of the ends of the time senes where the window will still be present. A simple way to handle the segmenting in order to satisfy Eq. (8.4) is to use a Hanning window with 50% overlap. This might not be the optimal way of doing this, but it is a good practical way of handling data.

The data segment can also be treated without direct windowing and instead using zero padding that indirectly will introduce a triangular window, see the details of overlap-add filtering in Oppenheim and Schafer [1].

8.2   Signal Classification

Once the basic quality of the signals has been enhanced and the unwanted DC drift has been removed, the next step is to evaluate the kind of signals we are dealing with; to what extend these signals are suitable for OMA, and to what extend we have unwanted components in the signals that we might want to remove.

8.2.1   Operating Condition Sorting

Since we do not expect to have the same modal parameters from different cases of loading and different environmental conditions, it is important to sort the data in terms of the load cases and sort the data according to different environmental conditions, such as temperature.

It might sometimes be difficult to find out which parameters to include in this sorting process, but it is important to understand which operating condition parameters have influence on the modal parameters and then include these operating condition parameters in the sorting procedure.

If only a single test is performed with the aim of getting just one set of modal parameters for the actual operating condition present during the testing, and where the load and the environmental conditions are not significantly influencing the modal parameters, then, there is no need of sorting. However, in cases where several sets of modal parameters are to be compared and where the influence from loads and

---

[7] This can be seen by evaluating Eq. (4.13) for zero frequency.
[8] See Section 7.1.3 for information about this quantity.

environment is significant, it is important to perform a sorting and to carefully consider how to perform this task. In order to illustrate the relatively big challenge that entails to find out what to include in the sorting procedure let us consider two cases, and discuss the problems involved; one case with a bridge with varying traffic, and one case with a wind turbine under varying wind loading. For the bridge case different load cases might be for instance

A. Excitation from micro tremors (no cars on the bridge).
B. Excitation from single small vehicles.
C. Excitation from single large vehicles.
D. Excitation from several small vehicles.
E. Excitation from heavy traffic.

Load cases A, D, and E should be defined in such a way that each data set is reasonably stationary. Load cases B and C, however, are event-based, and thus each event might not be long enough for an OMA analysis. In such case, several events of the same kind might be merged together to form a data record long enough for OMA. Each event should be gathered including a (small) silent period at the beginning and the end of each record. If large background noise is present, then a windowing should be applied in order to force the signal to zero at the beginning and the end of each event. In this case, it should be considered, for instance, if mass loading from the traffic might tend to decrease the natural frequencies of the structure. In the wind turbine case operating condition parameters influencing the modal parameters might be

A. Wind speed.
B. Wind turbulence.
C. Wind direction.
D. Pitch angle of the turbine blades.
E. Power production.

In such case where many of the operating condition parameters are varying at the same time and where the change of each parameters might significantly influence the OMA results, there exist no simple solutions to how the sorting and possible merging of the data should be performed.

Because of the central limit theorem, for a stationary data set, the response is expected to be approximately Gaussian. Thus, an important check for a good sorting naturally includes a check that each sorted pool of data is Gaussian distributed.

8.2.2   Stationarity

Basic assumptions in OMA are that the system is invariant and the response is stationary. The first assumption that the system is invariant is more oritical than the response is stationary because if the system is varying, then the response cannot be stationary. When the physical parameters of the system are changing, the response will also be changing.

In OMA practice, most systems are, in fact, invariant or close to be invariant, and lack of stationarity is mainly due to changes of the loading conditions. For instance, a building might be loaded mainly by wind; the turbulent wind excites the structure, however, as the velocity of the wind changes over time, the turbulence changes and thus the wind loads on the structure.

Even though the assumption of stationary response is a basic premise in OMA, it is not fair to say that this assumption needs to be fulfilled always in order to perform reliable OMA. It is more appropriate to say that even though this assumption is basic for all theory development, most of the theory works quite well under weakly nonstationary conditions. In practice it is often difficult to ensure or to prove that the data to be used for OMA can be considered stationary in a statistical sense.

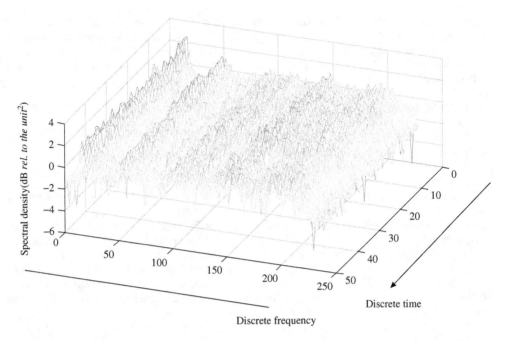

Figure 8.2 Time-frequency plot of channel one of the first data set of the Heritage Court Tower building test showing a relatively stationary loading case

In OMA, there is no tradition for making any big efforts on trying to design and conduct a test so that stationary datasets are obtained, or to use statistical tests for whether or not the data can be considered stationary. However, strong nonstationarity should be taken into account by sorting as described under the preceding section.

A simple way to roughly check the stationarity of the data is to make a time-frequency plot, as shown in Figure 8.2. A way of depicting the frequency content as a function of time is to segment the data as described earlier and then simply plot the spectral density of each segment as a function of time. The spectral density is not smoothed, but permits to directly compare the random variations from segment to segment with the tendency to systematic changes of the signal.

8.2.3 Harmonics

Harmonics[9] are almost always present in operating signals. Harmonics are usually caused by an engine operating somewhere on the structure or nearby that is transmitting a signal that becomes visible in the often very weak operating response obtained during the OMA testing.

---

[9] In this book, we will denote periodic signals as harmonics. It is common to denote the frequency of the periodic signal as the fundamental frequency and the overtones as harmonics. However, for simplicity here, we will just call them all harmonics.

Harmonics represent a challenge because they might be confused with modes as they appear as sharp peaks in spectral density estimates, or they may be a challenge because the estimation procedures are affected by this strong deviation from what is normally assumed that the loading is reasonably broad band.

Another problem is that sometimes the harmonics might be of much higher amplitude than the broad band operating forces. These harmonics may clip (saturate the ADC) and introduce a strong noise component in the analysis. There will be a challenge in having enough dynamic range in the measurement system to see the broad band signal below the level of the harmonics. Also the fact that over-harmonics often significantly increase the number of harmonics present in the data often makes this problem difficult to handle.

Normally harmonics cannot easily be removed. The main reason being that in many cases the harmonics have a slowly varying rotation frequency and thus not only the harmonics frequency and the amplitude and phase have to be identified in order to remove the harmonic, but also often a varying frequency must be identified. Therefore, we will not go further into the problem of removing harmonics from the signal.

The most important issue concerning harmonics is to identify their presence so that they are not confused with modes as explained earlier. Two simple ways of identifying harmonics are mentioned here. One relies on the statistics of a harmonic signal, the other one relies on the local increase in rank of the spectral density matrix at the frequency where a harmonic is present.

The first method is based on the fact that the probability density of a random signal from a structural mode is approximately Gaussian whereas a harmonic has a completely different density function with two peaks as shown in Figure 8.3. This is well-known in the literature; see for instance, Bendat and Piersol [1]. It was proposed to use this as an indicator for harmonics OMA in Brincker et al. [5]. The difference between the Gaussian distribution and the density of a harmonic appears most strongly in the kurtosis that is calculated as the fourth moment of the signal $y(t)$ using time averaging

$$\gamma = \frac{E[(y(t) - \mu)^4]}{\sigma^4} = \frac{1}{\sigma^4 T} \int_0^T (y(t) - \mu)^4 dt \tag{8.5}$$

where $y(t)$ is signal filtered with a notch filter to isolate the considered harmonic and where μ and σ are the mean and the standard deviation of the filtered signal. A Gaussian distribution has a kurtosis equal

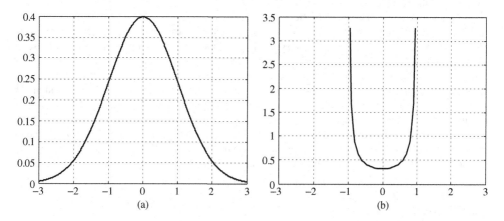

(a)   (b)

Figure 8.3 (a) Normalized probability density function of the response of a structural mode, and (b) Probability density function of harmonic component with unit amplitude

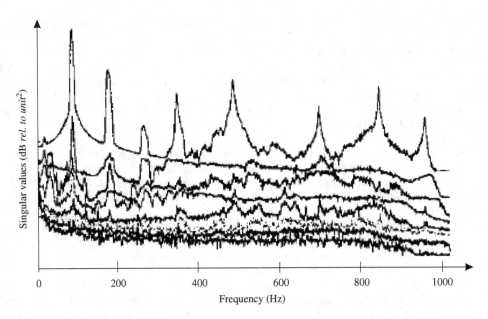

Figure 8.4 Plot of the singular values of the spectral matrix of the response of a plate loaded by random input and a harmonic around 90 Hz. The harmonic results in the first three peaks (the harmonic itself and its two over harmonics) and the random input results in the remaining peaks representing random response. As it appears the harmonics are visible in several singular values whereas the random response does not show this characteristic. From Brincker et al. [5]

to 3, thus values different from 3 indicate that the filtered signal might, in fact, represent a harmonic. See Jacobsen et al. [6] and Andersen et al. [7] for implementation and application of the kurtosis and examples of this technique.

The second method is based on the frequency domain decomposition (FDD) technique that gathers all spectral density functions in a single plot of the singular values of the spectral density matrix, see Figure 8.4. As it clearly appears from the spectral plot of the singular values of the example presented by Brincker et al. [5], the first three peaks in the FDD plot – that are in fact harmonics – look different than the remaining peaks which represent modal random response. The harmonics are clearly present not only in the first singular value but also in many of the lower singular values. This is a simple and important indicator of harmonics. See more about the property of the FDD technique in the chapter about frequency domain identification.

Finally it should be mentioned that if a harmonic has a constant frequency a simple way to remove it from a record is to make one big FFT of the considered signal (a periodogram) and then remove the harmonic peak and transform back to time domain by inverse FFT.

8.3 Filtering

In OMA, filtering mainly means to exclude some frequencies from the signal. In the past, this was normally done on the analog signals by different kinds of analog filters. However, since the only analog filters we are dealing with in OMA are the antialiasing filters that must be applied before the ADC, after the ADC everything is digital, and therefore only digital filtering is used in the subsequent signal processing.

8.3.1   Digital Filter Main Types

For digital filtering, two main types of filters exist. The finite impulse response (FIR) filters are of the form

$$y(n) = \sum_{k=0}^{nb-1} b(k)x(n-k) \tag{8.6}$$

where $x(n)$ is the input to the filter, $y(n)$ is the filtered signal, $b(k)$ are the filter constants and nb is the number of FIR filter coefficients.

The infinite impulse response (IIR) filters are of the form

$$y(n) = \sum_{k=1}^{na} a(k)y(n-k) \tag{8.7}$$

where na is the number of IIR filter coefficients and $a(k)$ are the filter constants. In this case, we need to add some external input, or at least specify some initial conditions, for the filter in order to obtain a response. As it appears from Eq. (4.124), the denominator polynomial of the filter transfer function has exactly na number of roots, and thus it is common to talk about na as the number of filter poles. In general, a digital filter is a combination of FIR and IIR terms such as Eq. (4.119) and is described by the two set of filter constants, $a(k)$, $b(k)$.

The FIR type is also called a filter of Moving Average (MA) type and has the advantage of often having a frequency independent phase shift (constant delay) and the quality of always being stable (as the FIR name indicate), but has the drawback of needing many coefficients in order to achieve a sharp filter cut-off.

The IIR type is also called a filter of the Auto Regressive (AR) type and has the advantage of being able to achieve a sharp filter cut-off with a limited number of coefficients, but has the drawback of having a frequency dependent phase shift. Further, since the AR coefficients turn into a numerator polynomial in the transfer function, frequencies close to the roots of this polynomial might cause the filter response to be unstable.

Both filter types might have problems with ripples close to the cut-off frequency. See more about how to formulate transfer functions and FRF's of general filters in the section on the z-transform in Chapter 4.

Classical filter types often used in OMA signal processing are as follows:

- High-pass filters that exclude the lower frequencies in the Nyquist band, these filters are characterized by cut-off frequency and slope.
- Low-pass filters that exclude the higher frequencies in the Nyquist band, these filters are characterized by cut-off frequency and slope.
- Band-pass filters that exclude all frequencies except the frequencies in a frequency band, these filters are characterized by two cut-off frequencies defining the frequency band and the slope.

Filter slope describes the steepness of the filter in the rejection area where the absolute value of the filter FRF reduces from unity to zero. The units of slope are either decibels per decade (dB/decade or decibels per octave (dB/octave). See Figure 8.5 for a graphical representation of frequency response functions (FRFs) of the main filter types and their slope.

In general, one can say that low-pass filters are typically used to remove DC drift,[10] high-pass filters are typically used to remove energy in the high-frequency region in relation to decimation, and band-pass

---

[10] De-trending as described in Section 8.1.3.

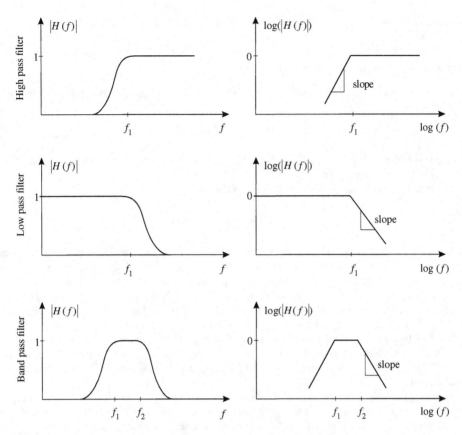

Figure 8.5 Frequency response functions $H(f)$ of the three commonly used filters: Top plots: High-pass filter with cut-off frequency f_1, middle plots: Low-pass filter with cut-off frequency f_1, and bottom plots: Band-pass filter with cut-off frequencies f_1 and f_2. Left plots have natural axes and right plots have double logarithmic axes making easier to see the slope (or steepness) of the filters around cut-off frequencies

filters are used in order to divide a frequency band with many modes into smaller frequency bands with a smaller number of modes in each band.

8.3.2   Two Averaging Filter Examples

A few simple examples of digital filters are given in this section. More information about different filter examples can be found in Oppenheim and Schafer [2] and Proakis and Manolakis [3].

We will consider two simple filters that can be used to remove the mean of a signal like the segmented version given by Eq. (8.2). An example of practical importance is the simple moving average filter

$$\mu(n) = \frac{1}{N} \sum_{k=0}^{N-1} x(n-k) \qquad (8.8)$$

We see from Eq. (8.6) that this filter is an FIR filter of order $nb = N$ with the constant coefficients $b(k) = 1/N$. As explained before in Section 8.3.1 removing the mean of a signal corresponds to removing the DC component of a signal, thus the signal

$$y(n) = x(n) - \mu(n) \tag{8.9}$$

must be equal to the signal $x(n)$ with the DC component removed. Let us prove this property of the filter using the theory that has been discussed. From the Transform chapter, we can use Eq. (4.122) to formulate the FRF for the FIR filter defined by Eq. (8.8) and sampling it in the frequency domain by setting $\omega = k2\pi/N\Delta t$

$$H(k) = \frac{1}{N}\sum_{n=0}^{N-1} e^{-i2\pi kn/N} \tag{8.10}$$

The result of the series given by Eq. (8.10) can be found by writing the corresponding discrete Fourier transforms. From the Transform chapter, we can write the discrete Fourier transform Eqs. (4.51/4.52) shifting the start of the summation to zero

$$y(n) = \sum_{k=0}^{N-1} Y(k)e^{i2\pi kn/N}$$

$$Y(k) = \frac{1}{N}\sum_{n=0}^{N-1} y(n)e^{-i2\pi kn/N} \tag{8.11}$$

Letting the time function $y(n)$ be equal to the discrete delta spike (or Dirac delta), we get function

$$\delta(n) = \begin{cases} 1, n = 0 \\ 0, n \neq 0 \end{cases} \tag{8.12}$$

We have from the lower part of Eq. (8.11) that the corresponding transform is $Y(k) = 1/N$. But from the upper part of Eq. (8.11), we see this also means that the delta function is given by

$$\delta(n) = \frac{1}{N}\sum_{k=0}^{N-1} e^{i2\pi kn/N} \tag{8.13}$$

Or since the delta function is symmetrical $\delta(n) = \delta(-n)$, we see from Eq. (8.10) that the FRF of the moving average filter given by Eq. (8.8) is in fact a delta spike at DC. Thus the FRF of removing the mean as given by Eq. (8.9) is exactly equal to unity except at DC, where the DC value is removed.

It should be noted that for practical reasons, it is better to use a noncausal version of the averaging filter that is symmetrical in future and past time such as

$$\mu(n) = \frac{1}{N-1}\sum_{k=-N/2+1}^{N/2-1} x(n-k) \tag{8.14}$$

because this filter does not introduce phase shift of the processed signal. Another simple averaging filter of practical importance is the mixed digital filter

$$y(n) = ay(n-1) + (1-a)x(n) \tag{8.15}$$

where $x(n)$ is the signal from which we want to remove the mean, and $y(n)$ is an estimate of the mean value. From Eqs. (8.6/8.7), we see that the filter has the single AR coefficient $a(1) = a$ and the single MA coefficient $b(0) = 1 - a$. From Eq. (4.121), we then have the transfer function

$$H''(z) = \frac{1-a}{1-az^{-1}} = \frac{(1-a)z}{z-a} \tag{8.16}$$

which from Table 4.2 can be shown to correspond to the impulse response

$$h(n) = (1-a)a^n \tag{8.17}$$

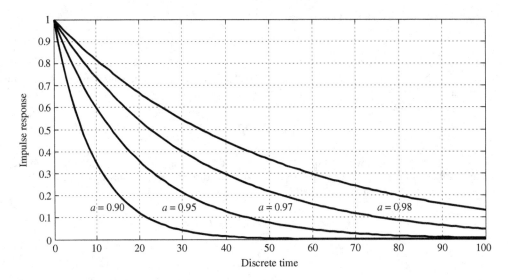

Figure 8.6 The relative impulse response function $h(n)/h(0)$ of the mixed filter given by Eq. (8.15) for different values of a

The impulse response is plotted in Figure 8.6. As shown in this figure, the values from the past are weighted with decreasing value, corresponding to a fading memory falling of exponentially. The larger the value of a, the longer the memory and the more values from the past are included in the averaging. Because of this, the filter (8.15) is also denoted exponential moving average. The impulse response given by Eq. (8.17) can also be found by inserting a delta spike $x(n) = \delta(n)$ into the filtering equation (8.15).

8.3.3   Down-Sampling and Up-Sampling

As mentioned in the Measurement chapter, we often sample the signals with a higher sampling frequency than actually needed for the final analysis. Thus, an important step in OMA signal processing will often be to resample the signal afterward to a smaller and more suitable sampling frequency. This task is denoted down sampling or decimation.

To illustrate the principle of decimation and the needed antialiasing filters to perform this task, let us consider the case where we need to reduce the sampling frequency by half. At first one might think that this could simply be done by omitting every second data point, because this would immediately reduce the sampling frequency by half.

However, this will introduce aliasing and thus a filter – an antialiasing filter – must be applied first. Due to the Parseval theorem, we cannot just omit every second sample point in order to decimate by a factor of two, because the energy in the signal between the new and the old Nyquist frequency still exists. This energy will jump into the frequency band below the new Nyquist frequency and pollute the signal. Thus, the energy above the new Nyquist frequency must be removed first by a low-pass filter, and then we can omit every second data point without any errors.

Up-sampling means to go from one sample frequency to a higher sampling frequency. For instance one could want to double the sampling frequency in order to obtain a time signal with double resolution. This procedure is based on the Shannon sampling theorem,[11] and actually enables us to do this without

---

[11] See the Shannon sampling theorem in Chapter 4.

any errors. For instance, it is possible first to decimate by a factor of two by discarding every second as described earlier and then later to reestablish the discarded sample points in the signal to their exact value. However, it should be noted that it is only possible to reestablish the original signal after the antialiasing filter has been applied, because the Shannon sampling theorem only holds for signals where the energy outside of the Nyquist band is zero.

8.3.4 Filter Banks

Just like it is sometimes practical to segment a time series in the time domain by time domain segmenting it as described in Section 8.1.3, it is sometimes practical to segment the data in the frequency domain.

Let us consider the signal $y(t)$ and its Fourier transform $Y(f)$. We might then apply a band-pass filter with center frequency f_k resulting in a filtered signal with the corresponding frequency domain function $Y_k(f)$.

A filter bank is used to decompose the signal into a number of band-pass filtered signals such that the individual components in the frequency domain $Y_1(f), Y_2(f), \ldots$ can be used to assemble the original signal by adding all the components

$$Y(f) = \sum_k Y_k(f) \tag{8.18}$$

The task of assembling the signal from its band-pass filtered contributions is often called synthesis. In order to reconstruct the original signal by Eq. (8.18) the first filter must be a high-pass and the last one must be a low-pass filter.

8.3.5 FFT Filtering

The classical digital filters presented in Section 8.3.1 are frequently used in all OMA, and their usefulness cannot be overestimated. However, these were developed for electrical engineering applications where many of these require real-time processing, and for such applications their strengths can be fully exploited. In OMA we do normally not need real-time processing as we conduct signal processing after testing has taken place, so other ways of doing filtering may be considered.

A good alternative to the classical digital filters is to use what is often denoted as FFT filtering. The idea of the FFT filtering is most easily derived from the fact that any digital filter has a FRF that can be expressed in terms of Eq. (4.122). Further, remembering that the effect of any FRF $H(\omega)$ on a signal $y(t)$ can be determined by multiplying the Fourier transform of the signal $y(t)$ $Y(\omega)$ with the FRF. The filtered signal $y_f(t)$ can then be obtained in the frequency domain as

$$Y_f(\omega) = H(\omega)Y(\omega) \tag{8.19}$$

Using the Fourier integral in Eq. (4.31) and writing the FRF in polar form as

$$H(\omega) = |H(\omega)|e^{i\theta(\omega)} \tag{8.20}$$

where $|H(\omega)|$ is the absolute value and $\theta(\omega)$ is the phase angle, we obtain the following expression for the filtered signal

$$\begin{aligned} y_f(t) &= \int_{-\infty}^{\infty} Y(\omega) \, |H(\omega)| \, e^{i\theta(\omega)} e^{i\omega t} d\omega \\ &= \int_{-\infty}^{\infty} Y(\omega) \, |H(\omega)| \, e^{i(\omega t + \theta(\omega))} d\omega \end{aligned} \tag{8.21}$$

We see from this expression that the phase $\theta(\omega)$ is in general, strongly dependent upon frequency and complicates the filter application if the purpose of the filter is to suppress some frequencies and amplify others just by using the amplification function $|H(\omega)|$.

Thus proper FFT filtering is simply implemented following these steps:

- Convert the considered signal to the frequency domain by Fourier transform.[12]
- Define a window $W(\omega)$ in the frequency domain that is similar to the preferred shape of the amplification function $|H(\omega)|$.
- Multiply the Fourier transform of the signal by the window.
- Convert the resulting signal back to the time domain.

The advantages of the FFT filtering are obvious. Phase distortion disappears, and the user can use any kind of window in the frequency domain. The user has full and direct control over the filter properties. For instance, the filter bank defined in Section 8.3.4 can easily be obtained using FFT filtering by segmenting the data in the frequency domain data as we did in Section 8.1.3 in the time domain and applying a window according to Eq. (8.3), hence

$$Y_k(f) = \begin{cases} Y(f)\,W(f-f_k)f \in [f_k - B/2;\ f_k + B/2] \\ 0 \quad ;\ f \notin [f_k - B/2;\ f_k + B/2] \end{cases} \tag{8.22}$$

where $Y(f)$ is the Fourier transform of the considered time signal $y(t)$ and $Y_k(f)$ is the filtered signal in frequency domain corresponding to the kth band pass filter with center frequency f_k and width B. A way to implement Eq. (8.18) is to define the window $W(f)$ as a Hanning window with 50% overlap. The width of the data segment B in the frequency domain (corresponding to the size T of the window in the time domain) should be chosen so that it contains a reasonable number of modes. In OMA, it does not make sense to work with band-pass filters smaller than the minimum bandwidth of one single mode. This bandwidth can be estimated as the half-power bandwidth $2\varsigma f_0$, where ς is the damping ratio and f_0 is the natural frequency of the considered mode. Thus the size of the frequency domain data segments should fulfill the requirement

$$B > \max(2\varsigma f_0) \tag{8.23}$$

where $\max(2\varsigma f_0)$ is calculated over all modes of interest.

If the signals to be filtered by an FFT approach have a size that makes it possible to do the FFT filtering with a one-shot FFT then it is preferable to do this since the leakage error introduced by the FFT algorithm reduces with data segment size. If the signals are too large to do the processing by one single FFT, the possibility of segmenting the data in the time domain should be considered.

Even though the FRF of the FFT filtering is just defined as a real window in most cases this cannot always be done. For instance, using FFT filtering to differentiate or integrate, it is necessary to use an imaginary window as explained in the section. In FFT filtering sometimes we need to work with complex FRFs.

8.3.6 Integration and Differentiation

In OMA integration and differentiation is often important. Good examples of such cases are when different sensors are used during the same OMA test. For instance when strain gauge measurements are to be compared with measurements from accelerometers, the accelerometer signals can be integrated twice to obtain displacements, and then the displacements can in many case be related directly to the considered

---

[12] In principle, one should apply zero padding here in order to avoid cyclic convolution. In practice, it is only important if the length of the signals are not getting very short.

strains,[13] or sometimes velocity sensors like geophones might be used together accelerometers and the velocity signals might then be differentiated so both sensor signals can be used in the same OMA.

Differentiation and integration of time series can be formulated as discrete filters of the FIR and IIR types. For instance, a differentiation filter can be formulated as an FIR filter with filter coefficients $b(k) = (1, -1)/\Delta t$ where Δt is the sampling step. Using Eq. (8.6) this leads to

$$y(n) = \frac{x(n) - x(n-1)}{\Delta t} \qquad (8.24)$$

We see that in this case the differentiation is performed by calculating the difference coefficient. Similarly, integration can be formulated as a combined FIR/IIR filter with the filter coefficients $a(k) = 1$; $b(k) = \Delta t$. Using Eqs. (8.6/8.7) we obtain

$$y(n) = y(n-1) + \Delta t x(n) \qquad (8.25)$$

which is a simple Newton integration where the signal between $x(n-1)$ and $x(n)$ is approximated by the constant value $x(n)$. These filters are only presented here for illustrative purposes and should not be used in practice because of the rather large numerical errors introduced by the simple approaches represented by Eqs. (8.24/8.25). In practical applications, one should use more accurate filters with a larger number of filter coefficients, see for instance Pintelon and Schoukens [8].

Differentiation and integration is one of the areas where the FFT filtering is of great value. From the differentiation property of the Fourier transform given by Eq. (4.39), it follows directly that the derivative $\dot{y}(t)$ of a signal $y(t)$ is given by

$$\dot{y}(t) \leftrightarrow Y(\omega)i\omega \qquad (8.26)$$

where $y(t) \leftrightarrow Y(\omega)$ is a Fourier transform pair. Thus differentiation in the time domain corresponds to multiplying $Y(\omega)$ by $i\omega$ in the frequency domain. Differentiation can be accurately performed by FFT filtering using the FRF[14]

$$H(\omega) = i\omega \qquad (8.27)$$

Similarly, integration in the time domain corresponds to division in the frequency domain by $i\omega$. However, in this case we cannot let the division by $i\omega$ extend toward DC because the pole that the FRF has at DC will amplify the signal in the low frequency band. The response of DC is normally not of any practical interest in OMA and normally we are not interested in the response below the lowest mode. Furthermore, the signal close to DC is often governed by noise, so using the integration filter all the way to DC will just amplify this noise. Therefore, the integration FRF should rather be approximated by a high-pass filter around DC, see Figure 8.7.

Normally a high-pass filter will have an amplification of unity in the high frequency range. However, the FRF of any low-pass filter can be multiplied by a constant a to adjust the cross-over frequency ω_1 between the low-pass filter and the integration filter, see Figure 8.7. If the FRF of the low-pass filter is $H_{lp}(\omega)$, then we define the scaled function $(a/i)W(\omega)$ to replace the integration filter close to DC, where $W(\omega)$ is the amplification function of the low-pass filter $W(\omega) = |H_{lp}(\omega)|$, and the integration filter is then implemented with the following FRF

$$H(\omega) = \begin{cases} 1/i\omega & ; \text{ for } \omega \geq \omega_1 \\ (a/i)\, W(\omega); & \text{ for } \omega < \omega_1 \end{cases} \qquad (8.28)$$

---

[13] See Section 12.8.2.

[14] Example 4.1 in the transforms chapter gives a detailed description of the implementation of differentiation in the frequency domain.

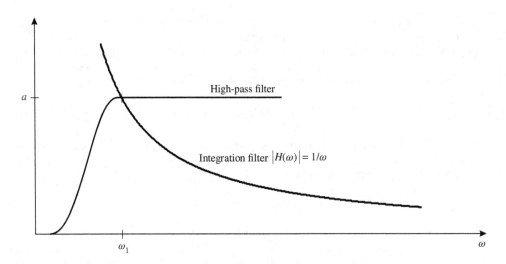

Figure 8.7 Integration in the time domain corresponds to division in the frequency domain by $i\omega$. However for frequencies close to DC the pole at DC must be removed and replaced by a high-pass filter

At the cross-over frequency ω_1 the constant a is determined from

$$(a/i)W(\omega_1) = 1/i\omega_1 \Rightarrow$$

$$a = \frac{1}{\omega_1 W(\omega_1)} \tag{8.29}$$

This equation can be used to determine the cross over frequency ω_1 for a selected high pass filter without the need to change the filter cut-off frequency. But an even simpler solution is to use a straight line in the frequency range between DC and the integration filter.

As a main rule, whenever differentiation and integration is performed, it should be considered to combine the differentiation and integration procedures with some kind of additional filtering to remove excessive noise at the boundaries of the Nyquist band. Differentiation filters often produce high frequent noise. Therefore, a low-pass filtering should be considered whenever differential filtering will be performed. Similarly, integration filters often produce low frequent noise[15] because of the amplification of noise close to DC as explained earlier. Thus, a high-pass filtering should be considered whenever integration will be performed.

8.3.7 The OMA Filtering Principles

There are two main principles in OMA that we should follow all the time. The first one states that if we perform the same filtering on all channels, then modal parameters are not affected. This principle has been formally proven by Ibrahim et al. [9].

The principle is the basis of the concept of a "loading filter" that creates the unknown forces acting on the structure as explained in the introduction chapter, and allows us to deal with colored force input.

The second principle states that it is in fact essential that whenever filtering is used, the same filtering must be applied to all channels of data. If this is not the case, then normally modal parameters will be affected.

---

[15] The suppression of low frequency noise has been taken care of when performing integration according to Eq. (8.28).

The two principles can be summarized by the following rule that must be respected in all OMA:

The same filtering must be applied to all channels of data

This rule will ensure that we do not disturb mode shapes characteristics by our signal processing. Sometimes our signal processing might introduce amplitude errors and phase errors as explained earlier, but this does not matter much as long as we respect this main rule. This main rule must be respected over all data sets in an OMA test.

A practical implication of this rule is that if we use the same kind of sensors in an OMA test, then even though phase distortion and amplitude distortion is present in the sensors, this will not affect the OMA results even though we do not correct for these errors as long as the distortion is the same in all sensors. It should be noted that distortion is never exactly the same from sensor to sensor, even for sensors of the same type. Therefore, it is always better to calibrate the sensors and correct for the individual distortion in each sensor. If several different sensor types are used, the distortion removal is a must, otherwise the different distortions might significantly affect the OMA results – especially for the mode shapes.

Example 8.1 Band pass filter, FIR, IIR, and FFT

This example considers the simple case of a signal that consists of three harmonics. The signal is band pass filtered to isolate the harmonic that is lying in between the two other harmonics. The frequencies of the three harmonics are $f_1 = 0.5825\,\text{Hz}$, $f_2 = 0.6472\,\text{Hz}$, and $f_3 = 0.7119\,\text{Hz}$ and the signal is generated with 8192 data points using a sample rate of $f_s = 2f_v = 2.5887\,\text{Hz}$ as

$$y(t) = \cos(2\pi f_1 t + \phi_1) + \cos(2\pi f_2 t + \phi_2) + \cos(2\pi f_2 t + \phi_2) \tag{8.30}$$

where the phases ϕ_1, ϕ_2, ϕ_3 are random numbers between 0 and 2π. The resulting auto spectral density for the signal is shown in Figure 8.8.

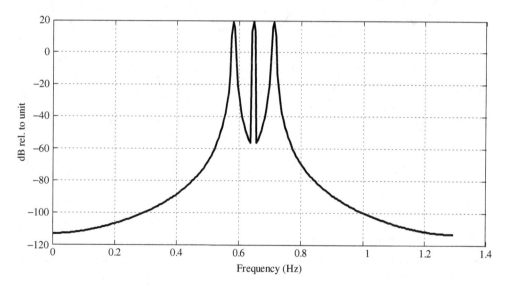

Figure 8.8 Auto spectral density of a signal consisting of three harmonics with frequencies at $f_1 = 0.5825\,\text{Hz}$, $f_2 = 0.6472\,\text{Hz}$, and $f_3 = 0.7119\,\text{Hz}$. Note that leakage is small on the mid harmonic and larger on the side harmonics

Three band-pass filters were designed, an FIR filter, an IIR filer and an FFT filter. All filters were designed with a center frequency that was 1.5% higher than f_2 (in order to reflect the fact that the exact frequency of a harmonic is never known in practice), and with a width of the pass band equal to 2% of the sampling frequency f_s, this defines the pass band from 0.6310 to 0.6828 Hz. The middle harmonic is clearly inside the pass band.

The FIR band pass filter was designed using the fdesign.m function in the Matlab signal processing toolbox with the "equiripple" option, 1 dB error in the pass band and −60 dB errors in the stop band. This defines an FIR filter with 199 moving average filter coefficients. The IIR filter was also designed using the fdesign.m function defining a Butterworth filter with 10 poles. The FFT filter was defined using a data segment size of 1024 data points, and the band pass window in the frequency domain was flat in the pass band and a Hanning tapering to each side was used over a band with a size equal to the half width of the pass band.

The results of the filtering are shown in Figure 8.9 where all signals are up-sampled by a factor 8 in order to give a clear plot. As it appears, all three filters manage to isolate the middle harmonic, but both the FIR and IIR filter introduce large phase errors (delay) on the result. This is no problem in practice because the filters can be applied forward and backward to completely remove the phase error. Amplitude errors of 2–3% are introduced by both digital filters due to the filter ripples and the fact the harmonic is not completely centering in the pass band. The FFT filter has no significant phase or amplitude errors.

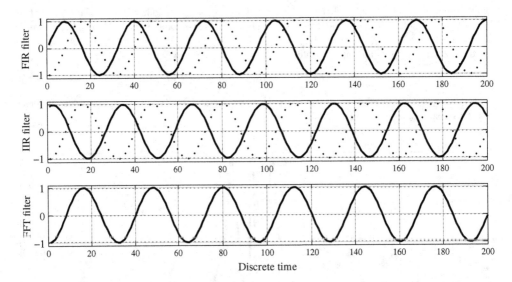

Figure 8.9 Result of band pass filtering to isolate the middle harmonic. The true harmonic is shown by the dotted line and filtering results are shown by the solid line

8.4 Correlation Function Estimation

Two of basic assumptions in OMA are related to correlation functions. One main assumption is that the correlation extracts all (or nearly all) the information about the modal characteristics from the random signal,[16] another one is that the correlation functions can be considered as free decays of the system.[17]

---

[16] See the discussion about this issue in Chapter 2, especially the Sections 2.2.1 and 2.3.2.
[17] See Chapter 6 for details concerning this assumption.

Therefore, estimation of correlation functions is central in OMA. We shall in the following consider the most important ways of estimating these key functions.

8.4.1 Direct Estimation

The simplest way of estimating unbiased correlation functions is to calculate them using Eq. (2.36). For a vector time series $\mathbf{y}(t)$ the unbiased correlation function (CF) matrix can be estimated using the formula

$$\widehat{\mathbf{R}}(\tau) = \frac{1}{T} \int_0^T \mathbf{y}(t)\mathbf{y}^T(t+\tau)dt \tag{8.31}$$

However, if the signal is only known in a certain time interval $t \in [0, T]$, where T is the total length of the time series, we cannot perform the integration as prescribed by Eq. (8.31) because then the integrant $t + \tau$ will exceed the interval whenever $\tau \neq 0$. The solution to this problem is to restrict the time lag to be a positive valued, and then reduce the integration interval to a maximum possible size, that is, $T - \tau$ so that for positive time lag the CF matrix is estimated by

$$\widehat{\mathbf{R}}(\tau) = \frac{1}{T-\tau} \int_0^{T-\tau} \mathbf{y}(t)\mathbf{y}^T(t+\tau)dt \tag{8.32}$$

For negative time lags the CF matrix can be estimated by the symmetry relation given by Eq. (2.35)

$$\widehat{\mathbf{R}}(-\tau) = \widehat{\mathbf{R}}^T(\tau) \tag{8.33}$$

For discrete time the integral in Eq. (8.32) becomes a summation and the infinitesimal time integration segment dt becomes the sampling time step Δt

$$\begin{aligned}\widehat{\mathbf{R}}(k) &= \frac{1}{(N-k)\Delta t} \sum_{n=1}^{N-k} \mathbf{y}(n)\mathbf{y}^T(n+k)\Delta t \\ &= \frac{1}{N-k} \sum_{n=1}^{N-k} \mathbf{y}(n)\mathbf{y}^T(n+k)\end{aligned} \tag{8.34}$$

where N is the total number of data points in the time series, k corresponds to the time lag $\tau = k\Delta t$ and $(N-k)\Delta t$ corresponds to $T - \tau$ in Eq. (8.32). We here using a discrete time formulation so that $\mathbf{y}(n)$ means $\mathbf{y}(n\Delta t)$. If we arrange the measured responses $\mathbf{y}(n)$ as columns in a data series matrix $\mathbf{Y}$, we can write

$$\mathbf{Y} = [\mathbf{y}(1), \mathbf{y}(2), \ldots] \tag{8.35}$$

then the direct correlation function matrix estimate given by Eq. (8.34) can be obtained by the matrix product

$$\widehat{\mathbf{R}}(k) = \frac{1}{N-k}\mathbf{Y}_{(1:N-k)}\mathbf{Y}^T_{(k+1:N)} \tag{8.36}$$

where $\mathbf{Y}_{(1:N-k)}$ results from removing k columns from the right of $\mathbf{Y}$ and $\mathbf{Y}_{(k+1:N)}$ results from removing k columns from the left of $\mathbf{Y}$.

The idea of performing correlation matrix estimation by matrix multiplications like in Eq. (8.36) can be further developed by gathering a family of matrices that are created by shifting the data matrix into one single block matrix like

$$
\mathbf{H} = \begin{bmatrix} \mathbf{Y}_{(1:N-2s)} \\ \mathbf{Y}_{(2:N-2s+1)} \\ \vdots \\ \mathbf{Y}_{(2s:N)} \end{bmatrix}
\tag{8.37}
$$

This matrix is a block Hankel matrix because it is constant along its anti-diagonals. It should be noted that it is this matrix that in the stochastic subspace technique is split in an upper and lower parts of equal size

$$
\mathbf{H} = \begin{bmatrix} \mathbf{Y}_{hp} \\ \mathbf{Y}_{hf} \end{bmatrix}
\tag{8.38}
$$

We can then calculate the matrix $\mathbf{HH}^T/C$ where C is the number of block columns in $\mathbf{H}$. From Eq. (8.37) $C = N - 2s$ and we see from Eq. (8.36) the resulting matrix contains the unbiased CF matrices

$$
\frac{1}{C}\mathbf{HH}^T = \frac{1}{C} \begin{bmatrix} \mathbf{Y}_{(1:N-2s)} \\ \mathbf{Y}_{(2:N-2s+1)} \\ \vdots \\ \mathbf{Y}_{(2s:N)} \end{bmatrix} \begin{bmatrix} \mathbf{Y}^T_{(1:N-2s)} & \mathbf{Y}^T_{(2:N-2s+1)} & \cdots & \mathbf{Y}^T_{(2s:N)} \end{bmatrix}
$$

$$
= \begin{bmatrix} \hat{\mathbf{R}}(0) & \hat{\mathbf{R}}(1) & \cdots & \hat{\mathbf{R}}(2s) \\ \hat{\mathbf{R}}(-1) & \hat{\mathbf{R}}(0) & \ddots & \vdots \\ \vdots & \ddots & \ddots & \hat{\mathbf{R}}(1) \\ \hat{\mathbf{R}}(-2s) & \cdots & \hat{\mathbf{R}}(-1) & \hat{\mathbf{R}}(0) \end{bmatrix}
\tag{8.39}
$$

We see that the resulting matrix is a block Toeplitz matrix because it is constant along its block diagonals.[18]

For cases with large amounts of data where the multiplication $\mathbf{HH}^T$ might lead to memory overflow, the Hankel matrix can be segmented like

$$
\mathbf{H} = [\mathbf{H}_1, \ \mathbf{H}_2, \ \cdots]
\tag{8.40}
$$

and the needed computer memory can be reduced by calculating the Toeplitz matrix $\mathbf{HH}^T$ as the summation

$$
\mathbf{HH}^T = \sum_n \mathbf{H}_n \mathbf{H}_n^T
\tag{8.41}
$$

With today's computers one can easily handle these computations. These ways of estimating CF information are preferable and should be considered whenever possible because of their ability to perform the unbiased estimation by simple means. However, in some cases where minimization of calculation time is essential then other estimation techniques might be considered.

---

[18] Block Hankel and block Toeplitz matrices are introduced in Section 3.2.5.

8.4.2   Biased Welch Estimate

One of the most popular ways of estimating correlation functions is to use the Welch method. This method uses data segmenting as described in Section 8.1.3 and subsequent Fourier transform of each data segment. The Welch method is based on Eq. (8.31) implemented under the assumption that all individual data segments are periodic.[19] If this is indeed the case, then we are no longer limited by the size of the data segment.

Under this assumption and generalizing the ideas in Section 2.2.4 to the case of vector signals, we arrive at the following equation that is a similar to Eq. (2.47) and expresses the convolution property of the CF matrix

$$\widehat{\mathbf{R}}(\tau) = \frac{1}{T} \int\limits_0^T y(t) y^T(t + \tau) dt$$

$$= y(-t) * y^T(t) \tag{8.42}$$

Similarly to what we did at the end of Section 6.1.2 using the definition of the spectral density given by Eqs. (6.12/6.16) and using the convolution property and the time reversal property of the Fourier transform, we arrive at the following expression for the spectral density (SD) matrix

$$\widehat{\mathbf{G}}(\omega) = \tilde{\mathbf{y}}(\omega)^* \tilde{\mathbf{y}}^T(\omega) \tag{8.43}$$

This is the vector equation corresponding to the scalar equation (6.26), $\tilde{\mathbf{y}}(\omega)$ being the Fourier transform of $\mathbf{y}(t)$. Using this expression for the data segmented in the time domain with the data segments $\mathbf{y}_1(t), \mathbf{y}_2(t), \dots, \mathbf{y}_S(t)$ the SD matrix can be averaged in the frequency domain

$$\widehat{\mathbf{G}}(\omega) = \frac{1}{S} \sum_{s=1}^S \tilde{\mathbf{y}}_s(\omega)^* \tilde{\mathbf{y}}_s^T(\omega) \tag{8.44}$$

where S is the number of data segments. If the segments satisfy the condition given by Eq. (8.4), that is, if the summation of all the data segments exactly constructs the original time series, then the estimate in Eq. (8.44) is in good agreement with the Parseval theorem.[20] However, if the condition given by Eq. (8.4) is not fulfilled, for instance, if a window is multiplied on nonoverlapping data segments, then the signal at the ends of each data segment is not presented in the spectral density, thus the energy of the original signal is not the area under the spectral density as claimed by the Parseval theorem, and therefore a correction factor must be introduced in Eq. (8.44).

See for instance Brandt [4] for the details on how to calculate the aforementioned correction factor and for guidelines on application of suitable windows and overlap. Here it should be remembered that for a given total amount of data, the data segmenting is always a trade-off between random error due to the limited number of averages S in Eq. (8.44) and bias error due to leakage.

The leakage bias is introduced by the wrong assumption of periodicity, and therefore, the smaller the discontinuities at the ends when periodicity is assumed, the smaller the leakage bias. For instance, if a signal is actually periodic as assumed, the bias has vanished; if a signal is close to be periodic; the bias is small; and if the discontinuities are suppressed by application of a window the bias is reduced.

---

[19] This is of course an assumption that is in conflict with our knowledge about random signals because normally such data does not at all fulfill this assumption.

[20] See Section 6.1.2 for information about the Parseval theorem.

Any window that acts as a tapering on the data segments will in average[21] reduce the bias, but cannot completely remove it.

The implication of the bias error introduced in the Welch method is that frequencies carrying high energy, for instance modal peaks in the spectral density or harmonics in the signal, will leak that energy to the surrounding frequencies. This bias is denoted "leakage error." See Brandt [4] for a detailed description of the leakage error and how to minimize it. The same reference also provides an overview of the qualities of commonly used time windows.

We conclude that even though the Hanning window with 50 % overlap is not the optimal solution, it is nevertheless a reasonable choice in all OMA cases. Since the data is partially corrupted by leakage errors due to the Welch method, it does seem reasonable to be concerned about a marginal improvement by, for instance, a more optimal window and an optimal overlap.

It is important to note that because leakage will blunt the modal peaks of the spectral density functions, in OMA the leakage bias will lead to an overestimation of the modal damping.

8.4.3   Unbiased Welch Estimate (Zero Padding)

The reason why the biased estimate introduced in the previous section is popular in practical applications is that estimating the correlation function using the Welch method is significantly faster than using the direct procedures mentioned in Section 8.4.1. However, it is possible to implement the Welch method using zero padding in such a way that the leakage error on the correlation functions vanish completely. The approach has been known since the sixties. In Bendat and Piersol [1] it is called "roundabout FFT" and is the recommended approach for FFT-based correlation function estimation.

The idea is most easily explained for scalar correlation function estimation; however, the generalization to vector signals is straight forward. We are considering the estimation of the correlation function as given by Eq. (2.33), which for limited data is an estimate indicated by the hat symbol

$$\hat{R}_{xy}(\tau) = \frac{1}{T} \int_0^T x(t)y(t+\tau)dt \tag{8.45}$$

Given the two signals $x(t), y(t)$ are only known inside the time interval $t \in [0, T]$, as explained earlier, in order not to exceed the definition set for the time, in practice, we are forced to use the form

$$\hat{R}_{xy}(\tau) = \frac{1}{T-\tau} \int_0^{T-\tau} x(t)y(t+\tau)dt \tag{8.46}$$

This is an unbiased estimate because the expectation of the estimator is equal to the definition given by Eq. (2.32)

$$E[\hat{R}_{xy}(\tau)] = \frac{1}{T-\tau} \int_0^{T-\tau} E[x(t)y(t+\tau)]dt = \frac{1}{T-\tau} \int_0^{T-\tau} R_{xy}(\tau)dt \tag{8.47}$$
$$= R_{xy}(\tau)$$

---

[21] Those segments with frequency components that are close to fulfill the condition of periodicity (a natural number of cycles is equal to the length of the data segment). Application of a window will increase the bias at these frequencies.

In practice, reduction of the upper integration limit can be introduced by zero padding using the following formulation

$$\widehat{R}_{xy}(\tau) = \frac{1}{T - \tau} \int_0^T x_0(t)y_0(t + \tau)dt \tag{8.48}$$

where the signals $x_0(t)$, $y_0(t)$ are the zero padded versions of the original signals $x(t)$, $y(t)$ defined as

$$x_0(t), y_0(t) = \begin{cases} x(t), y(t); & \text{for } t \in [0; \ T] \\ 0, \ 0; & \text{for } t \in \ T; \ 2T] \end{cases} \tag{8.49}$$

This forces the same reduction of the upper integration limit as in Eq. (8.46) because the integrand in Eq. (8.48) is zero as soon as $t + \tau > T$. Extending the data segments to double size defines the maximum possible time shift τ to be equal to T. The zero padding only solves a part of the problem, because in order to have the advantage of calculation speed, we want to perform the calculation using FFT taking advantage of the convolution property given by Eq. (2.48). Therefore, we need to modify Eq. (8.48) by defining

$$\widehat{R}_{0xy}(\tau) = \frac{1}{T} \int_0^T x_0(t)y_0(t + \tau)dt \tag{8.50}$$

The estimate $\widehat{R}_{0xy}(\tau)$ can be directly calculated using the FFT because by following the same ideas as in Section 2.2.4 we have from Eq. (2.48) that the correlation estimate can be calculated as a convolution

$$\widehat{R}_{0xy}(\tau) = x_0(-t) * y_0(t) \tag{8.51}$$

The correlation function estimate can be calculated segmenting the data in the time domain, estimating the corresponding spectral density as

$$\widehat{G}_{0xy}(\omega) = X_0(\omega)^* Y_0(\omega) \tag{8.52}$$

which is averaged in the frequency domain. Finally, we go back to the time domain as explained for the biased Welch method in Section 8.4.2. We see that the unbiased estimate $\widehat{R}_{xy}(\tau)$ defined in Eq. (8.47) and the estimate $\widehat{R}_{0xy}(\tau)$ defined in Eq. (8.49) are related by a triangular window

$$\widehat{R}_{0xy}(\tau) = \frac{T}{T - \tau} \frac{1}{T} \int_0^T x_0(t)y_0(t + \tau)dt = \frac{T}{T - \tau}\widehat{R}_{xy}(\tau) \tag{8.53}$$

The correlation function estimated by zero padding the time series becomes unbiased if the estimate is divided by a triangular time window of the form $w_t(\tau) = (T - \tau)/T$ also called the basic lag window.

One can say that in this procedure the leakage bias, often called "wrap around bias" due to the fact that the FFT algorithm let the integration be carried into the continuation of the data segment assuming periodicity, is removed completely by the zero padding and is replaced by a window bias from the basic lag window. The leakage bias is more or less random as it depends on frequency, whereas the bias from the basic lag window is well defined and therefore can be removed.

For the correlation function estimate, the division by the window approaching zero will amplify the noise, and therefore normally the outermost part of the correlation estimate is discarded before dividing

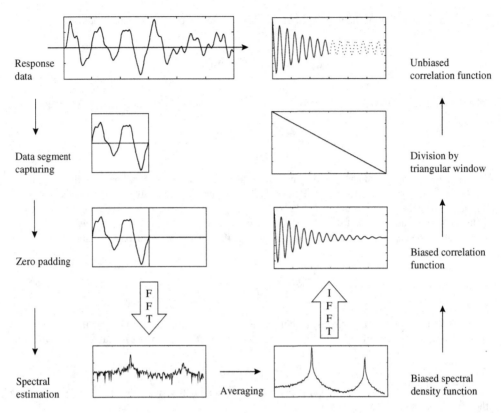

Figure 8.10 Unbiased estimation of correlation functions by the Welch method. After data segments have been defined, they are zero padded to double length and Fourier transformed, the spectral density is calculated according to Eq. (8.52) and smoothed by averaging the spectral density estimates from the different data segments. The resulting spectral estimate is transformed back to time domain by inverse FFT and the bias is removed dividing by the triangular window. The outer most part of the estimate is discarded in order not to amplify the noise by division with the basic lag window where it is approaching zero

by the basic lag window. The whole procedure of zero padding, averaging in the frequency domain, discarding the outermost part and finally dividing by the triangular window to obtain the unbiased Welch estimate of the correlation function is graphically described in Figure 8.10.

8.4.4 Random Decrement

The random decrement (RD) technique is a fast and simple technique for estimation of correlations functions for Gaussian processes by simple time averaging.

The RD technique was developed at NASA in the late sixties and early seventies by Cole [10–12], some years after the FFT technique had become generally accepted. In modal analysis, the technique has also been associated with the Ibrahim time domain identification technique, Ibrahim [13]. The basic idea of the technique is to estimate a so-called RD signature. Given the signals $x(t), y(t)$, the RD signature estimate $\hat{D}_{xy}(\tau)$ is formed by averaging S segments from the signal $x(t)$

$$\hat{D}_{xy}(\tau) = \frac{1}{S}\sum_{s=1}^{S} x(t_s + \tau)|C_{y(t_s)} \tag{8.54}$$

where the signal $y(t)$ at the time steps $t_s = t_1, t_2, \ldots, t_S$ satisfy the triggering condition $C_{y(t_s)}$ and S is the number of triggering points. The triggering condition might be for instance that $y(t_s) = a$ (the level triggering condition) or some similar condition. The algorithm is illustrated in Figure 8.11.

In Eq. (8.54), the cross signature $\hat{D}_{xy}(\tau)$ is estimated from the accumulated average calculation and the triggering condition applied on the two different signals $x(t)$ and $y(t)$. If instead the triggering condition is applied to the same signal $x(t)$ as the data segments are taken from, then the auto signature $\hat{D}_x(\tau)$ is estimated.

The meaning of the RD signature was unclear until it was shown by Vandiver et al. [14] that defining the auto RD signature for the level triggering condition as the conditional expectation

$$D_x(\tau) = E[x(t + \tau)|x(t) = a] \tag{8.55}$$

and then assuming a Gaussian signal, the RD signature is simply proportional to the corresponding correlation function

$$D_x(\tau) = \frac{R_x(\tau)}{\sigma_x^2}a \tag{8.56}$$

where σ_x^2 is the variance of the zero mean Gaussian signal $x(t)$ and a is the triggering level as defined by Eq. (8.56). Vandivers results were generalized by Brincker et al. [15] to the cross signature case. Let $x(t)$, $y(t)$ be stationary, zero mean Gaussian signals. The dependency between the signals is then completely described by the covariance matrix

$$E\left[\left\{ \begin{array}{c} x(t+\tau) \\ y(t) \end{array} \right\} \{x(t+\tau) \quad y(t)\} \right] = \left[\begin{array}{cc} \sigma_x^2 & R_{yx}(\tau) \\ R_{yx}(\tau) & \sigma_y^2 \end{array} \right] \tag{8.57}$$

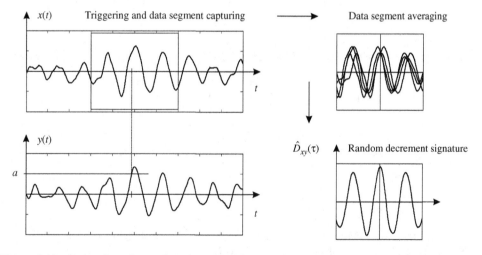

Figure 8.11 Estimation of a random decrement signature by defining a triggering condition (in this case the level triggering condition $y(t_s) = a$) capturing the data segments around the triggering points and averaging them to form the signature

where σ_x^2 and σ_y^2 are the variances of the two signals and where we have used the definition of the correlation function in Eq. (2.31). Using Eqs. (2.56/59) by taking the Gaussian vector $\mathbf{x}$ as the scalar $x(t + \tau)$ and the Gaussian vector $\mathbf{y}$ as the scalar $y(t)$ we get that the signals have zero mean, that is, $\mu_x = \mu_y = 0$

$$D_{xy}(\tau) = E[x(t + \tau)|y(t) = a] = \mu_{x|y} = \mu_x + \mathbf{B}(a - \mu_y)$$
$$= \frac{R_{yx}(\tau)}{\sigma_y^2} a \qquad (8.58)$$

We have used Eq. (2.58) to obtain the matrix $\mathbf{B}$ that in this case is just equal to the scalar $R_{yx}(\tau)/\sigma_y^2$. This result is limited to the level triggering condition. A more general definition of an RD signature is obtained by combining Vandivers' definition of the signature as a conditional expectation with the general triggering condition from Eq. (8.54)

$$D_{xy}(\tau) = E[x(t + \tau)|C_{y(t)}] \qquad (8.59)$$

At this point, it is important to note that the RD technique is unbiased because it follows directly from Eqs. (8.54/8.59) that the expectation of the estimator is equal to the definition of the RD signature

$$E[\hat{D}_{xy}(\tau)] = \frac{1}{S}\sum_{s=1}^{S} E[x(\tau + t_s)|C_{y(t_s)}] = D_{xy}(\tau) \qquad (8.60)$$

The general definition of the RD signature given by Eq. (8.59) can be used to derive a fundamental solution, forming a basis for application of different triggering conditions. Conditioning of $y(t)$ and its derivative $\dot{y}(t)$ is done by defining the Gaussian vector $\mathbf{x}$ as the scalar $x(t + \tau)$ and the Gaussian vector $\mathbf{y}$ as the vector $\mathbf{y} = \{y(t) \ \dot{y}(t)\}^T$. Using Eq. (2.59) and assuming that all signals are zero mean (we also assume that this is true for the derivative signal $\dot{y}(t)$), we have

$$D_{xy}(\tau) = E\left[x(\tau + t)\Big|\mathbf{y}(t) = \mathbf{y}_0 = \{a \ v\}^T\right] = \mu_{x|y} = \mu_x + \mathbf{B}(\mathbf{y}_0 - \mu_y)$$
$$= \mathbf{B}\mathbf{y}_0 \qquad (8.61)$$

The matrix $\mathbf{B}$ is obtained using Eq. (2.58) and by using the condition vector $\mathbf{y}_0 = \{y_0 \ \dot{y}_0\}^T$ we get

$$D_{xy}(\tau) = \mathbf{B}\mathbf{y}_0 = E\left[x(t + \tau)\{y(t) \ \dot{y}(t)\}\right] \left(E\left[\begin{Bmatrix} y(t) \\ \dot{y}(t) \end{Bmatrix}\{y(t) \ \dot{y}(t)\}\right]\right)^{-1} \begin{Bmatrix} y_0 \\ \dot{y}_0 \end{Bmatrix}$$

$$= \{R_{yx}(\tau) \ R_{\dot{y}x}(\tau)\}\begin{bmatrix} \sigma_y^2 & 0 \\ 0 & \sigma_{\dot{y}}^2 \end{bmatrix}^{-1}\begin{Bmatrix} y_0 \\ \dot{y}_0 \end{Bmatrix} = \{R_{yx}(\tau) \ -\dot{R}_{yx}(\tau)\}\begin{bmatrix} 1/\sigma_y^2 & 0 \\ 0 & 1/\sigma_{\dot{y}}^2 \end{bmatrix}\begin{Bmatrix} y_0 \\ \dot{y}_0 \end{Bmatrix}$$

$$= \frac{R_{yx}(\tau)}{\sigma_y^2}y_0 - \frac{\dot{R}_{yx}(\tau)}{\sigma_{\dot{y}}^2}\dot{y}_0 \qquad (8.62)$$

Here we have said that any signal $y(t)$, the signal is uncorrelated with its velocity $\dot{y}(t)$ and that $R_{\dot{y}x}(\tau) = -\dot{R}_{yx}(\tau)$. The first part of the statement is shown in Section 2.2.4. The second part of the statement follows directly from manipulations similar to the ones used for the derivation of the Eqs. (2.42/2.43).

From this fundamental solution, it is possible to explain the meaning of the RD signature for several triggering conditions of practical interest, see Table 8.1. The results of the table follow directly from the fundamental expression in Eq. (8.62).

Table 8.1 Commonly used RD triggering conditions and their effects

Condition name	Triggering condition	Effect on RD estimate
Level crossing	$y(t) = a$	$D_{xy}(\tau) \propto R_{yx}(\tau)$
Level band	$y(t) \in [a_1;\ a_2]$	$D_{xy}(\tau) \propto R_{yx}(\tau)$
Velocity crossing	$\dot{y}(t) = v$	$D_{xy}(\tau) \propto \dot{R}_{yx}(\tau)$
Positive point	$y(t) > a$	$D_{xy}(\tau) \propto R_{yx}(\tau)$
Positive velocity	$\dot{y}(t) > v$	$D_{xy}(\tau) \propto \dot{R}_{yx}(\tau)$

For digital time series, it is normally not possible to fulfill any of the triggering conditions completely. For instance, the level triggering condition $y(t) = a$ maynot be fulfilled, due to the discretization forced onto the signal by the analogue to digital converter. The signal $y(t)$ will either be smaller or larger than the level a. Therefore, at the triggering point the initial condition will in reality have a certain variation that must be observed. In practice, the average value of the initial conditions must be applied instead of the prescribed value.

It is important to note that due to the symmetry properties of the correlation function, see Eq. (2.32), it is possible to estimate both the correlation function and its derivate from the RD signature using a triggering condition on both a signal and its derivative. This is easily proved. Similar to Eq. (8.62), we have that

$$D_{yx}(-\tau) = \frac{R_{yx}(\tau)}{\sigma_x^2}x_0 + \frac{\dot{R}_{yx}(\tau)}{\sigma_{\dot{x}}^2}\dot{x}_0 \tag{8.63}$$

where for the first term we have used the symmetry relation Eq. (2.32), and for the last term we have used that from the derivate of the symmetry relation, $\dot{R}_{xy}(-\tau) = -\dot{R}_{yx}(\tau)$. Assuming that the triggering conditions are scaled equally, that is

$$\begin{aligned} e_0 &= \frac{y_0}{\sigma_y} = \frac{x_0}{\sigma_x} \\ \dot{e}_0 &= \frac{\dot{y}_0}{\sigma_{\dot{y}}} = \frac{\dot{x}_0}{\sigma_{\dot{x}}} \end{aligned} \tag{8.64}$$

then from Eqs. (8.62/8.63) we have

$$\begin{aligned} D_{xy}(\tau) &= \frac{R_{yx}(\tau)}{\sigma_y}e_0 - \frac{\dot{R}_{yx}(\tau)}{\sigma_{\dot{y}}}\dot{e}_0 \\ D_{yx}(-\tau) &= \frac{R_{yx}(\tau)}{\sigma_x}e_0 + \frac{\dot{R}_{yx}(\tau)}{\sigma_{\dot{x}}}\dot{e}_0 \end{aligned} \tag{8.65}$$

The correlation function and its derivative can then be expressed as

$$\begin{aligned} R_{yx}(\tau) &= \frac{\sigma_{\dot{x}}D_{yx}(-\tau) + \sigma_{\dot{y}}D_{xy}(\tau)}{(\sigma_{\dot{x}}/\sigma_x + \sigma_{\dot{y}}/\sigma_y)e_0} \\ \dot{R}_{yx}(\tau) &= \frac{\sigma_x D_{yx}(-\tau) - \sigma_y D_{xy}(\tau)}{(\sigma_x/\sigma_{\dot{x}} + \sigma_y/\sigma_{\dot{y}})\dot{e}_0} \end{aligned} \tag{8.66}$$

The RD signature has from the beginning been interpreted as a free response for the system. This is true only when considering the loading filter introduced in the Chapter 1 as a part of the system. The correlation from the loading filter will color the correlation functions of the response, and consequently will color the resulting RD functions as it appears from Eq. (8.62). Taking this into account, any vector RD signature

$$\mathbf{d}_r(\tau) = E[\mathbf{y}(\tau + t)|C_{y_r(t)}] \tag{8.67}$$

where we have applied the triggering condition $C_{y_r(t)}$ to the rth channel $y_r(t)$, is a free response of the system. We can therefore estimate as many free responses as we have components in the response vector $\mathbf{y}(t)$.

Here it is useful to note the fact that a signal is always uncorrelated to its derivative, and thus if we estimate a vector RD signature for a level triggering condition, and we have no condition on velocity, then if the velocity signal has zero mean, the initial condition of the free response corresponds to velocity zero at the time point where the triggering condition is applied. However, if we sort all data segments that fulfill the triggering condition in two sets, calculating one RD estimate corresponding to positive velocity and one corresponding to negative velocity, then we can obtain more information. This is because the average of the two estimates will form the result corresponding to pooling all date segments in one estimate regardless of velocity. Each of the two estimates, however, includes information about the impulse response due to the initial velocity component.

Following this idea, we see that we always have two independent RD estimates, one corresponding to initial velocity zero and one corresponding to initial value zero. If we arrange the RD vectors $\mathbf{d}_r(\tau)$ in the RD matrix

$$\mathbf{D}(\tau) = [\mathbf{d}_1(\tau), \mathbf{d}_2(\tau), \dots] \tag{8.68}$$

then the two independent estimates can be represented as the symmetrical part $\mathbf{D}(\tau)$ and the unsymmetrical part $\mathbf{D}_u(\tau)$ with the properties

$$\begin{aligned} \mathbf{D}_s(-\tau) &= \mathbf{D}_s^T(\tau) \\ \mathbf{D}_u(-\tau) &= -\mathbf{D}_u^T(\tau) \end{aligned} \tag{8.69}$$

so that

$$\mathbf{D}(\tau) = \mathbf{D}_s(\tau) + \mathbf{D}_u(\tau) \tag{8.70}$$

Combining Eqs. (8.69) and (8.70) we have

$$\begin{aligned} \mathbf{D}_s(\tau) &= (\mathbf{D}(\tau) + \mathbf{D}^T(-\tau))/2 \\ \mathbf{D}_u(\tau) &= (\mathbf{D}(\tau) - \mathbf{D}^T(-\tau))/2 \end{aligned} \tag{8.71}$$

Comparing Eqs. (8.66/8.71) we realize that the symmetrical part of the RD matrix $\mathbf{D}_s(\tau)$ is closely related to the transpose of the correlation function matrix (the difference is just some scaling factors on the columns of the RD matrix). The unsymmetrical part of the RD matrix $\mathbf{D}_u(\tau)$ is closely related to the transpose of the derivative of the correlation function matrix.

It is useful to note that we can apply triggering conditions with initial values at different levels, and obtain free responses corresponding to different amplitude levels. This means that we can use the RD technique to estimate modal parameters for different amplitudes. This is an important consideration when investigating the influence of nonlinearities on the random responses.

Finally it should be noted that Eq. (2.63) representing the conditional correlation matrix for a Gaussian vector can be used to express the uncertainty on the RD estimate. More information about the RD technique including information about uncertainty estimating of the RD signature can be found in Asmussen [16].

8.5   Spectral Density Estimation

The final topic in this chapter is to present the different ways of estimating spectral density functions. This is an important subject because no matter what methods are used for identification it is a common practice to always inspect visually the spectral density functions to see if clear modal peaks are present.

8.5.1   Direct Estimation

The simplest way of estimating spectral density functions is to estimate the corresponding correlation functions by the direct estimation as described in Section 8.4.1 and then take the discrete Fourier transform to estimate the spectral density.

This way of estimating the spectral density is simple and works well under one basic assumption:

That the correlation function has completely decayed to zero inside the considered time window

If this is not the case, a tapering window must be applied in order to force the correlation function to zero, see more about this issue in Section 8.5.5.

8.5.2   Welch Estimation and Leakage

Welch estimation is well known and described in nearly all literature on signal processing, and the basic idea is already described in Section 8.4.2. The reader is referred to the comprehensive literature on the subject. Concerning the influence of leakage and how to make the best trade-off between leakage and random error, the reader is referred to Brandt [4].

Example 8.2   Scaling and units in spectral estimation

It is often a problem to find a proper way of scaling spectral density functions; therefore, in this example, we will discuss this issue in relation to the Matlab preferences.

If we define a random signal in Matlab by generating a sequence of white noise using randn.m then the signal will approximately have a standard deviation equal to unity and so will any part of the signal. In this example, we have scaled the signal by multiplying it by a factor of 10. If we segment the signal as described in section 8.1.3 using the OMA toolbox function fftseg.m then we will have the signal represented by a variable $Y(r,c,s)$ in the frequency domain (one-sided representation) where r is the DOF index (in this case restricted to unity because the signal in one-dimensional), c is the frequency index, and s is the segment index. Implementing the loop:

```
for s=1:ns,
    G1 = G1 + conj(Y(:,:,s)).*Y(:,:,s);
    Ys = [Y(:,:,s); flipud(conj(Y(2:nr-1,:,s)))];
    ys = real(ifft(Ys))';
    V = V + ys*ys'/N;
end
```

where ns is the number of segments we have estimated the one-sided spectral density function G1. We have reestablished the full frequency domain representation Ys for each data segment, and taken this back to time domain by inverse FFT to find the corresponding time representation ys. Finally, we have added the variance estimates from each data segment ys*ys'/N (where N is the number of points in each data segment) to form the total variance V of all segments. Since a Hanning window W has been applied to each segment, the variance must be corrected for the influence of this window and also divided by the number of segments ns. We introduce the correction

$$V = V/(2 * ns * W' * W/N); \qquad (8.72)$$

to make the final estimation of the variance of the segmented signal. We are including the factor 2 because of the 50 % overlap of the segments. However, we can also use the Parseval theorem to calculate the variance from spectral density and we introduce the same correction of the spectral density as

$$G1 = G1/(2 * ns * W' * W); \qquad (8.73)$$

Here we do not include factor N for reasons gives later in this example. If we now flip the one-sided spectral density and take the inverse FFT, then we get the correlation function. According to Parseval, the initial value is equal to the variance. Comparing this value with the value estimated earlier we see that the two values are identical (the difference is of the order of the Matlab floating point precision).

Comparing the two estimates of the variance with the variance calculated for the original signal as y*y'/Np, where Np is the total number of points in the signal we see, however, that the result is not the same. Typical values from the example using

$$Np = 1024 * 16; \quad N = 1024; \qquad (8.74)$$

are 100.3260 and 99.6504 which are of the order that we expect because of the used standard deviation of 10, and thus the variance is then around 100. The first value is estimated as the initial value of the correlation function and the last one estimated as the variance of the total signal. There is no bias in between the two estimates, but a small random difference introduced by the different weighting of the middle of the signal and the beginning and the end, which are weighted differently due to the segmenting. If similar values for the variance are estimated using a constant signal where the influence of the different weighting is disappeared, the two resulting estimates are equal.

The results of the analysis mentioned above on a Gaussian white noise sequence are shown in Figure 8.12 for two different segment sizes. As it appears, when using this approach the spectral density becomes independent of the segment size, and the average value of the spectral density (see top plots of Figure 8.12) is equal to the variance of the considered signal and also equal to the initial value of the correlation function (bottom plots of Figure 8.12). The middle of Figure 8.12 shows the common logarithmic plots relative to the unit of the signal such that a unity standard deviation gives a value of 0 dB, a standard deviation of 10 gives a value of 20 dB, and so on. As it appears from the results shown in Figure 8.12, when a smaller data segment is being used, we have a larger number of averages, and as expected, we see a smaller scatter on the spectral density.

Using Eq. (4.53) but realizing that Matlab has placed the factor $1/N$ on the Fourier series instead of on the Fourier transform, we have for the relationship between the discrete spectral density $G(k)$ and the discrete correlation function $R(n)$ that

$$R(n) = \frac{1}{N} \sum_{k=1}^{N} G(k)e^{i2\pi(k-1)(n-1)/N} \qquad (8.75)$$

By calculating the correlation function at time lag zero, we obtain the discrete Parseval theorem

$$R(1) = \frac{1}{N} \sum_{k=1}^{N} G(k) \qquad (8.76)$$

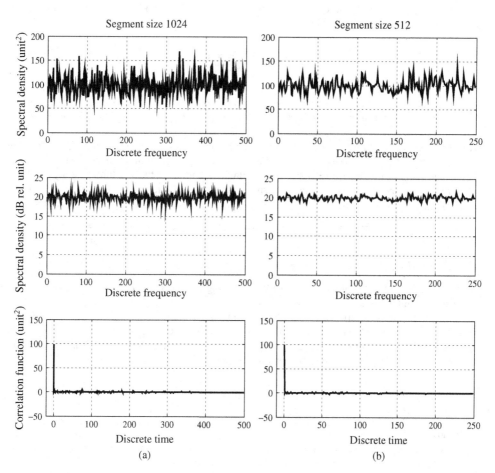

Figure 8.12 Examples of scaling spectral densities using different segment sizes for a white noise sequence with 16384 data points and a variance equal to 100. (a) Results for a segment size of 1024 and (b) for a segment size of 512. Using the guidelines presented in this example the spectral density becomes independent upon the segment size, and the average value of the natural spectral plots (top plots) is equal to the variance of the signal and like the initial value of the correlation function (bottom plots)

which states that the average value of the spectral bins is equal to the correlation function, which in turn is equal to the variance of the signal. This is a good way of scaling the spectral density. Doing it in this way – and this is why we skipped the N in Eq. (8.73) – we have ensured that the values of the spectral density are independent of the number of data point in the data segments, that the spectral values have a simple interpretation, and that the correlation function is the inverse FFT of the double sided spectral density function.

This way of representing the discrete spectral densities will define the units of measure of the spectral density as the same as the units of the correlation function. These are given as unit^2 where "unit" is the unit of measure of the signal. In this case, it is normal to denote the spectral density as a power spectrum.

It is also normal to use discrete spectral densities that are shown as unit^2/Hz, which makes sense in relation to the continuous spectral densities where the Parseval theorem is formulated in terms of an integral. This idea can be introduced from Eq. (8.76) by dividing both sides of the equation by the frequency increment and making use of Eq. (4.48)

$$\frac{1}{\Delta f}R(1) = \frac{1}{2f_v}\sum_{k=1}^{N}G(k) \tag{8.77}$$

We have divided the spectral density bins by the total spectral bandwidth that is equal to twice the Nyquist frequency. It makes sense to plot the spectral density $G(k)/2f_v$ that now has the same units as a continuous spectral density function. In this case, the spectral values are independent upon the number of data points in the data segment, but the spectral values have no simple interpretation. In order to obtain the correlation function, the double sided spectral density must be inversely Fourier transformed and then multiplied by the factor Δf in order to obtain the corresponding correlation function given in Eq. (8.75).

Therefore, it is recommended to stay with the simple definition of the spectral density $G(k)$. If it is needed to obtain spectral values in unit^2/Hz simply scale these values by the frequency increment as $G(k)/\Delta f$.

8.5.3 Random Decrement Estimation

The RD technique offers an alternative to the Welch averaging that should be considered in cases where speed and minimum leakage bias is of importance.

Even though the correlation functions can be estimated by the RD technique using Eq. (8.66), it is often preferable just to estimate a kind of spectral density function corresponding to the actual triggering condition. This can be done by taking the Fourier transform of the RD signature. For the vector RD signature defined in Eq. (8.67) the corresponding spectral densities can be obtained as

$$\tilde{\mathbf{d}}_r(\omega) = \frac{1}{2\pi}\int_{-\infty}^{\infty}\mathbf{d}_r(t)e^{-i\omega t}dt \tag{8.78}$$

The advantage of this approach is that because the RD signature naturally falls off to zero, the integration is limited to the time segment where the RD signature decays without any errors introduced. If the RD signature has decayed totally, the spectral density is leakage free.

However, this only gives us one column of the spectral density matrix. We often want to have a full spectral density matrix, so that singular decomposition can be applied on the spectral matrix. One would think that the full matrix could be obtained by estimating the different columns of the spectral density matrix using Eq. (8.78), but this is not straight forward process because the scaling between the columns is difficult to manage.

When we want to estimate a full spectral density matrix using the RD technique, it is easier to perform Welch averaging using each of the RD vector signatures as individual data segments[22] such that

$$\hat{\mathbf{G}}(\omega) = \frac{1}{R}\sum_{r=1}^{R}\tilde{\mathbf{d}}_r(\omega)^*\tilde{\mathbf{d}}_r^T(\omega) \tag{8.79}$$

where R is the number of RD vector signatures (number of free decays). A basic idea in this approach is to average in the time domain to obtain a decaying time function (the RD signature) so that we can perform the Welch averaging in the frequency domain without leakage error.

However, since the RD functions have noise tails just like the correlation function estimates, the same considerations concerning tapering windows as mentioned in Section 8.5.1 also apply to all Fourier transform of RD signatures. See more about this issue in Section 8.5.5.

---

[22] This approach can be used for any set of free decays.

8.5.4 Half Spectra

It is shown in Chapter 5, that neither the columns nor the rows of the spectral density (SD) matrix are proportional to the mode shapes. This makes it difficult to accurately extract the modal information from the SD matrix.

Therefore, it has become popular to use an alternate version of the SD matrix where the rows (or columns) of the matrix are directly proportional to the mode shapes. From Eqs. (6.90/6.91), we see that we can develop a modified version of the SD matrix by taking the Fourier transform of the positive part of the correlation function (CF) matrix. The modified SD matrix has the form

$$\mathbf{G}_{yy}(\omega) = \sum_{m=1}^{N} \left(\frac{\boldsymbol{\gamma}_m \mathbf{b}_m^T}{i\omega - \lambda_m} + \frac{\boldsymbol{\gamma}_m^* \mathbf{b}_m^H}{i\omega - \lambda_m^*} \right) \tag{8.80}$$

and the transpose of the SD matrix becomes

$$\mathbf{G}_{yy}^T(\omega) = \sum_{m=1}^{N} \left(\frac{\mathbf{b}_m \boldsymbol{\gamma}_m^T}{i\omega - \lambda_m} + \frac{\mathbf{b}_m^* \boldsymbol{\gamma}_m^H}{i\omega - \lambda_m^*} \right) \tag{8.81}$$

We see that each column of this matrix has the same form as a free decay in the frequency domain.

8.5.5 Correlation Tail and Tapering

As mentioned in Sections 8.5.1 and 8.5.3, when Fourier transforming correlations or free decays such as RD functions, it is important to consider if a tapering window should be applied. This issue is discussed in this section.

The reason for the application of a tapering window is to avoid what is called side lobe noise in the spectral density estimate. This kind of noise is avoided if the domain function has decayed to zero inside the considered time interval. However, this is normally not possible to achieve as one would expect just by increasing the size of the time interval, because a noise tail is always present on the correlation function estimate due to the limited information available. The noise tail will prevent the correlation estimate from decaying completely. Or even worse, increasing the time window will increase the random error and consequently just increase the problem of the noise tail. Therefore, before taking Fourier transform of correlation functions and free decays, such as RD functions, it is essential to apply a tapering window.

A classical solution to this problem is to apply an exponential window. See for instance Brandt [4] for more information on this issue. It should be noted that if an exponential window is applied, then this will influence the damping estimate and thus a correction must be included as part of damping estimation. This means that the exponential window will bias the spectral estimate, blunting the spectral peaks.

In OMA we always should consider to minimize all sources of bias, including cases where the bias can be corrected by application of the exponential window. It is worth considering using what we here call the flat-triangular window. This window follows from the considerations on the unbiased Welch estimate in Section 8.4.3. In that section, we showed that by padding zeros to the time domain data segment and using the Welch algorithm, we end up with a correlation estimate that is biased by the triangular window, but the bias can be removed by division with the same window. However, due to the noise tail problem we cannot divide by a triangular close to the boundaries of the time interval where the windows approaches zero, as this will blow up the noise tail. Therefore, as discussed in Section 8.4.3, we should only use the innermost part of the correlation function as shown in Figure 8.10.

However, a more natural solution is simply to use the triangular window for the innermost part of the correlation function, but keep the outermost part as it is. We want to keep the triangular window as it is on the outermost part, because we want to suppress the noise tail and to force the correlation function to be zero inside the considered time interval. Therefore, we can use a basic triangular window for a total time interval of size T,[23] which is equal to zero for times greater than $T/2$, and for the time interval $(0, T/2)$ is equal to

$$w_t(\tau) = \frac{T/2 - \tau}{T/2} \tag{8.82}$$

to define a window of the form

$$w_{tf}(\tau) = \begin{cases} \dfrac{T/2 - \tau}{T/2} & ; \quad \tau \leq \alpha T/2 \\ \dfrac{T/2 - \alpha T/2}{T/2} = (1 - \alpha); & \tau \geq \alpha T/2 \end{cases} \tag{8.83}$$

where α is the fraction of time interval that defines the transition from the lineal decay do the flat part of the window.

Now dividing the correlation function biased by the triangular window by this window we get the following window that is denoted as the flat-triangular window

$$w_{ft}(\tau) = w_t(\tau)/w_{tf}(\tau) = \begin{cases} 1 & ; \quad \tau \leq \alpha T/2 \\ \dfrac{1}{1 - \alpha} \dfrac{T/2 - \tau}{T/2}; & \tau \geq \alpha T/2 \end{cases} \tag{8.84}$$

As indicated above, this window is defined for $0 \leq \tau \leq T/2$, and equal to zero for $T/2 \leq \tau \leq T$.

Example 8.3   Side lobe noise on spectral densities

The effect of application of different tapering on the correlation functions of the first data set of the Heritage Court Tower building test[24] is studied in this example.

The case of no tapering is shown in the left plots of Figure 8.13, the top plot shows the autocorrelation function of the first measurement channel $R_{11}(\tau)$, and the bottom plot shows the corresponding spectral density estimated by direct FFT of the correlation function. The correlation function matrix was estimated using the direct method as given by Eq. (8.36) with a maximum time lag equal to 1024 times the sampling time step.

All spectral plots are in dB relative to the measurement unit show the singular values of spectral density matrix so that the essential information from all spectral functions is shown in one single plot.[25] The frequency band is reduced to include only the first three closely spaced modes of the Heritage court building.

As it appears from the case of no tapering (left side of Figure 8.13), there is significant side lobe noise on the spectral density functions. This is due to the fact that the correlation functions have a clear noise tail as it is seen on the left top plot of Figure 8.13.

A moderate tapering is then introduced by application of the flat-triangular window. In this case, a value of $\alpha = 0.5$ was used and the resulting tapered correlation function is shown in the middle top plot of Figure 8.13. As it appears from the corresponding spectral density below, this moderately reduces the side lobe noise of the spectral density.

---

[23] The triangular window introduced in Section 8.4.3 is defined for a total size of $2T$ because the size is doubled due to the zero padding.
[24] see Appendix B for more information on this case.
[25] See the section on the frequency domain decomposition (FDD) technique in Chapter 10.

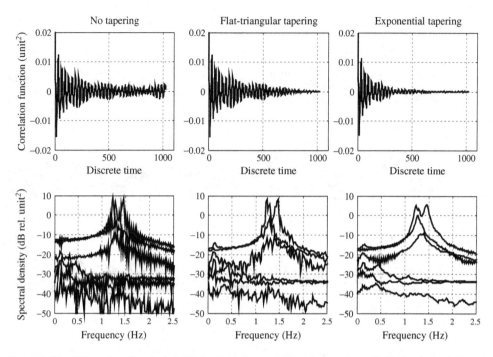

Figure 8.13 Effect of application of different tapering to the first data set from the Heritage Court Tower building. Top plots show the auto correlation function $R_{11}(\tau)$ for the three cases of tapering. From the left: First plot is without any tapering, middle plot is tapered by the flat-triangular window, and the third plot is tapered by an exponential window that is reduced to 0.05% at the boundaries. Bottom plots show the corresponding spectral density plots of the singular values of the spectral matrix plotted to emphasize the first three modes of Heritage Court building. As it appears, the side lobe noise is moderately reduced by the flat-triangular window, and significantly reduced by the exponential window, the latter at the expense of a clear bias introduced at the spectral peaks

Finally, a tapering is introduced by application of a classical exponential window. In this case, an exponential window was used that reduce the initial value to 5% at the boundaries and the resulting tapered correlation function is shown on the top right of Figure 8.13. As it appears from the corresponding spectral density, this significantly reduces the side lobe noise on the spectral density and provides a spectral plot that depicts the spectral peaks more clearly. However, it is also seen from Figure 8.13 that this is at the expense of a clear bias of the spectral peaks.

Example 8.4   Spectral estimation on Heritage Court data. Direct method, Welch, RD and half spectrum

The first data set from the Heritage Court Tower building[26] is analyzed to estimate spectral densities, and the results are shown in Figure 8.14. All plots are in dB relative to the measurement unit and show the

---

[26] See Appendix B for more information on this case.

Singular values of spectral density matrix

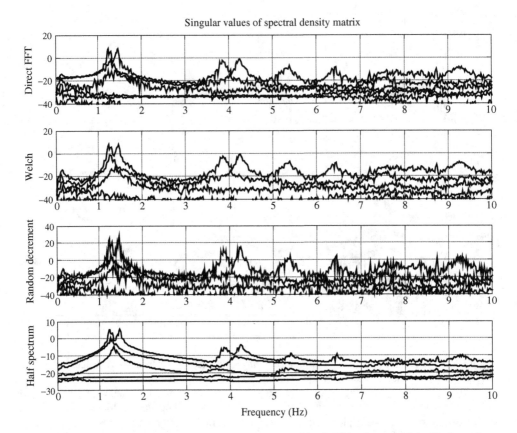

Figure 8.14 Singular values of the spectral density matrix for the first data set of the Heritage Court Tower. All plots are in dB relative to the measurement unit. Top plot shows the Fourier transformed direct estimate of the correlation functions; second from the top is the traditional Welch averaging with a Hanning window and 50% overlap; the third from the top is the RD estimate obtained by Eq. (8.79); and finally the bottom plot is the half spectrum based on the zero padded direct correlation function matrix estimate

singular values of spectral density matrix so that the essential information from all spectral functions is shown in one single plot. All estimations were performed using a maximum time lag of the correlation functions equal to 1024 times the sampling time step.

The top plot of Figure 8.14 shows the spectral density found from the direct estimates of the correlation function matrix given by Eq. (8.36) and then transformed by the FFT. Before the correlation functions were FFT'ed they were multiplied by the flat-triangular window with $\alpha = 0.5$ in order to suppress side lobe noise.

The second plot from the top is the traditional Welch averaging with a Hanning window and 50% overlap, obtained by segmenting as described in Section 8.1.3 and according to Eq. (8.44). The third plot from the top is the random decrement estimate obtained by Eq. (8.79) using band triggering with a symmetric band around zero and a width of $\pm 2\sigma$, where σ is the standard deviation of the triggering signal.

No additional windowing is used here because the natural decay of the RD functions removes the need for windowing.

The bottom plot is the half spectrum derived based in the zero padded direct correlation function estimate as given by Eq. (8.36) and then transformed by the FFT. In this case the correlation functions were multiplied by the flat-triangular window before computing the FFT to suppress side lobe noise.

We see that all four ways of estimating the spectral density basically give us the same information, and we see clearly in all four spectral representations the three closely spaced modes from 1.2 to 1.5 Hz. However, we should note that the scale of the RD spectral estimate is different due to the fact that the initial value of the RD function depends on the triggering condition, and that the half spectrum is smoother and thus gives a more clear picture of the desired physical information than any of the other estimates.

References

[1] Bendat, S.J. and Piersol, A.G.: *Random data. Analysis and measurement procedures.* 2nd edition. New York, John Wiley & Sons, 1986.

[2] Oppenheim, A.V. and Schafer, R.W.: *Discrete-time signal processing.* International edition. New Jersey, Prentice-Hall International, Inc., 1989.

[3] Proakis, G.P. and Manolakis, D.G.: *Digital signal processing. Principles, algoritms and applications.* 4th edition. New Jersey, Pearson Prentice Hall, 1996.

[4] Brandt, A.: *Noise and vibration analysis. Signal analysis and experimental procedures.* Chichester, Wiley, 2011.

[5] Brincker, R., Andersen, P. and Møller, N.: *An indicator for separation of structural and harmonic modes in output-only modal testing.* In Proceedings of IMAC 18 San Antonio, Texas, USA, February 7–10, p. 1649–1654, 2000.

[6] Jacobsen, N.J., Andersen, P. and Brincker, R.: *Eliminating the influence of harmonic components in operational modal analysis.* In proceedings of IMAC-XXIV, 2007.

[7] Andersen, P., Brincker, R., Ventura, C. and Cantieni, R.: *Modal estimation of civil structures subject to ambient and harmonic excitation.* In proceedings of IMAC-XXVI, 2008.

[8] Pintelon, R. and Schoukens, J.: *Real-time integration and differentiation of analog-signals by means of digital filtering. IEEE Trans. Instrum. Meas.,* V. 39. No. 6, p. 923–927, 1990.

[9] Ibrahim, S.R., Brincker, R. and Asmussen, J.C.: *Modal parameter identification from response of general unknown Forces.* In Proceedings of the International Modal Analysis Conference (IMAC), p. 446–452 1996.

[10] Cole, A.H.: *On-the-line analysis of random vibrations.* In AIAA/ASME 9th Structures, Structural and Materials Conference, Pam springs, California, April 1–3, Paper No 68–288, 1968.

[11] Cole, A.H.: *Failure detection of a space shuttle wing by random decrement.* NASA, TMX-62,041, May 1971.

[12] Cole, A.H.: *On-line failure detection and damping measurement of space structures by random decrement signatures.* NASA, CR-2205, March 1973.

[13] Ibrahim, S.R.: Random decrement technique for modal identification of structures. *J. Spacecraft Rockets,* 14(11), p. 696–700, 1977.

[14] Vandiver, J.K., Dunwoody, A.B., Cambell, R.B. and Cook, M.F.: A mathematical basis the random decrement vibration signature analysis technique. *J. Mech. Design,* V. 104, p. 307–313, 1982.

[15] Brincker, R., Krenk, S. Kirkegaard, P.H. and Rytter, A.: Identification of dynamical properties from correlation function estimates. *Bygningsstatiske Med.,* V. 63, No. 1, p. 1–38, 1992.

[16] Asmussen, J.C.: *Modal analysis based on the random decrement technique.* Application to civil engineering structures. Ph.D. Thesis, Department of Building Technology and Structural Engineering, Aalborg University, 1997.

9

Time Domain Identification

"If there are no closely spaced modes and no noise – everything works."

– Sam Ibrahim

In this chapter, we are at the core of OMA – the identification – where the modal parameters are estimated. There are two main tracks in this approach; one is to do the identification in the time domain (TD) – this is the subject for this chapter – the other main approach is to perform the identification in the frequency domain, which is the subject for the following chapter.

Even though this subject is the most central one, we shall not cover everything in detail. The reason is that a lot of material is in fact available on the different techniques; it is fair to say that in OMA, the description of the identification techniques is one of the subjects that are covered best. Further, there are many techniques available and each of them contains a lot of details that we cannot describe fully here. Good examples are the ARMA models that are described in details in, for instance, Ljung [1] and the stochastic subspace technique (SSI) that is covered in the book by Overschee and De Moor [2]. Further, many of the techniques are already well summarized in books such as Maia and Silva [3]. The reader should also be aware that in TD identification all the techniques that are used in classical modal analysis based on impulse response functions can be used in OMA on correlation functions. Therefore, the reader might want to check books on classical modal analysis such as Ewins [4] and He and Fu [5] for techniques that might also be useful in OMA.

In the time domain, we are dealing with free decays – in OMA normally estimated as correlation functions or similar functions such as random decrement functions. This means that all the modes that are present in the signal are present at any time during the considered free decay. Therefore, in the time domain, the identification problem has full rank so to speak because all modes contribute to the rank of the problem at any time. This is a drawback of the time domain identification. The advantage of the time domain identification is that it is more easy to obtain bias-free data (minimum bias on the free decays used in the identification) than it is in the frequency domain.

In this chapter, we will follow the tradition to refer to the input time functions as "free decays" and normally not refer to the input as correlation functions, even though in OMA normally the correlation functions are used as the free decays.

The aim of this chapter is to give a condensed overview of the most important concepts for each technique and illustrate how they work and that they in fact work. To this end, we will illustrate the techniques by estimating the first three modes of the Heritage Court Tower (HCT) building case, see Appendix B for an introduction to this case. It is not the intention of the presentation to present optimized identification results, thus we shall not try to come close to ideal conditions concerning model size or data

Introduction to Operational Modal Analysis, First Edition. Rune Brincker and Carlos Ventura.
© 2015 John Wiley & Sons, Ltd. Published 2015 by John Wiley & Sons, Ltd.
Companion Website: www.wiley.com/go/brincker

size, and we will not try to compare the results from the different techniques. Concerning the details of the technique, the reader is referred to the vast literature on the subject and we will refer to this literature as we move on through this chapter.

We will only present some of the most well-known techniques that have proven their value in practical OMA applications, thus many techniques are left out of the treatment in this chapter. Some of them because they are based on single input, which is known not to work well in OMA, or some are left out just because they have not been used much in practical OMA applications.

Other techniques are not explicitly covered in the following because they are closely related to the techniques presented herein. This is the case for the identification technique known as NeXT by James, Carne and coworkers [6–8], which is based on correlation functions. ERA has played an important role in the development of the OMA ideas as it was one of the first attempts to justify the application of correlation functions as free decays, James et al. [8]. Also the work by Zhang and coworkers [9, 10] is not explicitly presented even though it has played an important role in formulating a common theory of several of the techniques described in this chapter, such as the poly reference technique and the Ibrahim time domain technique.

As pointed out by the colophon of this chapter, if there are no closely spaced modes and no noise present in a specific case – then all the techniques work. Therefore, the example here – the first three modes of the HCT building – is carefully selected to show that we are dealing with a really difficult case. Any technique that can handle this case reasonably well is a good one. As it will be seen in the coming examples, all the techniques that are presented here can handle the case and identify the three closely spaced modes.

Concerning a more detailed overview of the different techniques and the vast literature on the subject, the reader is referred to the overview papers by Zhang et al. [11] and Masjedian and Keshmiri [12] and in general to the IMAC, ISMA, and IOMAC conference proceedings.

9.1   Common Challenges in Time Domain Identification

The idea in OMA is to obtain information physically-related to the structure from the correlation functions and spectral densities, thus in time domain identification this means to extract this physical information from the correlation functions. Normally this is done using some kind of fitting technique based on parametric model that uses regression as described in Chapter 3.

The major difference between the techniques is that they use different ways of formulating the regression problem. However, since not all of the techniques presented herein are formulated for direct application on the correlation functions, and since in OMA the correlation functions is our source of information, we shall start this chapter by looking at how the modal results of any technique can be fitted to the estimated correlation functions. The result of this step is to obtain the important information of the participation of each of the identified modes.

Further, since noise modes are present, ways of knowing which modes are physically-related to the structure and which are not becomes an important issue, and we shall briefly touch upon this subject before moving onto the details of the identification using the different techniques. For illustrations of how to use stabilization diagrams to deal with noise modes, the reader is referred to Chapter 11.

9.1.1   Fitting the Correlation Functions (Modal Participation)

Normally what we get from an identification technique is a set of estimated eigenvalues (poles) and mode shapes. However, in order to understand and illustrate the quality of the identification it is useful to perform a fit of the modal model to the correlation functions used for the identification process.

To accomplish this we use an identified set of poles arranged in the diagonal matrix $[\lambda_n]$ and the corresponding set of mode shapes arranged as columns in the matrix $\mathbf{A} = [\mathbf{a}_n]$.[1] Assuming the mode shape matrix $\mathbf{A}$ contains the complex conjugate pairs of the mode shapes and if in Eq. (6.92) we extent the summation to the set of $2N$ modal parameters, we can sample the CF matrix at the discrete time lags $\tau = k\Delta t$ to give

$$\mathbf{R}_y(k) = 2\pi \sum_{n=1}^{2N} \gamma_n \mathbf{a}_n^T e^{-\lambda_n k \Delta t} \tag{9.1}$$

This is a series of outer products that can be expressed by the matrix equation

$$\mathbf{R}_y(k) = 2\pi \mathbf{\Gamma} [\mu_n]^k \mathbf{A}^T \tag{9.2}$$

where the diagonal matrix $[\mu_n]$ contains the discrete time poles $\mu_n = e^{-\lambda_n \Delta t}$ and the matrix $\mathbf{\Gamma}$ contains the modal participation vectors $\mathbf{\Gamma} = [\gamma_n]$; both matrices are holding a set of complex conjugate pairs for each mode. Now, arranging the CF matrix in the single block row Hankel matrix

$$\mathbf{H} = [\mathbf{R}_y(0), \mathbf{R}_y(1), \mathbf{R}_y(2), \ldots] \tag{9.3}$$

and a similar arrangement for the modal information in the matrix is

$$\mathbf{M} = \left[[\mu_n]^0 \mathbf{A}^T, [\mu_n]^1 \mathbf{A}^T, [\mu_n]^2 \mathbf{A}^T, \ldots \right] \tag{9.4}$$

then Eq. (9.2) can be written as

$$\mathbf{H} = 2\pi \mathbf{\Gamma M} \tag{9.5}$$

This equation can be solved by regression, or by SVD as explained in chapter 3, to give an estimate of the modal participation vectors

$$\hat{\mathbf{\Gamma}} = \frac{1}{2\pi} \mathbf{H} \mathbf{M}^+ \tag{9.6}$$

And the resulting fit of the CF matrix is

$$\hat{\mathbf{R}}_y(k) = 2\pi \hat{\mathbf{\Gamma}} [\mu_n]^k \mathbf{A}^T \tag{9.7}$$

The participation vector can be used to determine a participation factor for each mode. This is important in practical identification because noise modes normally have a low modal participation factor, thus a high participation factor indicates that the corresponding mode might have physical importance. By taking one of the modal participation vectors $\hat{\gamma}_n$ in $\hat{\mathbf{\Gamma}}$ an absolute scalar measure p_n of the modal participation can be calculated as

$$p_n = \sqrt{\hat{\gamma}_n^H \hat{\gamma}_n} \tag{9.8}$$

and a relative modal participation factor can be calculated as

$$\pi_n = \frac{p_n^2}{\mathbf{p}^T \mathbf{p}} \tag{9.9}$$

where the column vector $\mathbf{p}$ contains the absolute modal participation factors as defined by Eq. (9.8) $\mathbf{p}^T = \{p_1, p_2, \ldots\}$. It follows from the definition that the factors π_n add up to 100% over all the considered modes.

---

[1] In this book, we are following the terminology that mode shapes from a model are denoted $\mathbf{B} = [\mathbf{b}_n]$ whereas mode shapes from a test are denoted $\mathbf{A} = [\mathbf{a}_n]$.

9.1.2   Seeking the Best Conditions (Stabilization Diagrams)

As it is well known – and as it appears from the following examples in the chapter – in all parametric modal estimation, noise modes are nearly always present and complicate the decision on which modes should be considered as the related to the structure (i.e., physical modes) ones.

Further, it is well known that noise modes influence the physical modes and that there is an optimal choice of model order where this influence is minimal. This means that we would like to search for this model because we would expect that this model provides the best modal estimate.

It is important to notice that similar considerations can be made on any parameter that can be varied in the estimation procedure, for instance, the amount of data used in the estimation, if filtering is used, the filtering parameters, and so on.

In the examples, we shall illustrate this problem, but not seek any optimization. The problem is further discussed in Chapters 11 and 12.

9.2   AR Models and Poly Reference (PR)

One of the simplest ways to perform OMA is to use AR models on the free decays by using the concept outlined in Section 5.3.5. These models are described in nearly any material concerning sampled time varying signals. Many of the initial references on this subject are from statistics, electrical engineering, and economics. Good references in this category include Harvey [13], Pandit and Wu [14], Norton [15], and Ljung [1].

The PR technique by Vold et al. [16, 17] is a technique that follows the same approach as the technique presented here. The main difference is that Vold uses impulse response functions, but we use correlation functions. In the following sections, we will refer to the technique presented here as AR/PR identification.

Using an AR model-based identification approach, the first step is to find the AR matrices in the homogenous (free decay) part of Eq. (5.203)

$$\mathbf{y}(n) - \mathbf{A}_1\mathbf{y}(n-1) - \mathbf{A}_2\mathbf{y}(n-2) - \cdots - \mathbf{A}_{na}\mathbf{y}(n-na) = 0 \tag{9.10}$$

given the free decay $\mathbf{y}(n)$; $n = 1, 2, \cdots np$ with np number of samples. To estimate the AR matrices, which in poly reference is denoted the polynomial coefficient matrices, we form a block Hankel matrix as described in earlier chapters with na block rows

$$\mathbf{H}_1 = \begin{bmatrix} \mathbf{y}(1) & \mathbf{y}(2) & \cdots & \mathbf{y}(np-na) \\ \mathbf{y}(2) & \mathbf{y}(3) & & \mathbf{y}(np-(na-1)) \\ \vdots & \vdots & \ddots & \vdots \\ \mathbf{y}(na) & \mathbf{y}(na+1) & & \mathbf{y}(np-1) \end{bmatrix} \tag{9.11}$$

and one block Hankel matrix with only a single block row

$$\mathbf{H}_2 = \begin{bmatrix} \mathbf{y}(na+1) & \mathbf{y}(na+2) & \cdots & \mathbf{y}(np) \end{bmatrix} \tag{9.12}$$

Equation (9.10) can then be formulated for all the possible $np - na$ values of n as

$$\mathbf{H}_2 - \mathbf{A}\mathbf{H}_1 = 0 \iff \mathbf{A}\mathbf{H}_1 = \mathbf{H}_2 \tag{9.13}$$

where $\mathbf{A}$ is a side-by-side collection of the AR matrices

$$\mathbf{A} = \begin{bmatrix} \mathbf{A}_{na}, \mathbf{A}_{na-1}, \dots \mathbf{A}_1 \end{bmatrix} \tag{9.14}$$

Equation (9.13), which in poly reference is called the least squares equation for estimation of the matrix polynomial, can be transposed to form an overdetermined problem for determination of the AR matrices and solved by least squares or by the SVD as explained in chapter 3 to obtain an estimate $\hat{\mathbf{A}}$ for the matrix $\mathbf{A}$ containing the AR matrices

$$\mathbf{H}_1^T \mathbf{A}^T = \mathbf{H}_2^T$$
$$\Updownarrow \qquad\qquad\qquad\qquad (9.15)$$
$$\hat{\mathbf{A}} = \left(\mathbf{H}_1^{T+}\mathbf{H}_2^T\right)^T = \mathbf{H}_2\mathbf{H}_1^+$$

It should be noted that the matrices $\mathbf{H}_1$ and $\mathbf{H}_2$ can easily be formed by constructing one block Hankel matrix with $na + 1$ block rows using all free response data and then taking $\mathbf{H}_1$ as the first na block rows from the top and $\mathbf{H}_2$ as the bottom block row.

If the response has nc number of channels, the response vector is $nc \times 1$, and the problem is overdetermined as long as $np - na > na \times nc$. However, since there is always some noise present the procedure will only work satisfactory if the problem is well overdetermined, thus we must require that

$$np - na \gg na \times nc \qquad\qquad (9.16)$$

Further, it should be noted that the procedure still holds in case of a gathering of several responses for a multiple input formulation

$$\mathbf{Y}(n) = \left[\mathbf{y}_1(n), \mathbf{y}_2(n), \ldots, \mathbf{y}_{nr}(n)\right] \qquad\qquad (9.17)$$

where $\mathbf{y}_1(n), \mathbf{y}_2(n), \ldots, \mathbf{y}_{nr}(n)$ is the nr number of responses and where $\mathbf{y}(n)$ is replaced by $\mathbf{Y}(n)$ in Eqs. (9.11–12). However, in this case the condition for the problem to be well overdetermined is

$$(np - na)\, nr \gg na \times nc \qquad\qquad (9.18)$$

The modal parameters are found by forming the companion matrix given by Eq. (5.206) and performing the eigenvalue decomposition (5.210). The modal model has in this case $na \times nc$ eigenvalues corresponding to $na \times nc/2$ modes.[2]

In the paper by Vold and Rocklin [17], the solving technique for the eigenvalue decomposition (here described by simply performing an eigenvalue decomposition of the companion matrix) is discussed in more detail.

Example 9.1   Estimating the first three modes of the HCT building case using AR models

In this example, we will identify the first three modes in the HCT building case. In order to isolate the first three modes, all four data sets were band-pass filtered using the `fftfilt.m` function in the OMA toolbox with a center frequency of 1.35 Hz, a flat band around the center frequency with a bandwidth of 0.4 Hz and roll-off bands on each side with a width of 0.4 Hz.

The number of AR parameters were chosen to be $na = 2$, so the first data set with six channels has $na \times nc/2 = 6$ modes while the other data sets that all have eight channels have $na \times nc/2 = 8$ modes. The CF matrix was estimated using the direct technique, the full CF matrix was transposed as described in Section 6.2.5 and all columns in the transposed CF matrix were then used for the identification using the AR models as described in the preceding section. For the first three data sets, 500 (discrete time lag)

---

[2] The number of eigenvalues $na \times nc$ might not be an odd number, thus the number of modes might not necessarily be a natural number; however, this just means that the "half mode" corresponding to only a single eigenvalue probably does not represent anything of physical interest.

data points were used for the CF matrix. For the last data where the modes are somewhat more difficult to identify, 950 data points were used.

After sorting the modes in ascending order according to frequency and only including the modes with positive damping, the results of the identification is shown in Table 9.1. As it appears, the first data set only has three modes with positive damping, and all three modes have a high modal participation, thus all three modes are chosen to be physical modes. The remaining data sets have some noise modes, but they are easily ruled out by comparing with the results from the first data set and we choose only the modes with the highest modal participation. The result of fitting the CF matrix is shown for the autocorrelation function of the first measurements channel in Figure 9.1.

Table 9.1 Identification results using AR/PR identification on the four data sets of the HCT building case, modes chosen as the physical modes are shown in bold

	Parameter	Mode 1	Mode 2	Mode 3	Mode 4	Mode 5	Mode 6	Mode 7	Mode 8
Data set 1	f_n (Hz)	**1.2271**	**1.2848**	**1.4515**					
	ς_n (%)	**0.96**	**1.19**	**1.11**					
	π_n (%)	**30.14**	**19.24**	**50.61**					
Data set 2	f_n (Hz)	1.2164	**1.2358**	1.2728	**1.2850**	1.4254	**1.4575**		
	ς_n (%)	0.22	**1.41**	0.33	**1.15**	0.41	**1.59**		
	π_n (%)	0.87	**9.82**	0.67	**40.45**	1.88	**46.31**		
Data set 3	f_n (Hz)	1.2203	**1.2381**	**1.2825**	1.3117	1.3834	**1.4556**	1.4768	
	ς_n (%)	0.75	**0.88**	**1.01**	2.36	2.33	**0.72**	0.45	
	π_n (%)	1.13	**4.03**	**6.32**	1.13	0.06	**87.20**	0.12	
Data set 4	f_n (Hz)	1.2120	**1.2457**	1.2814	**1.2985**	1.3683	1.4200	**1.4414**	1.4720
	ς_n (%)	0.77	**1.12**	0.37	**1.57**	1.09	0.17	**0.33**	1.76
	π_n (%)	4.74	**23.38**	0.26	**3.41**	0.38	5.64	**58.634**	3.56

Finally the components mode shapes corresponding to the modes shown in Table 9.1 were merged and the resulting mode shapes are shown in Figure 9.2.

9.3 ARMA Models

Instead of considering free decays – for instance, by first estimating the correlation functions – and using only the homogenous part of the Eq. (5.203), one can skip the need of using a two-stage approach and use the full equation. In this case, we identify the model parameters by directly fitting a model to the measured signal. Some of the first applications of ARMA models in the modal community are described in Zhao-qian and Yang [18], Shi and Stühler [19], Piombo et al. [20], Bonnecase et al. [21], and Giorcelli et al. [22].

For this technique, we will briefly go over the theory and only describe the main ideas. We will do this for two reasons. First, the model is a one-stage model that does not follow the main idea of this book by first extracting the correlation information and then fitting a free decay model to the correlation functions or spectral densities. The second reason is that this kind of models have never been widely used

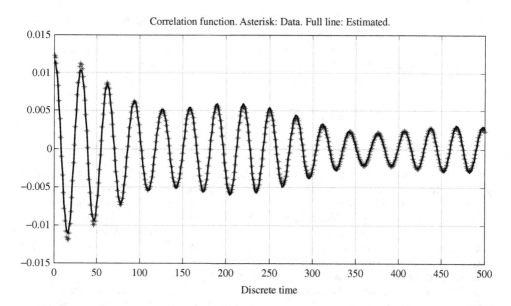

Figure 9.1 Result of using the AR/PR identification technique and fitting the correlation function matrix. The figure shows the autocorrelation function of the first measurement channel of the first data set of the HCT building case

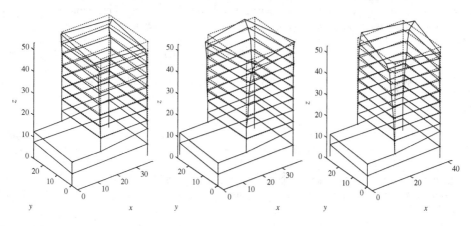

Figure 9.2 First three mode shapes of the HCT building found by AR/PR identification and merging the mode shapes components from the four data sets corresponding to the physical modes shown in Table 9.1

for OMA due to problems with calculation time and convergence of the large number of parameters in the associated nonlinear optimization problem.

The ARMA models are well described in the literature, the general theory which has mainly been developed for applications in electronics and economics is described in books such as [1, 13–15, 23]. Applications of the technique on structural systems go back to the eighties and nineties, and some of them can be found in the references [18–22, 24, 25].

We can explain the ARMA models in terms of (i) how the models relate to the fundamental assumptions in OMA that all physical information is hidden in the correlation functions and can be extracted from those functions, and (ii) how the identification process is implemented in practical applications.

First let us look at how the ARMA model is related to the dynamic models derived in Chapter 5. In Section 5.3.5, we have introduced the ARMA model as

$$\mathbf{y}(n) = \sum_{k=1}^{N} \mathbf{A}_k \mathbf{y}(n-k) + \sum_{k=0}^{M} \mathbf{B}_k \mathbf{x}(n-k) \Longleftrightarrow$$

$$\mathbf{y}(n) - \mathbf{A}_1 \mathbf{y}(n-1) - \cdots - \mathbf{A}_N \mathbf{y}(n-N) = \mathbf{B}_0 \mathbf{x}(n) + \cdots + \mathbf{B}_M \mathbf{x}(n-M)$$

(9.19)

Corresponding to the scalar models introduced in Section 4.4.4, we have seen in Section 5.3.5 how the physical properties, which are carried by the autoregressive matrices $\mathbf{A}_k$ can be extracted by performing an eigenvalue decomposition of the companion matrix given by Eq. (5.210). We have seen how physical properties given by the mass, stiffness, and damping matrices can be turned into a set of autoregressive matrices by forming the companion matrix by Eq. (5.222). By these earlier considerations, we have already proved the connection of the homogeneous part of Eq. (9.19) to the classical dynamic formulation in continuous time given by Eqs. (5.211/5.212).

To complete the explanation, we need to show that the right hand side of the lower part of Eq. (9.19) (the MA part) makes sense in relation to the same equations. The key to understanding the MA part of Eq. (9.19) is to realize that a part of the parameters comes directly from the continuous time dynamic equation and the remaining part comes from the fact that we need to ensure that the solution to Eq. (9.19) is covariance equivalent. That is, it has the same correlation functions as the continuous time models of the random response introduced in Chapter 6. This is somewhat tedious to prove and to do this is not justified here; the reader is referred to Kozin and Natke [23] and to Andersen et al. [24] for the theory of covariance equivalent ARMA models. However, let us briefly indicate the problem following Kozin and Natke [23]. We consider the classical simple 1-DOF oscillator as given in Eq. (5.5)

$$m\ddot{y}(t) + c\dot{y}(t) + ky(t) = x(t)$$

(9.20)

Using the state variable $\{y(t), \dot{y}(t)\}$ as $\{u_1(t), u_2(t)\}$ we can formulate the equation of motion in state space form as

$$\dot{u}_1(t) = u_2(t)$$

$$\dot{u}_2(t) = -\frac{c}{m} u_2(t) - \frac{k}{m} u_1(t) + \frac{1}{m} x(t)$$

(9.21)

The question is now what the appropriate discrete form is for this equation. If we sample the equation at the discrete times $t = k\Delta t; k = 0, 1, 2, \ldots$ and approximate the differential equation by the difference equation

$$u_1(k) - u_1(k-1) = \Delta t u_2(k)$$

$$u_2(k) - u_2(k-1) = -\Delta t \frac{c}{m} u_2(k) - \Delta t \frac{k}{m} u_1(k) + \Delta t \frac{1}{m} x(k)$$

(9.22)

We can then isolate $u_1(k) = y(k)$ in one equation as

$$y(k) - y(k-1) - (y(k-1) - y(k-2)) = -\Delta t \frac{c}{m} (y(k) - y(k-1)) - \Delta t \frac{k}{m} y(k) + \Delta t \frac{1}{m} x(k)$$

$$\Updownarrow$$

$$\left(1 + \Delta t \frac{c}{m} + \Delta t \frac{k}{m}\right) y(k) + \left(-2 - \Delta t \frac{c}{m}\right) y(k-1) - y(k-2) = \Delta t \frac{1}{m} x(k)$$

(9.23)

which is an ARMA model of the form

$$y(k) + a_1 y(k-1) + a_1 y(k-2) = b_0 x(k) \qquad (9.24)$$

We have already shown that the homogenous solution of this equation is equivalent to that of Eq. (9.20) if the autoregressive constants are chosen correctly, but it would be unexpected if the particular solution has the right properties if the rough approximation of Eq. (9.22) is taken into consideration. It can be shown that by adding one more term to the right hand side of Eq. (9.24) and scaling the input such that $e(k) = b_0 x(k)$ the resulting equation

$$y(k) + a_1 y(k-1) + a_2 y(k-2) = e(k) + b_1 e(k-1) \qquad (9.25)$$

will ensure that the correlation function of the solution becomes equivalent to the correlation function of Eq. (9.20), so that the sampled correlation function of Eq. (9.20) for an arbitrary white noise input is equal to the discrete time correlation function of Eq. (9.25). In Kozin and Natke [23], this is shown in two steps. First, the general solution given by Eq. (5.214) to the state space formulation in Eqs. (5.211/5.212) is sampled and it is shown that this results in a form according to Eq. (9.25). The second step is to show that the correlation functions are equivalent. Similarly, it can be shown that the N-DOF system given by Eq. (5.51) is equivalent to the ARMA model

$$\mathbf{y}(k) + \mathbf{A}_1 \mathbf{y}(k-1) + \mathbf{A}_2 \mathbf{y}(k-2) = \mathbf{e}(k) + \mathbf{B}_1 \mathbf{e}(k-1) \qquad (9.26)$$

which is often denoted as an ARMA(2, 1) model because it has two autoregressive matrices and one moving average matrix. It can be shown that adding one more extra term to the right-hand side to define the ARMA(2, 2) model

$$\mathbf{y}(k) + \mathbf{A}_1 \mathbf{y}(k-1) + \mathbf{A}_2 \mathbf{y}(k-2) = \mathbf{e}(k) + \mathbf{B}_1 \mathbf{e}(k-1) + \mathbf{B}_2 \mathbf{e}(k-2) \qquad (9.27)$$

improves the noise modeling properties. For details of the theory of ARMA models and for further information about the background and application of the models defined in Eqs. (9.26/9.27) for identification of structures, the reader is referred to Andersen [25].

The beauty of the ARMA models is the simple idea of estimation where the model is fitted directly to the time series. A predictor $\hat{\mathbf{y}}(k|k-1; \boldsymbol{\theta})$ is defined that predicts the response at time point k based on the parameter vector $\boldsymbol{\theta}$ containing all parameters in the AR and MA matrices and on the measured data up to the time point $k-1$. The idea of ARMA calibration is, therefore, to determine the parameter vector $\boldsymbol{\theta}$ in such a way that the prediction error defined as

$$\boldsymbol{\varepsilon}(k) = \mathbf{y}(k) - \hat{\mathbf{y}}(k|k-1; \boldsymbol{\theta}) \qquad (9.28)$$

is minimized. This is normally done by minimizing the scalar measure $\det(\mathbf{C}_\varepsilon)$ of the prediction error covariance matrix

$$\mathbf{C}_\varepsilon = \frac{1}{np} \sum_{n=1}^{np} \boldsymbol{\varepsilon}(k) \boldsymbol{\varepsilon}^T(k) \qquad (9.29)$$

where np is the number of data points in the time series. In this approach, the correlation functions are indirectly determined at the same time as the modal parameters of the physical system is determined and therefore it might seem that this approach is more sound than the two-stage approach advocated in this book. However, due to the practical problem of making the large multivariate nonlinear optimization problem converge effectively, the approach has never been used much for identification of structures.

9.4   Ibrahim Time Domain (ITD)

The Ibrahim time domain (ITD) is one of the first techniques developed in the modal community for identification of multiple output systems, that is, where advantage is taken of information from several measurements channels at the same time. The technique was introduced by Ibrahim in the late 1970s [26–28]. The idea of this was aimed at extracting OMA information from the random responses by the random decrement technique, Ibrahim [27].

The ITD technique operates with three time delays, Δt_1, Δt_2, and Δt_3; but with a certain (reasonable) choice of parameters we can give the classical ITD the following modern formulation. The background of the ITD is that any free response, thus also the correlation matrix[3] can be formulated as a linear combination of mode shapes and exponential decays

$$\mathbf{y}(t) = \mathbf{y}(k\Delta t) = c_1 \mathbf{a}_1 e^{\lambda_1 k\Delta t} + c_2 \mathbf{a}_2 e^{\lambda_2 k\Delta t} + \cdots = c_1 \mathbf{a}_1 \mu_1^k + c_2 \mathbf{a}_2 \mu_2^k \cdots \tag{9.30}$$

where $\mathbf{a}_1, \mathbf{a}_2, \ldots$ are the mode shapes, $\lambda_1, \lambda_2, \ldots$ are the continuous time poles, $\mu_1, \mu_1, \ldots$ are the discrete time poles, and $c_1, c_2, \ldots$ are the initial modal amplitudes defining the free decay at time zero

$$\mathbf{y}(t = 0) = c_1 \mathbf{a}_1 + c_2 \mathbf{a}_2 + \cdots \tag{9.31}$$

We now form a block Hankel matrix with four block rows and spilt it into two matrices at the middle

$$\mathbf{H} = \begin{bmatrix} \mathbf{y}(1) & \mathbf{y}(2) & \cdots & \mathbf{y}(np-3) \\ \mathbf{y}(2) & \mathbf{y}(3) & \cdots & \mathbf{y}(np-2) \\ \mathbf{y}(3) & \mathbf{y}(4) & \cdots & \mathbf{y}(np-1) \\ \mathbf{y}(4) & \mathbf{y}(5) & \cdots & \mathbf{y}(np) \end{bmatrix} = \begin{bmatrix} \mathbf{H}_1 \\ \mathbf{H}_2 \end{bmatrix} \tag{9.32}$$

In this two smaller block Hankel matrices the top matrix $\mathbf{H}_1$ consists of the two top block rows and the bottom matrix $\mathbf{H}_2$ consists of the two bottom block rows. The free decay has np number of points. Taking the free decay expression of Eq. (9.30) into account we can express $\mathbf{H}_1$ as

$$\mathbf{H}_1 = \mathbf{\Psi}\mathbf{\Lambda} \tag{9.33}$$

where $\mathbf{\Psi}$ is a matrix holding the mode shapes in the columns

$$\mathbf{\Psi} = \begin{bmatrix} \mathbf{a}_1 & \mathbf{a}_2 & \cdots \\ \mu_1 \mathbf{a}_1 & \mu_2 \mathbf{a}_2 & \cdots \end{bmatrix} \tag{9.34}$$

and $\mathbf{\Lambda}$ is a matrix holding the discrete time poles raised to different powers times the corresponding modal amplitudes

$$\mathbf{\Lambda} = \begin{bmatrix} c_1 \mu_1^0 & c_1 \mu_1^1 & \cdots & c_1 \mu_1^{np-3} \\ c_2 \mu_2^0 & c_2 \mu_2^1 & \cdots & c_2 \mu_2^{np-3} \\ \vdots & \vdots & & \vdots \end{bmatrix} \tag{9.35}$$

Similarly, the other block Hankel matrix $\mathbf{H}_2$ is just $\mathbf{H}_1$ delayed two time steps and thus can be expressed as

$$\mathbf{H}_2 = \mathbf{\Psi}[\mu_n]^2 \mathbf{\Lambda} \tag{9.36}$$

where $[\mu_n]$ is a diagonal matrix holding the discrete eigenvalues. From Eqs. (9.33) and (9.36) the matrix $\mathbf{\Lambda}$ can be eliminated to give

$$\mathbf{\Psi}^{-1}\mathbf{H}_1 = [\mu_n]^{-2}\mathbf{\Psi}^{-1}\mathbf{H}_2 \tag{9.37}$$

---

[3] This is only true for the rows of the correlation functions matrix, see the discussion about this part in Chapter 6.

On premultiplying both sides of Eq. (9.37) by $\mathbf{\Psi}\left[\mu_n\right]^2$ we get the equation

$$\mathbf{\Psi}\left[\mu_n\right]^2\mathbf{\Psi}^{-1}\mathbf{H}_1 = \mathbf{H}_2 \qquad (9.38)$$

Finally, defining the system matrix

$$\mathbf{A} = \mathbf{\Psi}\left[\mu_n\right]^2\mathbf{\Psi}^{-1} \qquad (9.39)$$

we have the ITD equation for estimation of the system matrix $\mathbf{A}$

$$\mathbf{A}\mathbf{H}_1 = \mathbf{H}_2 \qquad (9.40)$$

Transposing this equation, it is clear that we have an overdetermined system of equations for determination of the system matrix $\mathbf{A}$ as long as $np - 3$ is larger than $2nc$, where nc is the number of channels in the considered free decay. The overdetermined problem can be solved either by SVD or by regression as described in chapter 3. However, in ITD the traditional way of solving the equation is by regression. Premultiplying both sides of Eq. (9.40) by $\mathbf{H}_1^T$ provides the normally full rank matrix $\mathbf{H}_1\mathbf{H}_1^T$ on the left hand side of Eq. (9.40), which can be inverted to provide the estimate

$$\widehat{\mathbf{A}}_1 = \mathbf{H}_2\mathbf{H}_1^T\left(\mathbf{H}_1\mathbf{H}_1^T\right)^{-1} \qquad (9.41)$$

An alternative solution can be found by premultiplying by $\mathbf{H}_2^T$, which provides the normally full rank matrix $\mathbf{H}_1\mathbf{H}_2^T$ on the left hand side of Eq. (9.31), which can be inverted to provide the alternative estimate

$$\widehat{\mathbf{A}}_2 = \mathbf{H}_2\mathbf{H}_2^T\left(\mathbf{H}_1\mathbf{H}_2^T\right)^{-1} \qquad (9.42)$$

It is well known that both estimates for the $\mathbf{A}$-matrix are weakly biased, and that an average estimate is often used instead and referred to as the unbiased (or double least square) estimate

$$\widehat{\mathbf{A}} = \left(\widehat{\mathbf{A}}_1 + \widehat{\mathbf{A}}_2\right)/2 \qquad (9.43)$$

One of the problems with the classical formulation of the ITD as it was presented initially is that it was formulated for a single input, that is, only one free decay was allowed. However, this limitation can easily be removed by realizing that the matrices in Eq. (9.41)

$$\begin{aligned}\mathbf{T}_{11} &= \mathbf{H}_1\mathbf{H}_1^T \\ \mathbf{T}_{21} &= \mathbf{H}_2\mathbf{H}_1^T\end{aligned} \qquad (9.44)$$

and the matrices in Eq. (9.42)

$$\begin{aligned}\mathbf{T}_{12} &= \mathbf{H}_1\mathbf{H}_2^T \\ \mathbf{T}_{22} &= \mathbf{H}_2\mathbf{H}_2^T\end{aligned} \qquad (9.45)$$

are block Toeplitz matrices basically containing correlation information. This means that contributions from several free decays to these matrices can simply be added. Thus, if nr free decays are available then an MIMO version of ITD can be obtained by calculating the Toeplitz matrices containing information from all free decays

$$\begin{aligned}\mathbf{T}_{11} &= \sum_{r=1}^{nr}\mathbf{H}_{1,r}\mathbf{H}_{1,r}^T; \quad \mathbf{T}_{12} = \sum_{r=1}^{nr}\mathbf{H}_{1,r}\mathbf{H}_{2,r}^T \\ \mathbf{T}_{21} &= \sum_{r=1}^{nr}\mathbf{H}_{2,r}\mathbf{H}_{1,r}^T; \quad \mathbf{T}_{22} = \sum_{r=1}^{nr}\mathbf{H}_{2,r}\mathbf{H}_{2,r}^T\end{aligned} \qquad (9.46)$$

where $\mathbf{H}_{1,r}$ and $\mathbf{H}_{2,r}$ are the block Hankel matrices from the rth free response. Finally, the estimates of the system matrix corresponding to Eqs. (9.32/9.33) can be obtained by

$$\begin{aligned}\widehat{\mathbf{A}}_1 &= \mathbf{T}_{12}\mathbf{T}_{11}^{-1} \\ \widehat{\mathbf{A}}_2 &= \mathbf{T}_{22}\mathbf{T}_{21}^{-1}\end{aligned} \qquad (9.47)$$

The MIMO formulation of the ITD technique can also be obtained by stacking several free decays side-by-side in a response matrix according to Eq. (9.17), and using the response matrix when building the Hankel matrix in Eq. (9.32), see Fukuzono, [29]. The result is the same as adding the Toeplitz matrices for each free decay as per Eq. (9.46).

The final step in ITD is to take the eigenvalue decomposition of the estimated system matrix according to Eq. (9.39). The mode shapes are found from the columns of the eigenvector matrix, and the discrete time poles are found by taking the square root of the diagonal elements of the eigenvalue matrix.

The ITD has as many eigenvalues as the dimension of the system matrix. Since the system matrix is $2nc \times 2nc$, the ITD has $2nc$ eigenvalues or nc number of modes, two eigenvalues for each mode. The solution by regression is only robust when it is well overdetermined, and we must require that

$$np - 3 >> 2nc \tag{9.48}$$

For a single input ITD formulation and for the MIMO formulation with nr free decays we have the following requirement

$$(np - 3)\, nr >> 2nc \tag{9.49}$$

Since the model has a fixed number of modes that is defined by the number of channels in the free decay, it is normal to extend to more modes by adding pseudo measurements to the response vector without adding more physically-related information. This can be done by using some channels of data of the same free decays delayed in time.

Example 9.2   Estimating the first three modes of the HCT building case using ITD

The data was band-pass filtered as described in Example 9.1 and the CF matrix was also used as described in Example 9.1 where all columns of the transposed CF matrix were used as free decays in

Table 9.2 Identification results of using ITD identification on the four data sets of the HCT building case, modes chosen as the physical modes are shown in bold

	Parameter	Mode 1	Mode 2	Mode 3	Mode 4	Mode 5	Mode 6	Mode 7	Mode 8
Data set 1	f_n (Hz)	**1.2270**	**1.2848**	**1.4514**					
	ς_n (%)	**0.97**	**1.20**	**1.11**					
	π_n (%)	**30.20**	**19.29**	**50.52**					
Data set 2	f_n (Hz)	**1.2340**	**1.2851**	1.4262	**1.4575**				
	ς_n (%)	**1.39**	**1.14**	0.28	**1.56**				
	π_n (%)	**11.51**	**33.35**	1.86	**53.28**				
Data set 3	f_n (Hz)	1.2212	**1.2380**	**1.2824**	1.3121	1.3849	**1.4557**	1.4782	
	ς_n (%)	1.25	**0.82**	**0.97**	2.05	2.12	**0.72**	0.11	
	π_n (%)	0.82	**4.05**	**5.89**	1.16	0.06	**87.52**	0.06	
Data set 4	f_n (Hz)	1.2124	**1.2456**	1.2817	**1.2999**	1.3693	**1.4416**	1.4722	
	ς_n (%)	0.74	**1.05**	0.20	**1.40**	0.69	**0.31**	1.63	
	π_n (%)	6.09	**25.98**	0.28	**3.57**	0.61	**34.11**	29.37	

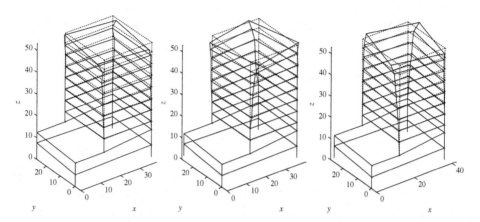

Figure 9.3 First three mode shapes of the HCT building found by ITD identification and merging the mode shapes components from the four data sets corresponding to the physical modes shown in Table 9.2

the identification averaging the Toeplitz matrices given by Eq. (9.46). It should be noted that the ITD formulation has the same number of modes as an AR model with $na = 2$, thus we have the same number of modes as in Example 9.1.

The modes were sorted in ascending order of frequencies including only the modes with positive damping. The results of the identification are shown in Table 9.2. The results are very similar to those of Example 9.1. The result of fitting the CF matrix is not shown here because there is no visible difference between Figure 9.1 and the plot based on the ITD data.

Finally the components of the mode shapes corresponding to the modes shown in Table 9.2 were merged and the resulting mode shapes are shown in Figure 9.3.

9.5 The Eigensystem Realization Algorithm (ERA)

The eigensystem realization algorithm (ERA) introduced in the mid-eighties by Pappa and coworkers [30–32] is related to control theory and thus the terminology of the original theory is somewhat different from the terminology of this book. Therefore, we shall not introduce the whole theory of ERA, but only introduce the main ideas from an OMA point of view.

As derived in Section 5.3.5, any free decay $\mathbf{y}(k)$ in discrete time $t_k = k\Delta t$, where Δt is the sampling time step, can be expressed using a state space formulation as in Eq. (5.217)

$$\mathbf{y}(k) = \mathbf{PD}^k\mathbf{u}_0 \qquad (9.50)$$

where $\mathbf{P}$ is the observation matrix, $\mathbf{D}$ is the discrete time system matrix

$$\mathbf{D} = \exp(\mathbf{A}\Delta t) \qquad (9.51)$$

where $\mathbf{A}$ is the continuous time system matrix in the state space formulation, see Eq. (5.212) and $\mathbf{u}_0$ is the state space initial conditions of the free decay. Having several free decays $\mathbf{y}_1(k), \mathbf{y}_2(k), \dots$ these can be placed side by side to form the common free decay matrix

$$\mathbf{Y}(k) = \begin{bmatrix} \mathbf{y}_1(k), & \mathbf{y}_2(k), & \cdots \end{bmatrix} \qquad (9.52)$$

and similarly we can place the corresponding initial conditions $\mathbf{u}_{01}, \mathbf{u}_{02}, \ldots$ side-by-side to form the initial condition matrix

$$\mathbf{U}_0 = \begin{bmatrix} \mathbf{u}_{01}, \mathbf{u}_{02}, \ldots \end{bmatrix} \qquad (9.53)$$

A common expression for all the free decays is then given by

$$\mathbf{Y}(k) = \mathbf{P}\mathbf{D}^k\mathbf{U}_0 \qquad (9.54)$$

Two block Hankel matrices are then formed with s block rows

$$\mathbf{H}(0) = \begin{bmatrix} \mathbf{Y}(0) & \mathbf{Y}(1) & \cdots \\ \mathbf{Y}(1) & \mathbf{Y}(2) & \cdots \\ \vdots & \vdots & \\ \mathbf{Y}(s-1) & \mathbf{Y}(s) & \cdots \end{bmatrix}; \quad \mathbf{H}(1) = \begin{bmatrix} \mathbf{Y}(1) & \mathbf{Y}(2) & \cdots \\ \mathbf{Y}(2) & \mathbf{Y}(3) & \cdots \\ \vdots & \vdots & \\ \mathbf{Y}(s) & \mathbf{Y}(s+1) & \cdots \end{bmatrix} \qquad (9.55)$$

As it appears the only difference between the two matrices is a time shift of one sample step in between them. Using Eq. (9.54), we can express the two matrices as

$$\mathbf{H}(0) = \mathbf{\Gamma}\mathbf{\Lambda}; \quad \mathbf{H}(1) = \mathbf{\Gamma}\mathbf{D}\mathbf{\Lambda} \qquad (9.56)$$

where the so-called observability matrix $\mathbf{\Gamma}$ and the controllability matrix $\mathbf{\Lambda}$ are given by

$$\mathbf{\Gamma} = \begin{bmatrix} \mathbf{P} \\ \mathbf{P}\mathbf{D} \\ \mathbf{P}\mathbf{D}^2 \\ \vdots \\ \mathbf{P}\mathbf{D}^{s-1} \end{bmatrix}; \quad \mathbf{\Lambda} = \begin{bmatrix} \mathbf{U}_0 & \mathbf{D}\mathbf{U}_0 & \mathbf{D}^2\mathbf{U}_0 & \cdots \end{bmatrix} \qquad (9.57)$$

The main idea in the ERA identification technique is that taking the SVD of the first block Hankel matrix in Eq. (9.55)

$$\mathbf{H}(0) = \mathbf{U}\mathbf{S}\mathbf{V}^T \qquad (9.58)$$

and by making use of Eq. (9.56) we estimate the observability and controllability matrices as

$$\hat{\mathbf{\Gamma}} = \mathbf{U}\sqrt{\mathbf{S}}$$
$$\hat{\mathbf{\Lambda}} = \sqrt{\mathbf{S}}\mathbf{V}^T \qquad (9.59)$$

The discrete time system matrix is then obtained using the right part of Eq. (9.56) estimated as

$$\hat{\mathbf{D}} = \hat{\mathbf{\Gamma}}^+\mathbf{H}(1)\hat{\mathbf{\Lambda}}^+ \qquad (9.60)$$

where $\hat{\mathbf{\Gamma}}^+$ and $\hat{\mathbf{\Lambda}}^+$ are the pseudo inverse of the estimated observability and controllability matrices respectively. A simple way to perform the pseudo inverse and the calculation specified by the right hand side of Eq. (9.60) is to restrict the diagonal matrix $\mathbf{S}$ to the first n singular values corresponding to $n/2$ modes, and reduce the matrices $\mathbf{U}\mathbf{S}\mathbf{V}$ and $\mathbf{U}\mathbf{S}\mathbf{V}$ to the corresponding singular vectors forming the restricted SVD matrices $\mathbf{U}_n$, $\mathbf{S}_n$, $\mathbf{V}_n$ and then calculate the discrete time system matrix as

$$\hat{\mathbf{D}} = \mathbf{S}_n^{-1/2}\mathbf{U}_n^T\mathbf{H}(1)\mathbf{V}_n\mathbf{S}_n^{-1/2} \qquad (9.61)$$

which in the ERA is denoted an n-order realization of the discrete time system matrix. The eigenvalues and the eigenvectors of the estimate of the discrete time system matrix are then obtained from

$$\hat{\mathbf{D}} = \begin{bmatrix} \boldsymbol{\varphi}_n' \end{bmatrix} \begin{bmatrix} \mu_n \end{bmatrix} \begin{bmatrix} \boldsymbol{\varphi}_n' \end{bmatrix}^{-1} \qquad (9.62)$$

The corresponding continuous time eigenvalues λ_n (of the continuous time system matrix $\mathbf{A}$) are then found by the relation given by Eq. (9.49)

$$\mu_n = \exp\left(\lambda_n \Delta t\right) \iff \lambda_n = \ln\left(\mu_n\right) / \Delta t \tag{9.63}$$

The eigenvectors in Eq. (9.62) are not direct estimates of the mode shapes of the structure because it can be shown that calculating the realization given by Eq. (9.61) leads to transformed mode shapes and thus it is needed to transform the mode shapes back to the physical coordinates using the transformation

$$\left[\boldsymbol{\varphi}_n\right] = \hat{\mathbf{P}}\left[\boldsymbol{\varphi}_n'\right] \tag{9.64}$$

where $\hat{\mathbf{P}}$ is the observation matrix of the realization estimated as

$$\hat{\mathbf{P}} = \mathbf{U}_r \mathbf{S}_r^{1/2} \tag{9.65}$$

The reader is referred to the original paper [31] and the ERA user's manual [33] for further information on this issue.

Example 9.3 Estimating the first three modes of the HCT building case using ERA

The data was bandpass filtered as described in Example 9.1 and the CF matrix was also used as described in Example 9.1 where all columns of the transposed CF matrix were used as free decays in the identification. It should be noted that using the number of block rows of the block Hankel matrices equal to $s = 2$ ERA has the same number of modes as an AR model with $na = 2$, thus using $s = 2$ as we have done here result in the same number of modes as in Example 9.1.

One of the advantages of the ERA technique is that when we know how many modes we want to estimate, the number of singular values in the reduced SVD matrices can be adjusted accordingly. Only the physical modes are estimated because the large singular values also corresponds to the modes with

Table 9.3 Identification results of using ERA identification on the four data sets of the HCT building case, modes chosen as the physical modes are shown in bold

	Parameter	Mode 1	Mode 2	Mode 3	Mode 4	Mode 5	Mode 6	Mode 7	Mode 8
Data set 1	f_n (Hz)	**1.2271**	**1.2848**	**1.4515**					
	ς_n (%)	**0.96**	**1.19**	**1.11**					
	π_n (%)	**30.14**	**19.25**	**50.61**					
Data set 2	f_n (Hz)	1.2164	**1.2358**	1.2728	**1.2850**	1.42554	**1.4575**		
	ς_n (%)	0.22	**1.41**	0.33	**1.15**	0.41	**1.59**		
	π_n (%)	0.87	**9.82**	0.67	**40.45**	1.88	**46.31**		
Data set 3	f_n (Hz)	1.2203	**1.2381**	**1.2825**	1.3117	1.3834	**1.4556**	1.4768	
	ς_n (%)	0.75	**0.88**	**1.01**	2.36	2.33	**0.72**	0.45	
	π_n (%)	1.13	**4.03**	**6.32**	1.13	0.06	**87.20**	0.12	
Data set 4	f_n (Hz)	1.2120	**1.2457**	1.2814	**1.2985**	1.3683	1.4200	**1.4414**	1.4720
	ς_n (%)	0.77	**1.12**	0.37	**1.57**	1.09	0.17	**0.33**	1.76
	π_n (%)	4.74	**23.39**	0.26	**3.41**	0.38	5.64	**58.63**	3.56

high modal participation. However, this works only in cases where the physical modes are well excited and where noise is not disturbing the identification too much. In this case, reducing the SVD matrices to the three modes that we know are present produces the right modes for the first three data sets, but not for the last one. For the last data set, it is needed to produce all possible eight modes to have a good estimate of the three physical modes present in the system and therefore, in this example the maximum number of modes was produced for all four data sets.

The modes were sorted in ascending order of frequencies including only the modes with positive damping and the results of the identification are shown in Table 9.3. The results are very similar to those of Example 9.1. The result of fitting the CF matrix is not shown here because there is no visible difference between Figure 9.1 and the plot based on the ERA data.

Finally the mode shapes components corresponding to the modes shown in Table 9.3 were merged and the result is shown in Figure 9.4.

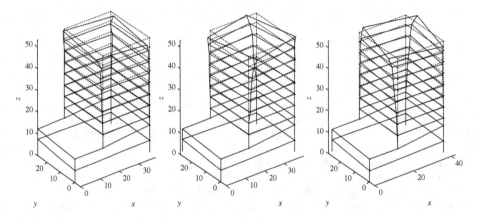

Figure 9.4 First three mode shapes of the HCT building found by ERA identification and merging the mode shapes components from the four data sets corresponding to the physical modes shown in Table 9.3

9.6   Stochastic Subspace Identification (SSI)

The Stochastic subspace identification (SSI) has been quite popular over the first decade of the millennium, mainly due to user-friendly implementations in the MACEC [33] and the ARTeMIS Extractor [34] software. The SSI is a collection of techniques that has a certain overall formulation in common as described in Overschee and De Moor [2]. Similarly to the ERA technique – and even to a higher degree – the SSI is formulated in a terminology dominated by concepts from control theory, and therefore somewhat difficult to include in the terminology used in this book. Therefore – and also because the theory is very well described in Overschee and De Moor [2] and also in Peeters [35/36] – we shall only briefly introduce the main ideas of the SSI here.

Two main types of SSI formulations exist, that is, the correlation driven SSI and the data driven SSI. The correlation driven SSI, see for instance Peters [35], is very similar to the ERA, and thus is not further discussed here. The data-driven SSI, however, is an approach where the estimation of the correlation functions is integrated into the identification algorithm, and the main ideas of the approach are outlined in the following. However, we will only present the main steps of the algorithm and not try to introduce the underlying theory. The first step of the data-driven SSI techniques is to form the block Hankel matrix based directly on the measured responses $\mathbf{y}(k)$ with np number of data points and $2s$ number of block rows

$$
\mathbf{H} = \begin{bmatrix}
\mathbf{y}(1) & \mathbf{y}(2) & \cdots & \mathbf{y}(np - 2s + 1) \\
\mathbf{y}(2) & \mathbf{y}(3) & \cdots & \mathbf{y}(np - 2s + 2) \\
\vdots & \vdots & & \vdots \\
\mathbf{y}(s) & \mathbf{y}(s + 1) & \cdots & \mathbf{y}(np - s) \\
\mathbf{y}(s + 1) & \mathbf{y}(s + 2) & \cdots & \mathbf{y}(np - s + 1) \\
\mathbf{y}(s + 2) & \mathbf{y}(s + 3) & \cdots & \mathbf{y}(np - s + 2) \\
\vdots & \vdots & & \vdots \\
\mathbf{y}(2s) & \mathbf{y}(2s + 1) & & \mathbf{y}(np)
\end{bmatrix} = \begin{bmatrix} \mathbf{H}_1 \\ \mathbf{H}_2 \end{bmatrix} \tag{9.66}
$$

As indicated in this equation, the matrix is split in the middle into the two block Hankel matrices $\mathbf{H}_1$ and $\mathbf{H}_2$ each with s block rows, $\mathbf{H}_1$ is the upper part also called "the past" and $\mathbf{H}_2$ is the lower part also called "the future" in the SSI theory. Based on the defined block Hankel matrices, the projection matrix is calculated and is defined as

$$
\mathbf{O} = \mathrm{E}\left[\mathbf{H}_2 \,|\, \mathbf{H}_1\right] \tag{9.67}
$$

where the expectation is operating on $\mathbf{H}_2$ that is considered to be stochastic and where $\mathbf{H}_1$ contains the deterministic conditions. In the SSI theory, it is shown that the projection can be calculated as

$$
\mathbf{O} = \mathbf{T}_{21}\mathbf{T}_{11}^{+}\mathbf{H}_1 \tag{9.68}
$$

where $\mathbf{T}_{21}$ and $\mathbf{T}_{11}$ are block Toeplitz matrices defined as

$$
\begin{aligned}
\mathbf{T}_{21} &= \mathbf{H}_2\mathbf{H}_1^{T} \\
\mathbf{T}_{11} &= \mathbf{H}_1\mathbf{H}_1^{T}
\end{aligned} \tag{9.69}
$$

It is worth noting the close similarity with Eq. (9.41) in the ITD. Further, it should be noted that Eq. (9.68) is easily implemented by assuming that the signals in the block Hankel matrices are zero mean Gaussian signals and by using the conditional mean as expressed by Eqs. (2.58/2.59). Further, since the projection is a conditional mean of the observed random signals, which is also the idea of forming a random decrement signature, the projection is basically the same as a random decrement signature, see Section 8.4.4. This also means that the projection consist of free decays, in this case free decays are organized in a block Hankel matrix.

However, in the SSI technique the projection is normally not performed by calculation the block Toeplitz matrices as described in Eq. (9.69) because this will be too time consuming and memory consuming in many practical applications. The projection is carried out using a QR decomposition of the transposed block Hankel matrix in Eq. (9.66) taking advantage of the fact that only a part of R factor matrix is needed for the projection. When the projection has been calculated and the free decays have been established in the projection matrix, and similarly to Eq. (9.56) in the ERA technique, it is possible to show that the projection matrix can be expressed as

$$
\mathbf{O} = \mathbf{\Gamma}\mathbf{X} \tag{9.70}
$$

where $\mathbf{\Gamma}$ is the observability matrix as given by Eq. (9.57) and $\mathbf{X}$ is a matrix of Kalman states that can be thought of as containing the initial conditions of the free decays in the projection matrix. Parallel to the solution idea in the ERA, we now take the SVD of the projection matrix

$$
\mathbf{O} = \mathbf{U}\mathbf{S}\mathbf{V}^{T} \tag{9.71}
$$

The diagonal matrix $\mathbf{S}$ is restricted to the first n singular values corresponding to $n/2$ modes reducing the matrices $\mathbf{U}$ and $\mathbf{USV}$ to the corresponding singular vectors, forming the restricted SVD matrices $\mathbf{U}_n$, $\mathbf{S}_n$, $\mathbf{V}_n$ and the two matrices in Eq. (9.70) are estimated as

$$\hat{\Gamma} = \mathbf{U}_n \sqrt{\mathbf{S}_n}$$
$$\hat{\mathbf{X}} = \sqrt{\mathbf{S}_n} \mathbf{V}_n^T$$

(9.72)

Finally, the discrete time system matrix $\mathbf{D}$ and the observation matrix $\mathbf{P}$ can then be found by solving a least squares problem. The poles and mode shapes are found similar to Eqs. (9.62–9.64). It should be noted that many different techniques can be formulated in the SSI using a generalized projection matrix by multiplying the real-valued weight matrices $\mathbf{W}_1$ and $\mathbf{W}_2$ on each side of the projection matrix and performing the singular value decomposition on the resulting matrix

$$\mathbf{W}_1 \mathbf{O} \mathbf{W}_2 = \mathbf{USV}^T$$

(9.73)

Different forms of the weight matrices form the SSI standard algorithms known as the principal components (PC) algorithm, the unweighted principal component (UPC) algorithm, and the canonical variate algorithm (CVA), see Overschee and De Moor [2].

Example 9.4   Estimating the first three modes of the HCT building case using SSI

The data was band-pass filtered as described in Example 9.1 and the CF matrix was only estimated to be used for determination of the participation factors as the modal mode shapes and poles are not based on a separate CF matrix estimate. The PC algorithm using the stochastic algorithm 1 from Overshcee and De Moor [36] has been used for the identification.

In contrast to the previously mentioned techniques in SSI there is a need for an oversized model. Thus, in this case it is not possible to use a small model with only two block rows (or a few block rows) as in the

Table 9.4   Identification results of using stochastic subspace identification (SSI) on the four data sets of the HCT building case, modes chosen as the physical modes are shown in bold

	Parameter	Mode 1	Mode 2	Mode 3	Mode 4	Mode 5	Mode 6	Mode 7	Mode 8
Data set 1	f_n (Hz)	**1.2271**	**1.2841**	**1.4497**					
	ς_n (%)	**0.53**	**0.58**	**0.63**					
	π_n (%)	**32.57**	**18.50**	**48.93**					
Data set 2	f_n (Hz)	**1.2302**	**1.2837**	**1.4573**					
	ς_n (%)	**0.74**	**0.68**	**0.80**					
	π_n (%)	**12.26**	**33.99**	**53.75**					
Data set 3	f_n (Hz)	**1.2307**	**1.2846**	**1.4567**					
	ς_n (%)	**0.48**	**0.50**	**0.45**					
	π_n (%)	**8.24**	**8.19**	**83.57**					
Data set 4	f_n (Hz)	1.0014	1.1160	**1.2147**	1.2786	1.2799	**1.4429**	1.5216	1.5941
	ς_n (%)	3.32	9.91	**1.92**	7.20	1.25	**0.89**	2.78	4.05
	π_n (%)	0.07	0.91	**16.32**	9.99	0.11	**68.95**	2.16	2.16

ERA technique. Therefore, in this example a large model with 80 block rows was used for the estimation. This corresponds to a model with 240 modes for the first data set and 320 modes for the remaining data sets. The SVD matrices in Eq. (9.72) were reduced to the first six singular values for the first three data sets (corresponding to a reduced model with only three modes) and to 18 singular values for the last data set (corresponding to a reduced model with nine modes). Nine modes were needed in the last data set in order to obtain a model including the three modes considered with a reasonable participation factor.

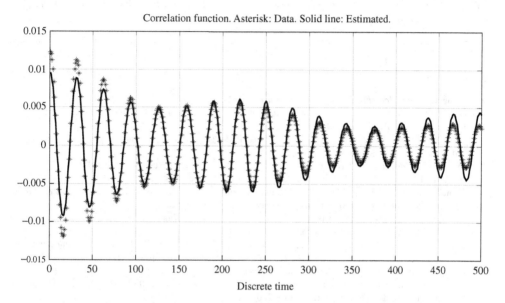

Figure 9.5 Result of using the stochastic subspace identification (SSI) technique and fitting the correlation function matrix. The figure shows the autocorrelation function of the first measurement channel of the first data set of the HCT building case

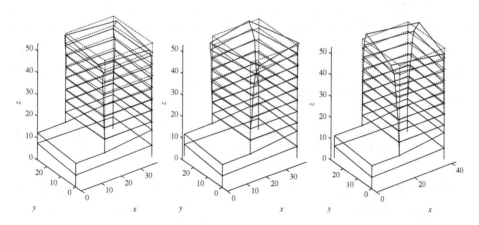

Figure 9.6 First three mode shapes of the HCT building found by the stochastic subspace identification (SSI) technique and merging the mode shapes components from the four data sets corresponding to the physical modes shown in Table 9.4

The modes were sorted in ascending order of frequencies including only the modes with positive damping. The results of the identification are shown in Table 9.4 where only the first eight modes of the nine estimated modes for data set 4 are shown (the remaining mode has only a participation factor of 0.28%).

The result of fitting the CF matrix is shown in Figure 9.5 and as it appears the fit is somewhat worse than the similar plot for the AR estimate in Figure 9.1. This is due to the fact that the damping estimated by the SSI seems to be too small.

Finally the mode shapes components corresponding to the modes shown in Table 9.4 were merged and the result are shown in Figure 9.6.

References

[1] Ljung, L.: *System Identification, theory for the user*. Prentice-Hall, Inc, 1987.

[2] Overschee, P. and De Moor, B.: *Subspace identification for linear systems, theory, implementation, application*. Kluwer Academic Publishers, 1996.

[3] Maia, N.M.M. and Silva, J.M.M. (Editors): *Theoretical and experimental modal analysis*. Research Studies Press Ltd., John Wiley & Sons, 1997.

[4] Ewins, D.J.: *Modal testing, theory practice and application*. 2nd edition. Research Studies Press Ltd., 2000.

[5] He, J. and Fu, Z.: *Modal analysis*. Oxford, U.K, Butterworth-Heinemann, 2001.

[6] Carne, G.C., Lauffer, J.P. and Gomez, A.J.: Modal testing of a very flexible 110m wind turbine structure. In Proceedings of International Modal Analysis Conference (IMAC), 1988.

[7] James, G.H., Carne, T.G. and Veers, P.S.: *Damping measurements using operational data*. SED-Vol 11, 10th ASME Wind Energy Symposium, ASME 1991.

[8] James, G.H., Carne, T.G. and Lauffer, J.P.: The natural excitation technique (NExT) for modal parameter extraction from operating structures. *Int. J. Anal. Exp. Modal Anal.*, V. 10, No. 4, p. 260–277, 1995.

[9] Zhang, L, Brincker R, Andersen, P.: *A unified approach for two-stage time domain modal identification*. In Proceedings of the 3rd International Conference on Structural Dynamics Modelling - Test, Analysis, Correlation and Validation, Madeira Island, Portugal, June 2002.

[10] Zhang, L.: *On the Two-stage Time Domain Modal Identification*. In Proceedings of the International Modal Analysis Conference (IMAC), Los Angeles, Feb 4–7, 2002.

[11] Zhang, L., Brincker, R. and Andersen, P.: *An overview of operational modal analysis: Major developments and issues*. In Proceedings of the 1st International Operational Modal Analysis Conference (IOMAC), Apr. 26–27, Copenhagen, 2005.

[12] Masjedian, M.H. and Keshmiri, M.: *A review on operational modal analysis researches: classification of methods and applications*. In Proceedings of the 3rd. Int Operational Modal Analysis Conference (IOMAC), May 4–6, Portonovo (Ancona), 2009.

[13] Harvey, A.C.: *Time series models*. Philip Allan Publishers Ltd., 1981.

[14] Pandit, S.M. and Wu, S.M.: *Time Series and system analysis with applications*. John Wiley and Sons, 1983.

[15] Norton, J.P.: *An introduction to Identification*. Academic Press, 1986.

[16] Vold, H., Kundrat, J., Rocklin, G.T. and Russell, R.: *A multi-input modal estimation algorithm for mini-computers*. SAE Paper Number 820194, 1982.

[17] Vold, H. and Rocklin, G.T.: *The numerical implementation of a multi-input modal estimation method for mini-computers*. In Proceedings of the International Modal Analysis Conference (IMAC), p. 542–548, 1982.

[18] Zhao-qian, T. and Yang, Q.: *Modal analysis using the time series analysis method (AR or ARMA model) and its computer program design*. In Proceedings of the 2nd International Modal Analysis Conference (IMAC), Feb 6–9, Orlando, Florida, p. 559–565, 1984.

[19] Shi, D and Stühler, W.: *Modal analysis with AR(ARMA) model and unknown excitation*. In Proceedings of the 5th International Modal Analysis Conference (IMAC), Apr 6–9, London, p. 1171–1176, 1987.

[20] Piombo, B., Giorcelli, E., Garibaldi, L., Fasana, A.: *Structure identification using ARMAV models*. In Proceedings of the 11th 8th International Modal Analysis Conference (IMAC), Orlando, FL, p. 588–592, 1993.

[21] Bonnecase, D., Prevosto, M. and Benveniste, A.: *Application of a multidimensional ARMA model to modal analysis under natural excitation*. In Proceedings of the 8th International Modal Analysis Conference (IMAC), Jan 29–Feb 1, Kissimmee, Florida, p. 382–388, 1994.

[22] Giorcelli, E., Fasana, A., Garibaldi, L and Riva, A.: *Modal analysis and system identification using ARMAV models*. In Proceedings of the 12th International Modal Analysis Conference (IMAC), Jan 31–Feb 3, Honolulu, Hawaii, p. 676–680, 1994.

[23] Kozin, F. and Natke, H.G.: System identification techniques. *Struct. Saf.*, 3:269–316, 1986.

[24] Andersen, P., Brincker, R. and Kierkegaard, P.H.: *Theory of covariance equivalent ARMAV models of civil engineering structures*. In Proceedings of the 14th International Modal Analysis Conference (IMAC). Feb 12–15, Dearborn, Michigan, p. 518–525, 1996.

[25] Andersen, P.: *Identification of civil engineering structures using vector ARMA models*. Ph.D. thesis, Department of Building Technology and Structural Engineering, Aalborg University, 1997.

[26] Ibrahim, S.R. and Milkulcik, E.C.: *A method for direct identification of vibration parameters from the free response. Shock Vibr. Bull.*, 47, p. 183–196, 1977.

[27] Ibrahim, S.R.: *Random decrement technique for modal identification of structures. J. Spacecr. Rockets*, V. 14, p. 696–700, 1977.

[28] Ibrahim, S.R.: Modal confidence factor in vibration testing. *J. Spacecr. Rockets*, V. 15, p. 313–316, 1978.

[29] Fukuzono, K.: *Investigation of multiple-reference Ibrahim time domain modal parameter estimation technique*. M.Sc. Thesis, Department of Mechanical and Industrial Engineering, University of Cincinnati, 1986.

[30] Juang, J.N. and Pappa, R.S.: An eigen system realization algorithm for modal parameter identification and modal reduction. *J. Guidance*, V. 8, No. 5, p. 620–627, 1985.

[31] Pappa, R.S., Elliott, K.B. and Schenk, A.: Consistent-mode indicator for the eigensystem realization algorithm. *J. Guidance Contr. Dyn.*, V. 16, No. 5, p. 852–858, 1993.

[32] Pappa, R.S.: *Eigensystem realization algorithm, user's guide for VAX/VMS computers*. NASA Technical Memorandum 109066, 1994.

[33] Van den Branden, B., Laquiere, A., Peeters, B, De Roeck, G. Reynders, E. and Schevenels, M.: *MACEC 3.2: A MATLAB Toolbox for Experimental and Operational Modal Analysis*. Structural Mechanics division of K.U.Leuven.

[34] ARTeMIS Extractor. Structural Vibration Solutions A/S.

[35] Peeters, B. and De Roeck, G.: Reference-based stochastic subspace identification for output-only modal analysis. *Mech. Sys. Signal Process.*, V. 13, No. 6, p. 855–878, 1999.

[36] Peeters, B.: *System identification and damage detection in civil engineering*. Ph.D. Thesis. Faculteit Toegepaste Wetenschapen, Katholieke Universiteir Leuven, 2000.

10

Frequency-Domain Identification

"Time and space are modes by which we think and not conditions in which we live."

– Albert Einstein

In the preceding chapter, and in this chapter, we are at the center of OMA – the identification – where the modal parameters are estimated. There are two main tracks in this approach: one is to do the identification in the time domain (TD) – this is the subject of the preceding chapter – the other approach is to perform the identification in the frequency domain, which is the subject of this chapter.

As for the chapter on Time-domain identification – and for the same reasons – we shall not cover everything in detail, but only present the main ideas and use examples to illustrate how the techniques work. Since we are only presenting what has been widely used in OMA and is known to work in practice, we shall only present the most well-known techniques, that is the classical frequency-domain approach, the frequency-domain decomposition technique (FDD), and the AR models in frequency domain, also known as frequency-domain poly reference (FDPR)[1] or PolyMax.[2]

While in the time domain we are dealing with free responses that are present over the full considered time span, in the frequency domain each mode has a small frequency band where a mode dominates, and thus, in the frequency domain we have the advantage of a "natural modal decomposition" simply by considering the different frequency bands where different modes dominate. This is the main advantage of the frequency-domain approach, and is the main idea of the classical approach introduced in Section 10.2. However, even though the FDD and FDPR are not based on an assumption of "natural modal decomposition," the techniques still take advantage of the fact that simply considering a smaller frequency band, the influence of modes outside of this band has a reduced influence on the results.

A drawback of the frequency domain is that the spectral density estimates suffer from some kind of bias, and thus it is not as easy to justify as in the time domain modal extraction based on information that is bias free or nearly bias free.

As we discussed in the preceding chapter, we will illustrate the techniques on estimating the first three modes of the Heritage Court Tower (HCT) building case, as an introduction to this method. Identification of the first three modes of the HCT building case is a really difficult case, and any technique that can handle this case reasonably well is a good one. As it will be seen in the coming examples, both the FDD and the FDPR techniques can handle the case and identify all three modes.

---

[1] In the following, we shall use this name for the technique.
[2] PolyMax is a trademark of LMS International [19].

Introduction to Operational Modal Analysis, First Edition. Rune Brincker and Carlos Ventura.
© 2015 John Wiley & Sons, Ltd. Published 2015 by John Wiley & Sons, Ltd.
Companion Website: www.wiley.com/go/brincker

Concerning a more detailed overview of the different techniques and the vast literature on the subject of OMA identification, the reader is referred to the overview papers by Zhang et al. [1] and Masjedian and Keshmiri [2] and in general to the IMAC, ISMA, and IOMAC conference proceedings.

10.1   Common Challenges in Frequency-Domain Identification

The idea in OMA is to obtain physical information from the correlation functions and spectral densities, thus in frequency-domain identification this means to extract the physical information from the spectral density functions. Even though fitting is not as common in the frequency domain as in the time domain, this can always be done using some kind of fitting technique using a parametric model and regression as described in Chapter 3.

We shall start this chapter by looking at how the modal results of any technique can be fitted to the estimated spectral density functions. The result of this step is to obtain important information of the participation of each of the identified modes.

Further, since noise modes are present, ways of knowing which modes are physical and which are not physical becomes an important issue, and we shall briefly touch upon this subject before moving onto the details of the identification using the different techniques. For a more in-depth treatment of this subject of dealing with noise modes, the reader is referred to Chapter 11 and Section 12.4 in Chapter 12.

10.1.1   Fitting the Spectral Functions (Modal Participation)

Normally what we get from an identification technique is a set of estimated eigenvalues (poles) and mode shapes; however, in order to understand and illustrate the quality of the identification it is useful to perform a fit of the modal model to the spectral density (SD) matrix that produced this information.

One way to do this is to estimate the modal participation vectors in the time domain based on the empirical correlation function (CF) matrix as described in Section 9.1.1 and then transform the empirical and the estimated CF matrix to frequency domain by a discrete Fourier transform. However, a similar estimation of the modal participation vectors can also be carried out directly in the frequency domain as follows.

We assume an identified set of poles arranged in the diagonal matrix $[\lambda_n]$ and the corresponding set of mode shapes arranged as columns in the matrix $\mathbf{A} = [\mathbf{a}_n]$.[3] As explained in Section 6.2.5 using the definition of the correlation functions in this book, in the SD matrix neither the columns nor the rows are proportional to the mode shapes and thus the SD matrix itself is difficult to use for this purpose. A more simpler approach is to fit the modal information to a version of the SD matrix that only contains information from the corresponding correlation functions for positive time lags – the so-called half spectrum, see Section 8.5.4. Assuming that the half spectrum version of the SD matrix contains the complex conjugate pairs of both mode shapes and poles as given by Eq. (8.80), we can extend the summation to the set of $2N$ modal parameters and the SD matrix can be sampled at the discrete frequencies $\omega = k\Delta\omega$ to give

$$\mathbf{G}_y(k) = \sum_{n=1}^{2N} \frac{\gamma_n \mathbf{a}_n^T}{ik\Delta\omega - \lambda_n} \tag{10.1}$$

This is a series of outer products that can be expressed by the matrix equation

$$\mathbf{G}_y(k) = \mathbf{\Gamma}\left[ik\Delta\omega - \lambda_n\right]^{-1}\mathbf{A}^T \tag{10.2}$$

---

[3] In this book we are following the terminology that mode shapes from a model are denoted $\mathbf{B} = [\mathbf{b}_n]$ whereas mode shapes from a test are denoted $\mathbf{A} = [\mathbf{a}_n]$.

where the diagonal matrix $\left[ik\Delta\omega - \lambda_n\right]^{-1} = \left[1/\left(ik\Delta\omega - \lambda_n\right)\right]$ and the matrix Γ contains the modal participation vectors $\Gamma = \left[\gamma_n\right]$. Arranging the SD matrix in the single block row Hankel matrix for the frequency interval $k = k_1, k_1 + 1, \ldots, k_2$

$$\mathbf{H} = \left[\mathbf{G}_y\left(k_1\right), \mathbf{G}_y\left(k_1 + 1\right), \ldots, \mathbf{G}_y\left(k_2\right)\right] \tag{10.3}$$

and similarly arranging the modal information in the matrix

$$\mathbf{M} = \left[\left[ik_1\Delta\omega - \lambda_n\right]^{-1}\mathbf{A}^T, \left[i\left(k_1 + 1\right)\Delta\omega - \lambda_n\right]^{-1}\mathbf{A}^T, \ldots, \left[ik_2\Delta\omega - \lambda_n\right]^{-1}\mathbf{A}^T\right] \tag{10.4}$$

then Eq. (10.2) can be written as

$$\mathbf{H} = \Gamma\mathbf{M} \tag{10.5}$$

This equation can be solved by regression or by SVD as explained in chapter 3 to give an estimate of the modal participation vectors

$$\hat{\Gamma} = \mathbf{H}\mathbf{M}^+ \tag{10.6}$$

And the resulting fit of the half spectrum SD matrix is

$$\hat{\mathbf{G}}_y\left(k\right) = \hat{\Gamma}\left[ik\Delta\omega - \lambda_n\right]^{-1}\mathbf{A}^T \tag{10.7}$$

Or the estimated modal participation vector can be used to calculate the estimated classical SD matrix as given by Eq. (6.91) by the corresponding matrix equation

$$\hat{\mathbf{G}}_y\left(k\right) = \mathbf{A}\left[-ik\Delta\omega - \lambda_n\right]^{-1}\hat{\Gamma}^T + \hat{\Gamma}\left[ik\Delta\omega - \lambda_n\right]^{-1}\mathbf{A}^T \tag{10.8}$$

where the first term corresponds to the negative part of the CF matrix and the last term corresponds to the positive part of the CF matrix. It should be noted that only the modes that play a role for the SD matrix in the considered frequency band should be included in the modal matrix $\mathbf{M}$.

As for the time-domain identification, the participation vector can be used to determine relative participation factors π_n for the modes. Eqs. (9.8/9.9) are also valid for the frequency domain.

10.1.2   Seeking the Best Conditions (Stabilization Diagrams)

As described in the previous chapter, whenever parametric models are used, they must typically be oversized, and thus noise modes are introduced.

At this point we will not go further into the subject of optimizing the modal identification. But in the subsequent examples, we shall illustrate this problem, without seeking any optimization. The problem is further discussed in Chapter 11 and in Section 12.4 in Chapter 12.

Example 10.1   Modal participation of the first three modes of the HCT building case

In this example, we will illustrate how to find and document the modal participation in the frequency domain of the first three modes of the first data set of the HCT building case. The modes were found as described in Example 9.1 with the exception that 1024 (discrete time lag) data points were used for the CF matrix. This slightly changes the results and introduces two noise modes as shown in Table 10.1, which also shows the modal participation factors π_n calculated in the time domain as described in the previous chapter.

The results of estimating the participation factors π_n based on the half spectral densities as described earlier is shown in Table 10.2. As it appears, the values are somewhat different than the values

Table 10.1 Identification results using AR/PR identification on the first data set of the HCT building case, modes chosen as the physical modes are shown in bold, the participation factors were calculated as described in Example 9.1

Parameter	Mode 1	Mode 2	Mode 3	Mode 4	Mode 5
f_n (Hz)	**1.2266**	1.2598	**1.2896**	1.3804	**1.4510**
ς_n (%)	**0.72**	0.19	**1.01**	0.34	**0.96**
π_n (%)	**32.32**	0.79	**16.98**	0.12	**49.78**

Table 10.2 Participation factors calculated based on the half spectral densities as described in Section 10.1.1

Parameter	Mode 1	Mode 2	Mode 3	Mode 4	Mode 5
π_n (%)	**24.02**	0.49	**21.47**	0.22	**53.81**

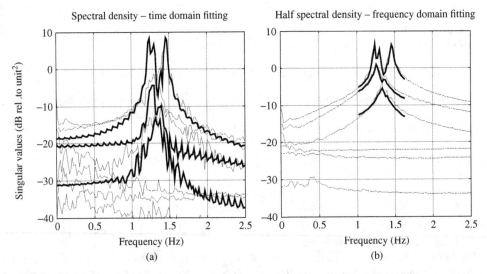

Figure 10.1 Spectral densities showing the empirical spectral density in dotted line and a parametric model fit to obtain the modal participation in solid line. The spectral plots are in dB relative to the measurement unit showing the singular values of the spectral density matrix. (a) shows the result of the modal participation fit in the time domain taking the empirical and synthesized spectral densities to frequency domain by discrete Fourier transform (corresponding to participation factors in Table 10.1), and (b) shows the result of fitting the half spectral densities in the frequency domain (corresponding to participation factors in Table 10.2)

estimated in the time domain; however, the participation factors are helpful in deciding which modes are physical and which modes are to be considered as noise modes.

The left side of Figure 10.1 shows the result of estimating the modal participation vectors in the time domain based on the empirical CF and then transforming the empirical and the estimated CF matrix to frequency domain by a discrete Fourier transform to obtain the spectral densities. The so-obtained spectral densities were estimated using a moderate tapering on the correlation functions by application of the flat-triangular window (value of $\alpha = 0.5$) in order to reduce side lobe noise.

The right side of Figure 10.1 shows the result of direct fitting in the frequency domain as described earlier using the half spectral densities. The fitting was performed in the frequency band with a center frequency of 1.35 Hz and a bandwidth of 0.60 Hz, and the synthesized half spectral densities are shown in the same frequency band.

10.2   Classical Frequency-Domain Approach (Basic Frequency Domain)

The classical frequency-domain approach – also known as the "basic frequency-domain" approach – has been used for modal identification for many decades. A good introduction to this early approach is given by Felber [3], but is also described in classical books such as Bendat and Piersol [4]. The main idea of this approach is that any structural mode that is reasonably lightly damped is only influencing the response of the structure in a narrow frequency band around the natural frequency of the considered mode.

Considering the structural response $\mathbf{y}(t)$ in a frequency band dominated only by a single mode, the response is given by

$$\mathbf{y}(t) = \mathbf{a}q(t) \tag{10.9}$$

where $\mathbf{a}$ is the mode shape and $q(t)$ is the modal coordinate of the mode. Using the definition of the correlation function (CF) matrix, see for instance Eq. (2.35), the CF matrix is given by

$$\mathbf{R}(\tau) = \mathrm{E}\left[\mathbf{y}(t)\,\mathbf{y}(t+\tau)^T\right] = \mathbf{a}\mathrm{E}\left[q(t)\,q(t+\tau)\right]\mathbf{a}^T = R_q(\tau)\,\mathbf{a}\mathbf{a}^T \tag{10.10}$$

where $R_q(\tau)$ is the autocorrelation function of the modal coordinate. The spectral density matrix $\mathbf{G}_y(f)$ of the response is (see Eq. (6.12))

$$\mathbf{G}_y(f) = G_q(f)\,\mathbf{a}\mathbf{a}^T \tag{10.11}$$

where $G_q(f)$ is the auto spectral density function of the modal coordinate. As it appears, this gives us a spectral density matrix of rank one, and, any row or column of the spectral matrix is proportional to the mode shape vector. This means that for this case, we can take any column $\mathbf{u}_c$ in the spectral density matrix $\mathbf{G}(f) = \left[\mathbf{u}_1, \mathbf{u}_2, \dots\right]$ as an estimate of the mode shape.

$$\hat{\mathbf{a}} = \mathbf{u}_c \tag{10.12}$$

This gives an estimate for the mode shape for the considered mode based on channel c. This can be done for all channels (all columns in the spectral matrix), and mode shape estimates might then be averaged over channels.

Cases where this approach can be used effectively are when only one mode is active, which is quite rare, because in practice nearly always many modes are active. However, when a considered mode is well separated from all other modes, then the response is mainly dominated by the considered mode, and thus, the influence from the surrounding modes can be neglected.

The mode can be considered "well separated" from other modes when the minimum frequency distance to any other mode is much larger than the bandwidth B of the considered mode. In Chapter 8, we have defined the bandwidth of one single mode as the half-power bandwidth defined as $B = 2\varsigma f_0$, where ς is the damping ratio and f_0 is the natural frequency.

The advantage of the approach is its simplicity, and if the above assumptions are fulfilled, then it will be not only simple to use, but also quite accurate. The problem here is that the approach does not give us any clear indication if the assumptions are not fulfilled. By using this approach alone, there are no easy ways to know if a closely spaced mode is actually present in the signal. Often closely spaced modes are in fact present, and often their presence will not be visible in a spectral density function influenced by the noise that is always present in any OMA problem.

Therefore, a more general approach is needed in order to be able to treat problems with closely spaced modes and random noise.

From a user point of view it is also a problem that, even when dealing with a limited number of sensors, the number of spectral densities that need to be inspected by the user in order to determine the number of modes present in the considered problem is large. For instance, for a case where eight sensors have been used, the total number of spectral density functions is 64. Therefore, in practice, the basic frequency-domain technique is not reliable because the user might miss the presence of closely spaced modes, and the amount of data to deal with is most of the time overwhelming.

10.3   Frequency-Domain Decomposition (FDD)

The frequency-domain decomposition (FDD) technique that is based on singular value decomposition (SVD) of the spectral density (SD) matrix takes the classical frequency-domain approach some steps further and makes it possible to analyze cases with closely spaced modes. Further, the FDD approach makes the frequency-domain technique more user friendly because it concentrates all information in one single plot; that is the plot of singular values of the SD matrix.

The technique was introduced by Brincker et al. [5–7] and has been widely used for OMA mainly due to its user friendliness and the implementation in the ARTeMIS Extractor software [8]. The technique is closely related to the complex modal indicator function (CMIF) introduced by Brown and co-workers [9] which was based on an SVD of the frequency response function (FRF) matrix and presentation of the singular values as function of frequency. However, it should be noted that there are some important differences between the CMIF and the FDD techniques. The important difference is that the decomposition of the FRF matrix is obtained as a modal decomposition that is closely related to the SVD as it appears from the general expression of the FRF matrix, see for instance Eq. (5.133), whereas the modal decomposition of the spectral matrix follows a different scheme, see Eqs. (6.77/6.91).

10.3.1   FDD Main Idea

The idea of the technique is most easily introduced by expressing the considered response $\mathbf{y}(t)$ in normal modes and modal coordinates[4]

$$\mathbf{y}(t) = \mathbf{a}_1 q_1(t) + \mathbf{a}_2 q_2(t) + \cdots = \mathbf{A}\mathbf{q}(t) \tag{10.13}$$

where $\mathbf{A}$ is the mode shape matrix $\mathbf{A} = [\mathbf{a}_1, \mathbf{a}_2, \ldots]$ and $\mathbf{q}(t)$ a column vector of modal coordinates $\mathbf{q}^T(t) = \{q_1(t), q_2(t), \ldots\}$ Using the definition of the CF matrix $\mathbf{R}_y(\tau)$, see Eq. (2.34), we obtain the following expression for the matrix

$$\mathbf{R}_y(\tau) = \mathrm{E}\left[\mathbf{y}(t)\mathbf{y}^T(t+\tau)\right]$$

$$= \mathbf{A}\mathrm{E}\left[\mathbf{q}(t)\mathbf{q}^T(t+\tau)\right]\mathbf{A}^T \tag{10.14}$$

$$= \mathbf{A}\mathbf{R}_q(\tau)\mathbf{A}^T$$

where $\mathbf{R}_q(\tau)$ is the CF matrix of modal coordinates. Taking the Fourier transform of both sides of this equation, we obtain the corresponding SD matrix

$$\mathbf{G}_y(f) = \mathbf{A}\mathbf{G}_q(f)\mathbf{A}^T \tag{10.15}$$

---

[4] This is more or less the same derivation as has been carried out in Section 6.3.2.

If we now assume that the modal coordinates are uncorrelated, that is, the off-diagonal elements of the CF matrix of modal coordinates $\mathbf{R}_q(\tau)$ is zero, then the SD matrix $\mathbf{G}_q(f)$ of the modal coordinates is both diagonal and positive valued.[5] Further, since we know that the SD matrix is Hermitian and that some complexity might be present in the mode shapes, we need to use the Hermitian instead of the transpose. The final form of Eq. (10.15) is

$$\mathbf{G}_y(f) = \mathbf{A} \left[\; g_n^2(f)\right] \mathbf{A}^H \tag{10.16}$$

where $g_n^2(f)$ are the auto spectral densities (diagonal elements) of $\mathbf{G}_q(f)$. A decomposition such as that given by Eq. (10.15) can be performed by taking an SVD of the SD matrix, which for a complex, Hermitian, and positive definite matrix, takes the form

$$\mathbf{G}_y(f) = \mathbf{U}\mathbf{S}\mathbf{U}^H$$
$$= \mathbf{U} \left[s_n^2 \right] \mathbf{U}^H \tag{10.17}$$

and it follows directly that the singular values s_n^2 in the diagonal matrix $\mathbf{S}$ should be interpreted as the auto spectral densities of the modal coordinates, and the singular vectors – that is the columns in $\mathbf{U} = \left[\mathbf{u}_1, \mathbf{u}_2, \ldots \right]$ – should be interpreted as the mode shapes.[6]

10.3.2 FDD Approximations

Because the SVD given by Eq. (10.17) does not correspond completely to the theoretical decomposition of the SD matrix, see Eqs. (6.77/6.91), the FDD is always an approximate solution.

In time domain techniques, like the Poly reference, see Section 9.2, where the model just assumes that the correlation functions are free decays, and since this is in fact always true for the case where the noise is vanishing, the poly reference is an unbiased technique.[7] Because the FDD is based on the approximate decomposition given by Eq. (10.17) (which is not completely true), the FDD will introduce bias.

The approximation can be studied from two different perspectives. One is the general decomposition in the frequency domain under the assumption of the white noise input, and the other perspective is the assumption of uncorrelation modal coordinates as introduced earlier.

The general decomposition in the frequency domain under the assumption of the white noise input is given in Chapter 6 (Sections 6.2.3 and 6.2.5) where it is shown that the SD matrix is given by

$$\mathbf{G}_y(\omega) = \sum_{n=1}^{N} \left(\frac{\mathbf{a}_n \boldsymbol{\gamma}_n^T}{-i\omega - \lambda_n} + \frac{\mathbf{a}_n^* \boldsymbol{\gamma}_n^H}{-i\omega - \lambda_n^*} + \frac{\boldsymbol{\gamma}_n \mathbf{a}_n^T}{i\omega - \lambda_n} + \frac{\boldsymbol{\gamma}_n^* \mathbf{a}_n^H}{i\omega - \lambda_n^*} \right) \tag{10.18}$$

The first two terms are from the negative part of the CF matrix and the last two terms are from the positive part of the CF matrix, $\mathbf{a}_n$ are the mode shapes,[8] $\boldsymbol{\gamma}_n$ are the modal participation vectors and λ_n are the corresponding poles. Considering this expression, it is clear that it is not possible to express it in the form given by Eq. (10.17); not even in the case where we use the general SVD formula

$$\mathbf{G}_y(\tau) = \mathbf{U}\mathbf{S}\mathbf{V}^H \tag{10.19}$$

---

[5] Because auto spectral density functions are always positive valued, this follows from the Parseval theorem, see Chapter 6.
[6] This interpretation of the SVD is protected by a patent, see for instance Brincker and Andersen [10].
[7] Asymptotically unbiased for the noise approaching zero.
[8] Now denoted by $\mathbf{a}_n$ because we are dealing with experimentally obtained mode shapes and not by $\mathbf{b}_n$ that we use when referring to mode shapes from a model.

In Section 6.2.6, it is shown that the modal participation vectors γ_n are a weighted average of the mode shapes. This results in participation vectors that are nearly proportional to the mode vectors in the case of reasonably well separated modes

$$\gamma_n \cong c_n^2 \mathbf{a}_n \tag{10.20}$$

For this case and taking the two midterms in Eq. (10.18) that are dominant in the case of lightly damped structures, the expression for the SD matrix reduces to

$$\mathbf{G}_y(\omega) \cong \sum_{n=1}^{N} \left(\frac{c_n^2 \mathbf{a}_n^* \mathbf{a}_n^H}{-i\omega - \lambda_n^*} + \frac{c_n^2 \mathbf{a}_n \mathbf{a}_n^T}{i\omega - \lambda_n} \right)$$
$$\cong \sum_{n=1}^{N} 2c_n^2 \mathrm{Re}\left(\frac{\mathbf{a}_n \mathbf{a}_n^T}{i\omega - \lambda_n} \right) \tag{10.21}$$

This is a decomposition of the form given by Eq. (10.17). The positive constant c_n^2 is proportional to the inner product between the mode shape over the input spectral density and inversely proportional to natural frequency, damping, and modal mass; see Eq. (6.102).

Following the idea of uncorrelated modal coordinates, the FFD approximation can be analyzed as follows. In Section 6.3.3, it is shown that assuming general complex mode shapes and uncorrelated modal coordinates the modal decomposition of the SD matrix is given by

$$\mathbf{G}_y(\omega) = \mathbf{A}\left[Q_n(-\omega)Q_n(\omega)\right]\mathbf{A}^T + \mathbf{A}\left[Q_n(-\omega)Q_n^*(\omega)\right]\mathbf{A}^H$$
$$+ \mathbf{A}^*\left[Q_n^*(-\omega)Q_n(\omega)\right]\mathbf{A}^T + \mathbf{A}^*\left[Q_n^*(-\omega)Q_n^*(\omega)\right]\mathbf{A}^H \tag{10.22}$$

where $Q_n(\omega)$ are the modal coordinates in the frequency domain. However, approximating the terms of the diagonal matrices based on an assumption of light damping in Eq. (10.18) as shown in section 6.3.3 we arrive at

$$\mathbf{G}_y(\omega) \cong \sum_{n=1}^{N} \left(\frac{(g_{1n}\mathbf{a}_n + g_{3n}\mathbf{a}_n^*)\mathbf{a}_n^T}{i\omega - \lambda_n} + \frac{(g_{2n}\mathbf{a}_n + g_{4n}\mathbf{a}_n^*)\mathbf{a}_n^H}{-i\omega - \lambda_n^*} \right) \tag{10.23}$$

which corresponds to the two dominant midterms of Eq. (10.18) with the difference that the modal participation vector is proportional to the mode shape vector.

Therefore, we can conclude that in the case of well-separated modes the SD matrix approximately decomposes as given by Eq. (10.17) and we can also conclude that this is true when the modes are not separated as long as the modal coordinates are uncorrelated.

However, it is clear from the aforementioned considerations that the complexity of the mode shapes that is estimated by the FDD technique most definitely is heavily biased and thus should not be used for physical interpretations.

One of the advantages of the FDD is that it separates the noise from the physics and provides the user with SVD plots that clearly show what is noise and what is structural-related information helping the user to decide how many modes are present. This is of special value in the case of closely spaced modes. The noise level is indicated by the first singular value that becomes flat compared to the modal response. On the other hand, in any frequency band the number of singular values that clearly raises above the noise level is most likely a modal response, at least if it has a peak as a modal response.[9]

The concept is clearly illustrated in the Heritage Court Tower case; see Example 8.4 and Figure 8.14. In this figure, the singular values of the SD matrix of the first data set are shown in the frequency band from DC to 10 Hz. Some peaks are present in the frequency bands from 1 to 1.5 Hz and from 3.5 to

---

[9] Harmonics might also appear as a peak in the SVD plot, see the comments on this issue in Section 8.2.4.

4.5 Hz. The question is now: how many modes are present in the two mentioned frequency bands? In the first band, the fourth singular value is flat, while the first three are peaking in the interval. This defines the fourth singular value as representative for the noise level[10] and thus three modes are present. Following the same kind of arguments for the upper frequency band similarly we see that two modes are present in this band.

It should be noted that the advantage of this feature of the FDD can only be used for cases where the SD matrix has full rank, that is, if the rank is not reduced by single input loading and/or by other means; see the discussion about the "rank of the problem" in Section 7.2.2.

10.3.3   Mode Shape Estimation

In FDD the mode shapes are estimated from the singular vectors of the SVD of the spectral density matrix. However, we have an SVD for each frequency where the SD matrix is known because the SVD can be carried out for all known frequencies. This also means that if we have the same number of modes as we have sensors, then in principle all mode shapes can be found at one single frequency line. This is of course not a correct way to perform the estimate and we need to consider more carefully how the estimate is obtained in a correct way.

To this end, we will consider a case of two closely spaced modes, mode 1 and mode 2 at the natural frequencies f_1 and f_2, see Figure 10.2. In the following discussion, we will refer to them as the left and right modes. The spectral plot in the figure shows the two first singular values of the SD matrix as a function of frequency. Two peaks are situated at the natural frequencies, and at these frequencies we have the singular values $s_1^2(f_1) > s_2^2(f_1)$ at the frequency f_1 and similarly the singular values $s_1^2(f_2) > s_2^2(f_2)$ at the frequency f_2. Assuming that the remaining singular values in the considered band can be neglected, at the peaks we have the SVD of the spectral matrix

$$\mathbf{G}_y(f) = [\mathbf{u}_1, \mathbf{u}_2] \begin{bmatrix} s_1^2 & 0 \\ 0 & s_2^2 \end{bmatrix} \begin{bmatrix} \mathbf{u}_1^H \\ \mathbf{u}_2^H \end{bmatrix}$$
$$= s_1^2 \mathbf{u}_1 \mathbf{u}_1^H + s_2^2 \mathbf{u}_2 \mathbf{u}_2^H \tag{10.24}$$

where the singular values, as well as the singular vectors are varying from frequency line to frequency line. Similarly, we have from Eq. (10.16) assuming that only the two closely modes are present in the considered band

$$\mathbf{G}_y(f) = [\mathbf{a}_1, \mathbf{a}_2] \begin{bmatrix} g_1^2(f) & 0 \\ 0 & g_2^2(f) \end{bmatrix} \begin{bmatrix} \mathbf{a}_1^H \\ \mathbf{a}_2^H \end{bmatrix}$$
$$= g_1^2(f) \mathbf{a}_1 \mathbf{a}_1^H + g_2^2(f) \mathbf{a}_2 \mathbf{a}_2^H \tag{10.25}$$

In Eq. (10.24) both the singular values and the singular vectors vary from frequency line to frequency line, whereas in Eq. (10.25) the mode shapes are constant vectors and frequency dependency is only present at the modal coordinate spectral densities $g_1^2(f)$ and $g_2^2(f)$. This fact, and the similarity between these two equations make it important to make the right decision on how the mode shapes vectors are estimated from the SVD.

Here it is critical to acknowledge that the SVD always forces the singular vectors to be orthogonal and that in structural dynamics the eigenvectors are not geometrically orthogonal, but orthogonal with respect to the mass matrix (see Chapter 5). As indicated in Figure 10.2, we can imagine that the mode shape vectors $\mathbf{a}_1$ and $\mathbf{a}_2$ are not being geometrically orthogonal. Therefore, we have to obtain the estimate of the

---

[10] Including residues from adjacent modes.

eigenvectors at a frequency line where the bias introduced by forcing the singular vectors to orthogonality has the smallest possible influence on the result.

Actually what we do when we take the SVD of the SD matrix is that we take the SVD of the modal decomposition as given by Eq. (10.25). This can also be seen as the outer product contribution to the SD matrix from the two vectors $g_1(f)\,\mathbf{a}_1$ and $g_2(f)\,\mathbf{a}_2$ because Eq. (10.25) can also be written as

$$\mathbf{G}_y(f) = \left(g_1(f)\,\mathbf{a}_1\right)\left(g_1(f)\,\mathbf{a}_1\right)^H + \left(g_2(f)\,\mathbf{a}_2\right)\left(g_2(f)\,\mathbf{a}_2\right)^H \tag{10.26}$$

As indicated on the left side of Figure 10.2, when the left mode is dominating, that is, $g_1(f) > g_2(f)$ then the vector $g_1(f)\,\mathbf{a}_1$ is also dominating the SVD, and even though the SVD is forcing the corresponding two vectors $s_1(f_1)\,\mathbf{u}_1$ and $s_2(f_1)\,\mathbf{u}_2$ to be orthogonal, the influence is minimum on the vector $s_1(f_1)\,\mathbf{u}_1$. We should perform the estimation of the mode shape at such a point defining

$$\hat{\mathbf{a}}_1 = \mathbf{u}_1 \tag{10.27}$$

We can conclude that the left mode shape $\mathbf{a}_1$ should be estimated at a frequency where the ratio s_1/s_2 is as large as possible, and similarly the right mode shape $\mathbf{a}_2$ should be estimated at a (another) frequency where s_1/s_2 is as large as possible and at this point we obtain the corresponding estimate

$$\hat{\mathbf{a}}_2 = \mathbf{u}_1 \tag{10.28}$$

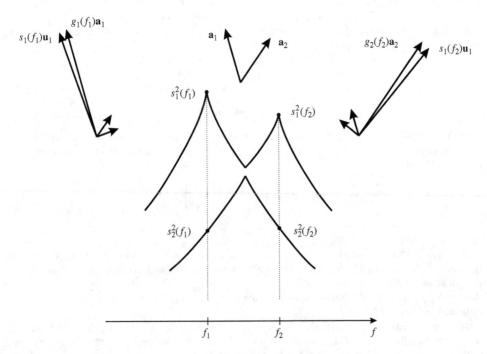

Figure 10.2 Minimizing the bias from forcing the singular vector to geometrical orthogonality. When the left mode is dominating $g_1(f) > g_2(f)$ then the vector $g_1(f)\,\mathbf{a}_1$ is dominating the modal contribution and the corresponding SVD estimate $s_1\mathbf{u}_1$ is close to the vector $g_1(f)\,\mathbf{a}_1$ even though the two mode shapes are not geometrically orthogonal. Therefore, any mode shape in FDD is estimated as the first singular vector at any frequency line where the corresponding singular value is dominating

Or we can state the following main rule in FDD mode shape estimation: *Any mode shape in FDD should be estimated as the first singular vector at a frequency line where the corresponding singular value is dominating.*

From the aforementioned analysis, it appears that at some point in between the two considered natural frequencies where the singular values are equal or nearly equal, the bias from the orthogonality is equally distributed at the two singular vectors.

However, if the two modes are very closely spaced the individual mode shape estimates are not interesting as such because in the limit where the poles are repeated, only the subspace defined by the two eigenvectors is related to the physical aspects of the problem (see Section 5.3.3). In this case, the actual representation of the two mode shapes can be freely chosen as long as the mode shapes serve as basis for the subspace. In this case, it makes sense to represent the subspace by two mode shapes that are geometrically orthogonal, and therefore for this case it might in fact be considered to use the two singular vectors at the midpoint (or where the first and second singular value are approximately equal) in between the two peaks to be representative for the mode shapes of the two closely spaced modes.

10.3.4 Pole Estimation

The simplest version of the FDD only estimates the mode shapes as explained in the preceding section and then provides an estimate of the natural frequency, normally the frequency of the frequency line corresponding to the peak where the mode shape were estimated. This process is also denoted peak-picking FDD and was the first version of the FDD introduced by Brincker et al. [5, 7].

The peak-picking FDD provides a rough estimate of the natural frequency because it is limited by the frequency resolution of the spectral density estimate and is influenced by noise. Actually, if noise is present on the spectral estimate – which is normally the case – the uncertainty on the estimate of the natural frequency can easily be higher than the frequency resolution. Further, the peak-picking FDD does not provide any estimate of the modal damping.

However, for many practical applications where damping is not an issue, the peak-picking FDD is to be considered due to its close relation to intuition and its user friendliness.

If higher accuracy on the natural frequencies and/or damping estimates is needed, then a better approach must be used. The first attempt in this direction was accomplished by the enhanced FDD (EFDD), Brincker et al. [6] where the identification was performed using information not only at the peak of a mode, but also including the information from the SVD around the considered peak.

In the classical EFDD, a primary mode shape estimate is defined as explained in the preceding section as the first singular vector at the considered peak. At the spectral lines around the peak, all the singular values were searched to find the singular value at each frequency line where the corresponding singular vector has the highest MAC value with the primary mode shape estimate, and this singular value is then associated with the single degree-of-freedom system auto spectral density function.

This procedure has some difficulties in isolating the modal coordinates in cases of closely spaced modes, and therefore, in the following sections a more efficient estimation procedure is explained. This procedure was first introduced by Zhang et al. [11] and later further developed by Brincker and Zhang [12]. The idea in this version of the FDD is to isolate the modal coordinates by modal filtering. Considering a certain frequency band with a set of modes dominating in this band, we have similarly to Eq. (10.25) the following expression

$$\mathbf{G}_y(f) = g_1^2(f)\,\mathbf{a}_1\mathbf{a}_1^H + g_2^2(f)\,\mathbf{a}_2\mathbf{a}_2^H + \cdots \tag{10.29}$$

We now define a matrix $\mathbf{V} = \begin{bmatrix} \mathbf{v}_1, & \mathbf{v}_2,, & \cdots \end{bmatrix}$ such that

$$\mathbf{V}^H\mathbf{A} = \mathbf{I} \tag{10.30}$$

It follows directly from Eq. (10.30) that the matrix $\mathbf{V}$ can be defined as the Hermitian of the pseudo inverse of the mode shape matrix

$$\mathbf{V} = \left(\mathbf{A}^+\right)^H \tag{10.31}$$

We see that the set of vectors $\mathbf{v}_1, \mathbf{v}_2, \ldots$ is orthogonal to the mode shapes $\mathbf{a}_1, \mathbf{a}_2, \ldots$. Therefore, performing the inner product of any of the vectors from the orthogonal set over the SD matrix isolates the spectral density of the corresponding modal coordinate

$$\mathbf{v}_n^H \mathbf{G}_y(f)\, \mathbf{v}_n = g_n^2(f) \tag{10.32}$$

The frequency-domain function can then be taken back to time domain and frequency and damping can be estimated by simple means interpreting the correlation function as a free decay of a corresponding SDOF system. This can be done, for instance, as in the original paper by Brincker et al. [6] by performing a linear regression on the logarithm of the extremes and on the crossing times of the free decay to estimate the logarithmic decrement and the period, respectively, of the SDOF oscillator. It can be done by performing some kind of fit of an SDOF model to the isolated modal coordinate as described in Zhang [11] or it can be done simply by using a time-domain identification technique with one DOF, for instance, by using ITD on each SDOF free decay as it is done in Example 10.2.

It should be noted that it is not possible to perform modal filtering using more modes than the number of sensors used in the test. Further, due to estimation errors the number of modes should be somewhat smaller than the number of sensors in order to have an over determined system that makes the modal filtering of Eq. (10.32) robust to noise and estimation errors. On the other hand, in order to take noise in the SD matrix and residues from surrounding modes into account some more modes than just the modes that need to be estimated should normally be included in the modal filtering. Thus, the number of modes included is a trade-off between these two different kinds of errors.

Finally, it should be noted that in the FDD the problem of noise modes is completely omitted because the user has decided the number of modes to be estimated, and has also determined approximately where each natural frequency is placed. This choice is made by deciding at which frequency lines the mode shapes are estimated according to Eqs. (10.27/10.28).

Example 10.2   Estimating the first three modes of the HCT building case using FDD

To estimate the first three modes of the HCT building case we consider the frequency band from DC up to 2.5 Hz, see Figure 10.3. The first step is to evaluate how many modes we have in the considered band and in FDD this choice is made by looking at a plot of the singular values of the SD matrix.

Figure 10.3 shows this plot for the first data set with the SD matrix estimated by the direct technique and depicting the first four singular values. As it appears, three singular values are peaking inside the considered interval while the fourth singular value is flat. The fourth singular value defines the noise floor and the first three singular values describe the physical aspects of the system. Therefore, it can be concluded that three modes are present.

All four data sets were analyzed taking as reference the same three frequency lines that are indicated in Figure 10.3 by asterisks – in the following discussion denoted as the initial frequencies. It should be noted that the initial frequency lines does not fall exactly at the peak of each of the three spectral peaks in Figure 10.3. This is because the three initial frequencies must be chosen in such a way that in all four data sets we obtain a reasonable modal decomposition. The modal decomposition were carried out using Eq. (10.32) with the three mode shapes estimated as the first singular vector at the initial frequencies over a frequency band centered at the initial frequencies and with a bandwidth of 1 Hz. As it appears in this figure the FDD clearly separates the three modes.

Each modal coordinate was then taken back to time domain by inverse discrete Fourier transform, and the poles were estimated from the corresponding correlation function using ITD on each single channel

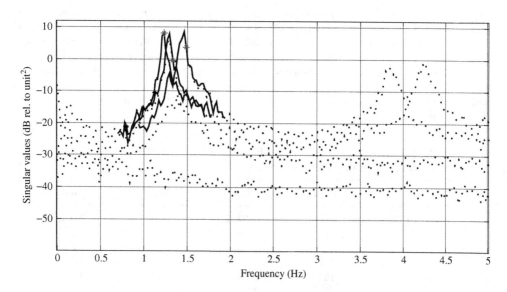

Figure 10.3 Results of using the FDD to identify the first three modes on the first data set of the HCT case. The singular values of the SD matrix are shown as dotted lines. In the plot, the frequency lines where the mode shape vectors are estimated are indicated by an asterisk and the corresponding modal decomposition using Eq. (10.32) is shown by the solid lines

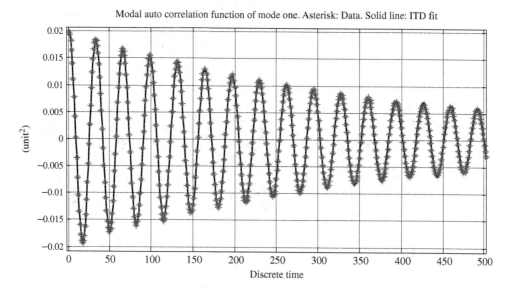

Figure 10.4 Results of using the FDD to identify the first mode on the first data set of the HCT case. The plot (asterisks) shows the autocorrelation function found by taking the modal coordinate estimated in the frequency domain by modal filtering according to Eq. (10.32) back to time domain by inverse discrete Fourier transform. The fit on the SDOF system is shown by the solid line

SDOF free decay (this forces the algorithm to return only one pole). The correlation function of the first mode of the first data set is shown in Figure 10.4. This plot also shows the free decay estimated by ITD.

The results of the identification are shown in Table 10.3 together with the modal participation factor estimated as described in Section 10.1.1. As mentioned before, in FDD we do not have to deal with noise modes because the user has decided how many modes are present by determining the initial frequencies. It should be noted that in data set 4, the second mode has a modal participation of less than one percent, which is significantly smaller than in the other data sets. This explains why the fourth data set is more difficult to handle.

Table 10.3 Identification results using FDD on the HCT building case, participation factors calculated as described in Section 10.1.1

	Parameter	Mode 1	Mode 2	Mode 3
Data set 1	f_n (Hz)	1.2252	1.2867	1.4500
	ς_n (%)	1.29	1.41	1.59
	π_n (%)	27.86	17.37	54.77
Data set 2	f_n (Hz)	1.2364	1.2862	1.4491
	ς_n (%)	2.58	1.42	1.94
	π_n (%)	15.61	32.20	61.18
Data set 3	f_n (Hz)	1.2332	1.2888	1.4557
	ς_n (%)	1.36	1.30	0.94
	π_n (%)	7.23	6.97	85.80
Data set 4	f_n (Hz)	1.2346	1.2929	1.4499
	ς_n (%)	2.72	2.45	1.27
	π_n (%)	35.52	0.72	63.76

Finally the mode shape components corresponding to the modes shown in Table 10.3 were merged and the resulting mode shapes are shown in Figure 10.5.

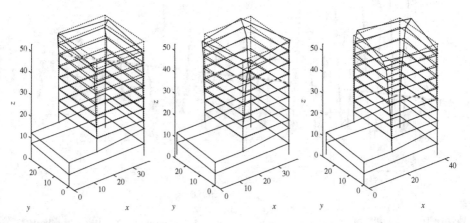

Figure 10.5 The first three mode shapes of the HCT building found by FDD identification merging the mode shape components from the four data sets

10.4   ARMA Models in Frequency Domain

The general problems of performing identification in the frequency domain using this kind of modeling are discussed in Pintelon and Schoukens [13] and in Guillaume et al. [14]. One of the first formulations for use in modal analysis is given in Zhang et al. [15] and some of the important references related to applications in OMA are Parloo [16] and Peeters et al. [17, 18].

In time domain, we can exclude the initial conditions, or the impulse that initiated the response, and only deal with the free response after the structure has been excited as we do it in a free response formulation such as Eq. (9.10), which we repeat here for convenience

$$\mathbf{y}(n) - \mathbf{A}_1 \mathbf{y}(n-1) - \mathbf{A}_2 \mathbf{y}(n-2) - \cdots - \mathbf{A}_{na} \mathbf{y}(n-na) = 0 \qquad (10.33)$$

In order to take this equation to frequency domain, a straightforward tool to do this is to use the Z-transform; see Section 4.4. However, as we have explained in Section 4.4.1, the Z-transform includes everything that has happened in the time domain, thus we cannot exclude the initial conditions and/or impulses from the analysis; everything that happened in the time domain must be included in the frequency-domain model.

This is the big drawback of the frequency-domain modeling: simple models such as Eq. (10.33) cannot be directly applied. One can say that the advantage of the frequency-domain approach is that the modes naturally decompose and thus only have an influence in a limited frequency band, but the drawback is that whatever happened at any time is mixed into every data point in the frequency domain.

Therefore, a right-hand side non-zero term must be added to Eq. (10.33). Thus in principle, we need a full ARMA model, like the one introduced in Eq. (9.27), which in generalized form can be expressed as, see Eq. (5.203)

$$\mathbf{y}(k) = \mathbf{A}_1 \mathbf{y}(k-1) + \cdots \mathbf{A}_{na} \mathbf{y}(k-na) + \mathbf{e}(k) + \mathbf{B}_1 \mathbf{e}(k-1) + \cdots \mathbf{B}_{nb} \mathbf{e}(k-nb) \iff$$

$$\mathbf{y}(k) - \sum_{n=1}^{na} \mathbf{A}_n \mathbf{y}(k-n) = \mathbf{e}(k) + \sum_{n=1}^{nb} \mathbf{B}_n \mathbf{e}(k-n) \qquad (10.34)$$

As discussed in Section 4.4.4, we can take the Z-transform of this difference equation using the time shift property (4.106) to obtain

$$\left(\mathbf{I} - \sum_{n=1}^{na} \mathbf{A}_n z^{-n}\right) \tilde{\mathbf{y}}(z) = \left(\mathbf{I} + \sum_{n=1}^{nb} \mathbf{B}_n z^{-n}\right) \tilde{\mathbf{e}}(z) \qquad (10.35)$$

where $\tilde{\mathbf{y}}(z)$ and $\tilde{\mathbf{e}}(z)$ are the Z-transforms of the response and the input, respectively. The transfer function that relates the input and the output $\tilde{\mathbf{y}}(z) = \mathbf{H}''(z)\tilde{\mathbf{e}}(z)$ can then be obtained

$$\mathbf{H}''(z) = \left(\mathbf{I} - \sum_{n=1}^{na} \mathbf{A}_n z^{-n}\right)^{-1} \left(\mathbf{I} + \sum_{n=1}^{nb} \mathbf{B}_n z^{-n}\right) \qquad (10.36)$$

If we have the data represented as a measured frequency response function $\mathbf{H}(f)$, then we can use it to evaluate Eq. (10.36) on the unit circle

$$z = e^{iw(k)} \qquad (10.37)$$

where the dimensionless frequency $w(k)$ goes from zero to π when the frequency $f = f(k)$ goes from zero (DC) to the Nyquist frequency. The Z-transform becomes the one-sided Fourier transform, see Eqs. (4.92/4.100), and using this in Eq. (10.36) we have that

$$\mathbf{H}(f(k)) = \left(\mathbf{I} - \sum_{n=1}^{na} \mathbf{A}_n e^{-inw(k)}\right)^{-1} \left(\mathbf{I} + \sum_{n=1}^{nb} \mathbf{B}_n e^{-inw(k)}\right) \qquad (10.38)$$

The coefficient matrices in this equation can then be found by nonlinear optimization. In principle, we have the same problems as mentioned when using ARMA models in the time domain. The problems with convergence of the nonlinear optimization are also present here, although to a smaller degree because only a few modes need to be estimated at the same time if a smaller frequency band is considered.

In OMA, the estimation can be performed on the half spectrum because this spectrum can be interpreted as free responses, and this is also the recommended solution in PolyMax; see for instance Peeters et al. [17]. In this reference, it is also explained how the nonlinear problem can be linearized. Once this is done, the problem can be solved by a least squares formulation of what is called a reduced set of normal equations.

A simple, but approximate identification can be performed assuming that the right-hand side polynomial in Eq. (10.35) is a constant matrix. This can be considered as an approximate AR model or poly reference model for the frequency domain; therefore, in the following sections we will denote this model as FDPR. Taking the response to be equal to the half spectrum transposed $\mathbf{Y}(f) = \mathbf{G}_y^T(f)$ and following the same arguments as for Eqs. (10.37/10.38), we have

$$\left(\mathbf{I} - \sum_{n=1}^{na} \mathbf{A}_n e^{-inw(k)} \right) \mathbf{Y}(f(k)) = \mathbf{Y}_0 \tag{10.39}$$

or we can use the alternative expression found from Eq. (10.39) by multiplying by the inverse of the constant matrix $\mathbf{Y}_0$

$$\sum_{n=0}^{na} \mathbf{A}'_n e^{-inw(k)} \mathbf{Y}(f(k)) = \mathbf{I} \tag{10.40}$$

thus the constant matrix is

$$\mathbf{Y}_0 = \mathbf{A}'^{-1}_0 \tag{10.41}$$

and the autoregressive matrices are given by

$$\mathbf{A}_n = -\mathbf{A}'^{-1}_0 \mathbf{A}'_n \tag{10.42}$$

Equation (10.40) can also be written as

$$[\mathbf{A}'_{na}, \ \cdots \ , \mathbf{A}'_1, \mathbf{A}'_0] \begin{bmatrix} \mathbf{Y}(f(k)) \, e^{-i(na)w(k)} \\ \vdots \\ \mathbf{Y}(f(k)) \, e^{-i(1)w(k)} \\ \mathbf{Y}(f(k)) \, e^{-i(0)w(k)} \end{bmatrix} = \mathbf{I} \tag{10.43}$$

$$\Updownarrow$$

$$\mathbf{A}\mathbf{Y}_{na}(f(k)) = \mathbf{I}$$

All unknown matrices $\mathbf{A}'_{na}, \ \cdots \ , \mathbf{A}'_1, \mathbf{A}'_0$ can then be found by solving the following overdetermined set of linear equations defined over the frequency band $B = [f(k_1) \, ; \, f(k_2)]$

$$\mathbf{A}\left[\mathbf{Y}_{na}(f(k_1)), \mathbf{Y}_{na}(f(k_1+1)), \ \cdots \ , \mathbf{Y}_{na}(f(k_2))\right] \mathbf{Y}_{na}(f(k)) = [\mathbf{I}, \mathbf{I}, \cdots, \mathbf{I}]$$

$$\Updownarrow \tag{10.44}$$

$$\mathbf{A}\mathbf{H}_1 = \mathbf{H}_2$$

As described in Chapter 3, the solution to the overdetermined set of equations is given by

$$\hat{\mathbf{A}} = \left(\mathbf{H}_1^{T+} \mathbf{H}_2^T \right)^T = \mathbf{H}_2 \mathbf{H}_1^+ \tag{10.45}$$

The modal parameters can then be found by forming the companion matrix based on the autoregressive coefficient matrices found from Eq. (10.42) as given by Eq. (5.206) and performing the eigenvalue decomposition (5.210). The modal model has in this case $na \times nc$ eigenvalues corresponding to $na \times nc/2$ modes[11] where nc is the number of channels in the measured response (the half spectrum matrix is an $nc \times nc$ matrix).

It should be noted that since Eq. (10.39) is an approximation, the identification is biased, and therefore in order to reduce the bias as much as possible the need for oversized models is critical and the model is mainly suited for narrow band identification. By using the full formulation in Eq. (10.35) the solution of the nonlinear problem defines a model that can include all important terms. Therefore, such model can be applied for broad band identification. Further, it should be noted that in such model the modal participation vectors are determined as a part of the solution.

Example 10.3   Estimating the first three modes of the HCT building case using FDPR

As in Example 10.2, we consider the first three modes of the HCT case. The identification was carried out based on the half spectrum matrix estimated with 1025 frequency lines. In all four data sets, the autoregressive matrices were found using Eq. (10.44) over a the frequency band with a center frequency of 1.35 Hz and a band width of 0.6 Hz. The first three data sets were identified using a model order of $na = 2$ (corresponding to six modes for the first data set and eight modes for data set two and three), however, for the last data set in order to identify all three modes with a reasonable modal participation, a model order of $na = 8$ was used (corresponding to 32 modes).

In the frequency domain, it is common to find many noise modes with high damping, thus in this case not only the modes with negative damping have been excluded, but also all modes with a damping ratio higher than 5% have been excluded from the results. The results of the identification are shown in Table 10.3 together with the modal participation factor estimated as described in Section 10.1.1.

Table 10.4   Identification results using FDPR on the HCT building case, the participation factors were calculated as described in Section 10.1.1

	Parameter	Mode 1	Mode 3	Mode 3	Mode 4	Mode 5	Mode 6
Data set 1	f_n (Hz)	**1.2275**	**1.2896**	**1.4543**			
	ς_n (%)	**1.32**	**1.46**	**1.36**			
	π_n (%)	**29.71**	**19.03**	**51.26**			
Data set 2	f_n (Hz)	**1.2317**	**1.2909**	1.3401	1.4500		
	ς_n (%)	**2.11**	**1.48**	3.46	1.46		
	π_n (%)	**14.78**	**33.18**	0.21	51.81		
Data set 3	f_n (Hz)	**1.2366**	**1.2866**	**1.4554**			
	ς_n (%)	**1.26**	**1.25**	**0.88**			
	π_n (%)	**7.42**	**7.47**	**85.10**			
Data set 4	f_n (Hz)	0.9113	**1.2129**	**1.3006**	**1.4458**	1.7866	
	ς_n (%)	2.72	**2.52**	**2.62**	**1.08**	0.93	
	π_n (%)	1.2411	**31.81**	**0.21**	**63.62**	3.12	

[11] The number of eigenvalues $na \times nc$ might not be an even number, thus the number of modes might not necessarily be a natural number; however, this just means that the "half mode" corresponding to only a single eigenvalue probably does not represent anything of physical interest.

As it appears for the first three data sets, the three first modes were easily identified with a modal participation that corresponds closely to the results of Example 10.2. For the last data set, the middle mode is again very weakly excited and thus difficult to identify. Finally the mode shapes corresponding to the modes listed in Table 10.4 were merged and the resulting mode shapes are shown in Figure 10.6.

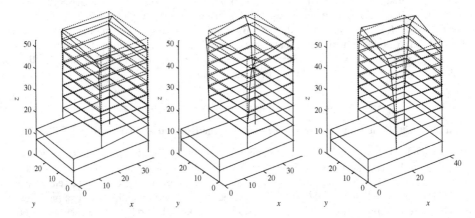

Figure 10.6 The first three mode shapes of the HCT building found by FDPR identification merging the mode shape components from the four data sets

References

[1] Zhang, L., Brincker, R. and Andersen, P.: *An overview of operational modal analysis: Major developments and issues.* In Proceedings of the 1st. Int Operational Modal Analysis Conference (IOMAC), Apr. 26-27, Copenhagen, 2005.

[2] Masjedian, M.H. and Keshmiri, M.: *A review on operational modal analysis researches: Classification of methods and applications.* In Proceedings of the 3rd. Int Operational Modal Analysis Conference (IOMAC), May 4–6, Portonovo (Ancona), 2009.

[3] Felber, A.J.: *Development of a hybrid bridge evaluation system.* PhD. thesis, University of British Columbia, Vancouver, Canada, 1993.

[4] Bendat, J.S. and Piersol, A.G.: *Engineering applications of correlation and spectral analysis.* 2nd edition. John Wiley & Sons, Inc, 1993.

[5] Brincker R., Zhang L., and Andersen P.: *Modal identification from ambient responses using frequency domain decomposition.* In Proceedings of IMAC 18, the International Modal Analysis Conference, p. 625–630, San Antonio, TX, USA, February 2000.

[6] Brincker, R., Ventura, C.E. and Andersen, P.: *Damping estimation by frequency domain decomposition.* In proceedings of the 19th International Modal Analysis Conference (IMAC), February 5–8, Kissimmee, Florida, p. 698–703, 2001.

[7] Brincker, R., Zhang, L. and Andersen, P.: *Modal identification of output-only systems using frequency domain decomposition.* Smart Mater. Struct., V. 10, p. 441–445, 2001.

[8] ARTeMIS Extractor and Modal software. Structural Vibration Solutions A/S, Denmark.

[9] Shih C.Y., Tsuei Y.G., Allemang R.J. and Brown D.L.: *Complex mode indication function and its application to spatial domain parameter estimation.* Mech. Sys. Signal Process., V. 2, No. 4, p. 367–377, 1988.

[10] Brincker, R. and Andersen, P.: *Method for vibration analysis.* United States patent no. US 6,779,404 B1, August 24, 2004.

[11] Zhang, L., Wang, T. and Tamura, Y.: *A frequency-spatial decomposition (FSDD) technique for operational modal analysis*. In Proceedings of the first International Operational Modal Analysis Conference (IOMAC), April 26–27, Copenhagen, p. 551–559, 2005.

[12] Brincker, R. and Zhang, L.: *Frequency domain decomposition revisited*. In Proceedings of the 3rd International Operational Modal Analysis Conference (IOMAC), May 4–6, Ancona, p. 615–626, 2009.

[13] Pintelon, R. and Schoukens, J.: *System Identification, a frequency domain approach*. IEEE Press, 2001.

[14] Guillaume P., Verboven P., Vanlanduit, S., Van der Auweraer, H. and Peeters, B.: *A poly-reference implementation of the least-squares complex frequency domain-estimator*. In Proceedings of the 21st International Modal Analysis Conference, Kissimmee (Florida), February 2003.

[15] Zhang, L., Kanda, H., Brown, D.L. and Allemang, R.J.: *Polyreference frequency method for modal parameter identification*. American Society of Mechanical Engineers, ASME Design Engineering Technical Conference, 1985.

[16] Parloo, E.: *Application of frequency-domain system identification techniques in the field of operational modal analysis*, PhD thesis, Department of Mechanical Engineering, Vrije Universiteit Brussels, Belgium, 2003.

[17] Peeters, B., Van der Auweraer, H., Guillaume, P. and Leuridan, J.: *The PolyMAX frequency-domain method: a new standard for modal parameter estimation? Shock Vibr.*, V. 11, p. 395–409, 2004.

[18] Peeters, B., Van der Auweraer, H., Vanhollebeke, F. and Guillaume, P.: *Operational modal analysis for estimating the dynamic properties of a stadium structure during a football game. Shock Vibr.*, V. 14, p. 283–303, 2007.

[19] PolyMax software. LMS, Siemens PLM software. m.plm.automation.siemens.com

11

Applications

"If experience was so important, we'd never have had anyone walk on the moon"

– Doug Rader

In this chapter, we discuss about practical applications of OMA. Without practical applications of OMA, this technology would not be of much interest to the engineering community. The modal parameters of any structure are not necessarily of great interest to many engineers; it is what we do with the modal parameters that make them useful to us. It is the variety of applications of modal parameter estimated by OMA that justifies this book. On the other hand, since the technology is relatively new, current applications are mainly limited to academic and scientific cases, or limited by our ability to "think outside the box" and find new areas where this technology can be applied. As indicated by the colophon of this chapter, our ability to move forward and try new things – in this case new OMA applications – is essential in all innovation, and we hope that applications making use of OMA will grow rapidly in the future. This also means that this chapter will be a much larger part of the next editions of this book.

We discuss first some practical issues related to OMA, such as the concept of the modal assurance criterion (MAC), how to assemble modes shapes from the results of the data analysis and what are the important issues to consider when this is being done. Next we address a very important issue that has gained significant attention in the last few years: model validation and updating using OMA results. We show how this can be done in practice using OMA test results. The third section of this chapter deals with the application of OMA in the field of structural health monitoring (SHM). In recent years, the focus of SHM has been on damage detection and localization, but SHM is more than that and we discuss the areas of SHM where OMA can be of practical value. For illustrations of how to deal with the subject of noise modes, the reader is also referred to Chapter 12 where stabilization diagrams are also discussed.

Finally, we present three case studies that illustrate how OMA has been used to determine the dynamic properties of large structures.

11.1 Some Practical Issues

The idea in OMA is to obtain physical information from the measured data, expressed generally in terms of spectral densities in the frequency domain, or in terms of correlation functions in the time domain. Normally this is done using some kind of fitting technique using a parametric model and regression as described in Chapters 3, 9, and 10. Time–frequency domain techniques can also be used for this purpose, but the treatment of data using these types of techniques is outside the scope of the present edition of this book.

Introduction to Operational Modal Analysis, First Edition. Rune Brincker and Carlos Ventura.
© 2015 John Wiley & Sons, Ltd. Published 2015 by John Wiley & Sons, Ltd.
Companion Website: www.wiley.com/go/brincker

11.1.1 Modal Assurance Criterion (MAC)

When performing modal tests in OMA, it is often useful to compare results from different tests on the same structure or from identification algorithms using the same collection of data sets. Comparing natural frequencies and damping values is a straightforward procedure because in this case we are dealing with single values. If the difference between two sets of values is less than a certain threshold value, then we consider the results satisfactory. This threshold value is generally defined by self-experience and the expected uncertainties associated with the test, or by prescribed values from standards of vibration tests, or by the owner or operator of the structure being tested.

When we are comparing mode shapes, the many degrees of freedom at each measured location make it very difficult to compare these quantities and in this case a correlation measure is normally used. Considering two different vectors $\mathbf{a}$ and $\mathbf{b}$, that are to be considered as two different estimates of the same experimental mode shape vector, we can calculate the correlation between the two vectors according to the MAC normally formulated as (see Allemang [1])

$$\mathrm{MAC}(\mathbf{a}, \mathbf{b}) = \frac{|\mathbf{a}^H\mathbf{b}|^2}{(\mathbf{a}^H\mathbf{a})(\mathbf{b}^H\mathbf{b})} \tag{11.1}$$

We can see that in the denominator of the expression above the terms $\mathbf{a}^H\mathbf{a}$ and $\mathbf{b}^H\mathbf{b}$ are the lengths of the vectors squared. By comparing this equation with Eq. (3.21), we see that the MAC value is equal to $\cos^2(\theta)$, where θ is the generalized angle between the two vectors.

If the shape vectors are very similar the MAC value is close to unity. In cases where the deviation between the two vectors is small, it is often useful to consider the value of the generalized angle θ as a measure of the deviation of the MAC value from unity.

In the case of closely spaced modes, the MAC value between the corresponding mode shape vectors does not make much sense. In this case the vectors are very sensitive to small perturbations that might not be of any physical significance (this is explained and illustrated in detail in Section 5.3.3). When this happens, the S2MAC value discussed in Section 12.1.1 should be considered. Alternatively, more general methods based on the subspaces of mode shape vectors should be considered for a more appropriate estimation of the correlation between two mode shapes of closely spaced modes.

Equation 11.1 can also be used to compare the modal estimates from two different OMA techniques using the same data sets. In such case vectors $\mathbf{a}$ and $\mathbf{b}$ correspond to the same mode, at the same frequency, but have been obtained from two different techniques. For instance, vector $\mathbf{a}$ would be the result obtained using the FDD technique and vector $\mathbf{b}$ would have been obtained using one of the SSI techniques.

11.1.2 Stabilization Diagrams

In general, a stability diagram is a plot of dynamic properties estimates, such as frequencies, for different curve fitting model sizes. A stability diagram has two advantages: (i) it is useful to determine how many modes are really contained in a frequency band; and (ii) by displaying a stable pole estimate regardless of the model size, it can be used to confirm that the estimate is correct.

In practice, different threshold values are defined to check the consistency of the modal properties. For instance, the following relations for frequency, damping and mode shapes can be used to develop a stabilization diagram:

$$|f_1 - f_2| < \Delta f_{\mathrm{th}}, |\varsigma_1 - \varsigma_2| < \Delta \varsigma_{\mathrm{th}}, \quad \text{and} \quad |f_1 - f_2| < f_{\mathrm{th}}, \mathrm{MAC}(\mathbf{a}_1, \mathbf{a}_2) > \mathrm{MAC}_{\mathrm{th}} \tag{11.2}$$

where f_1, ς_1, and $\mathbf{a}_1$ are modal frequency, damping ratio, and mode shape for a certain mode estimator or model order, and f_2, ς_2, and $\mathbf{a}_2$ are modal frequency, damping ratio, and mode shape for another mode estimator, or next model order. The thresholds Δf_{th}, $\Delta \varsigma_{\mathrm{th}}$, and $\mathrm{MAC}_{\mathrm{th}}$ are levels for frequency, damping ratio, and mode shape, respectively, that need to be specified by the analyst; like, 0.02 Hz,

0.01, and 0.9, respectively. A further discussion on generalizing the stabilization diagrams is presented in Section 12.4.2.

11.1.3 Mode Shape Merging

If the number of measurements,[1] that is, the number of components N_c in the response vector $\mathbf{y}(t) = \{y_r(t)\}; r = 1, 2, \dots , N_c$ is smaller than the needed number of components N_s to define the mode shape vector estimated in the OMA test $\mathbf{a} = \{a_r\}; r = 1, 2, \dots , N_s$, that is, if $N_c < N_s$, then we need to do repeated testing in order to measure all DOF's of interest on the structure. Each response measurement in this procedure is denoted as a data set[2] or a setup.

As explained earlier, normally we do not obtain scaled mode shapes, but each part of the mode shape has an arbitrary constant present on the mode shape vector, and thus, we cannot fit the different mode shape pieces together without doing something first to assemble a meaningful overall mode shape.

To be able to merge the different mode shape pieces together make sure that the different data sets have a common (overlapping) set of points. This set of points is denoted as the reference points, and the corresponding measurements are denoted as the "reference measurements."[3]

A simple and novel technique to assemble mode shapes from a series of test setups using a frequency domain approach was introduced by Felber in the early 1990s [2]. In this approach, the natural frequencies of the structure are determined first from each of the individual test setups. At each identified frequency, the spectral ratio between each roving location and the reference location are determined first. The amplitude is used to define the amount of displacement between the degree of freedom at the roving location and the reference location, and the direction of the displacement is determined by the phase angle obtained from the spectral ratio calculation. The coherence between the roving location and the reference location is also used to verify the linearity between the two points. Felber introduced the concept of potential modal ratio (PMR) to present in one single parameter amplitude, phase angle, and coherence between roving points and reference points. The mode shapes are then assembled by forming the vectors based on the geometry of the measured locations and the values of the PMRs for each of the setups. Merging of the mode shapes is based on the fact that all the spectral values of the reference points are normalized to unity. The details of this procedure are given in Felber [2].

This method is very efficient for systems that exhibit mainly 2-D modes shapes, such as beams and bridges. For more complex structures that exhibit 3-D mode shapes and that have closely spaced modes this technique is not very effective and assembling and merging mode shapes becomes a difficult task. The technique described next is more efficient in such cases, and is due to Peeters [3].

Let there be D data sets, and the different segments of an arbitrary mode shape $\mathbf{a}$ be denoted $\mathbf{a}_1, \cdots \mathbf{a}_D$ each of which can be divided into two parts as follows:

$$\mathbf{a}_i = \left\{ \begin{matrix} \mathbf{a}'_i \\ \mathbf{a}''_i \end{matrix} \right\} \tag{11.3}$$

where the part of the mode shape vector $\mathbf{a}'_i$ is defined over the common (overlapping) set of reference DOF's (or reference sensors), and $\mathbf{a}''_i$ is defined over the remaining set of DOF's. Taking the first mode shape part $\mathbf{a}'_1$ as the origin, we define the scaling between them as

$$\mathbf{a}'_1 = \alpha_{1i}\mathbf{a}'_i \tag{11.4}$$

---

[1] Also the same as the number of sensors being used in each data set.
[2] The applications of data sets are also discussed in Section 7.2.5.
[3] This is not to be confused with term "references" in traditional modal testing. In that case "references" means the number of independent inputs.

If the number of common points is larger than one for each dimension involved (1 for 1D problems, 2 for 2D problems, and 3 for 3D problems), the scaling is found by the least square solution

$$\hat{\alpha}_{1i} = (\mathbf{a'}_1^T \mathbf{a'}_1)^{-1} \mathbf{a'}_i \tag{11.5}$$

and the total mode shape is obtained by merging components as

$$\mathbf{a} = \begin{Bmatrix} \mathbf{a'}_1 \\ \mathbf{a''}_1 \\ \hat{\alpha}_{12}\mathbf{a''}_2 \\ \vdots \\ \alpha_{1D}\mathbf{a''}_D \end{Bmatrix} \tag{11.6}$$

Note that this leaves us with a merged mode shape with an arbitrary scaling factor, which of course can then be scaled to a target value. However, if the considered mode shape has small components in the common set of reference DOF's, then the scaling factor is prone to noise. Because it is difficult to ensure that all mode shapes have reasonable large values in just one common point, it is strongly recommended to use an over specified number of DOF's in the reference set. This ensures an overdetermined estimation of the scaling factor as given by Eq. (11.5) and further, the uncertainty on the mode shape can be evaluated from the different values of mode shape over the common set of DOF's.

11.2 Main Areas of Application

In this section, we briefly discuss the use of OMA results for model updating and validation and for SHM of existing structures. It is not the intent to present a detailed discussion of these topics as this is outside the scope of this book.

11.2.1 OMA Results Validation

In order to have high degree of confidence on the results obtained from OMA, it is always advisable to use two or more different identification techniques to confirm the results. Modal frequencies and damping can be compared on a one-on-one basis and the results are validated if the difference between pairs of values is within a predefined threshold value. The mode shape estimates can be compared using correlation indexes such as the MAC in Eq. (11.1) or the higher order S2MAC in Eq. (12.2) for closely spaced modes.

Another option to validate the OMA results is to repeat the tests under the same or similar conditions and compare the results. This, of course, is most likely to be the case for experiments in the laboratory, as it may not be practically possible to repeat the tests on a large structure such as a bridge or a container ship.

It should be kept in mind that in some cases one OMA technique may identify clearly a frequency, a damping value, or a mode shape, while another technique applied to the same data may not be able to identify one or more of these values. This is not uncommon when using experimental data from field tests, and judgment, experience, and a good understanding of the structure being tested are necessary to make a decision of whether or not the results should be used for further analysis. Typical causes of a situation such as this are the way the data was collected, the signal processing that has been used on the data, the length of the data sets, and the sources of excitation present when the data was obtained. See Chapters 7 and 8 for further discussion appropriate method of measuring data and signal procession.

11.2.2   Model Validation

Validation is the task of demonstrating that the model is a reasonable representation of the real structure, and that it reproduces the behavior of the structure with enough fidelity to satisfy the predefined analysis objectives. The development of a finite element model of a structure will be influenced by the objectives of the study being conducted, so the model validation will also be influenced by these objectives. As a result, the model may have different levels of validity for different components of the structure. For most models, there are three separate aspects, which should be considered during model validation:

- the assumptions being made about the structure being studied;
- the selection of values and their distribution for the input parameters; and
- the desired output values and expected results.

In practice, it is very difficult to achieve such a full validation of the model, except for the simplest structures, and mostly for structures being tested in a laboratory under well-controlled conditions. In general, initial validation attempts will be focused on the output of the model, and only if that validation is not satisfactory, a more detailed validation will be undertaken.

In OMA applications, a successful model validation will be based on:

- expert intuition and sound engineering judgment;
- real system measurements; and
- valid theoretical results/analysis.

Verification is like debugging – it is intended to ensure that the model does what it is intended to do. Therefore, all techniques that can help achieve this goal should be used as part of the arsenal of tools for successfully using OMA results to validate a finite element model of a structure. One of these techniques is model updating.

11.2.3   Model Updating

Model updating involves the controlled adjustment of parameters in a finite element (FE) model in order to yield analytical results that provide an improved match with experimental ones. The differences between experimental and analytical natural frequencies and mode shapes can provide insight into the types of changes necessary in the FE model to obtain a better agreement with the observed experimental response. Two different model updating techniques can be used: manual updating and automated updating.

The manual updating technique begins with the identification of differences in experimental and analytical modal parameters, such as natural frequencies and mode shapes. Next, based on this inspection of results, the analyst must select parameters in the model to change and adjust them in a way that will improve the match between the model results and the experimental ones. This process may be repeated several times until the desired correspondence between experimental and analytical modal characteristics is achieved. There are significant limitations and difficulties in trying to obtain a good general correlation between experimental and analytical modal properties for a large civil engineering structure using this model updating technique. However, if the objective of the model updating process is just to obtain a model that matches one or two experimental modes, then manual updating may suffice.

The automated updating technique follows the same general procedures as in manual updating except that the process is computer aided. The analyst can perform an in-depth inspection of the FE model through the use of a sensitivity analysis. The computer program can determine the sensitivity of the dynamic characteristics of the model to a large number of parameters, which can provide valuable information as to which parameters should be selected for updating. Once the user has completed the parameter selection, the computer program runs a number of iterations to match the experimental and analytical modal properties by changing the selected model parameters.

There are two main advantages of using an automated updating technique. When using a computer program to perform model updating, a wide range of parameters can be varied concurrently in order to achieve the best match between experimental and analytical natural periods and mode shapes. In addition, the model updating process can be made more effective by utilizing the results of the sensitivity analysis performed by the computer program in order to target specific parts of the model for parameter changes and in the parameter selection process.

A computer model of the structure is generally created using the information provided in the design drawings to establish geometry and material properties. At this point, the analyst needs to start making decisions related to the model. For instance, a building with a parking structure in the basement can be modeled with an assumed "fixed base" at ground level, as the experimental results may have indicated that the motions in the underground parking structure were negligible compared to the motions of the upper floors. The main structural elements of the building should, of course, be included in the model, and all setbacks and structural section changes throughout the height of the building should be taken into account.

A rational approach can be taken when selecting parameters for the initial FE model updating. The analyst will have to make a decision on which parameters related to the geometry and structural characteristics are to be selected for updating. In the case of a building, these can include, for instance, the following:

- The Young's modulus, E, of the beams, columns, shear walls, floor slabs, and cladding panels.
- The mass density, ρ, of the elements described above.
- The moment of inertia, I, of the columns.
- The thickness, H, of the cladding panels or of the slabs.

By permitting independent variations in Young's modulus, E, for the different groups of structural elements, it is possible to have a sense how these affect the overall stiffness of the structure, as there is always a degree of uncertainty about the actual material properties of elements and the most realistic representation of stiffness in the model. A variation of the mass density, ρ, of each group of elements makes it possible to establish a realistic mass distribution of structural and nonstructural elements in the model. The moment of inertia, I, and as a consequence the lateral stiffness, of the columns is one of the most uncertain parameters to model in concrete frame structures. It is dependent on the choice of concrete section to be used (cracked or uncracked) and degree of composite action of the steel reinforcement. The lateral stiffness of the columns is also related to the effective length of the members. Modeling the envelope of the structure can be rather complicated. Cladding panels can be modeled as plates of thickness H, and this parameter can be selected for updating since these elements will provide the additional stiffness and mass of the envelope of the building. In practical structural analyses of buildings, very seldom is the influence of cladding taken into account in the structural model, but numerous studies of vibration records from buildings have shown that these do have an influence on the dynamic properties of the structure and should be included in the model. The greatest influence in buildings is the value of the rotational mass moment of inertia of each floor.

Using a sensitivity analysis approach for automated updating, the relative and normalized sensitivities of response of the model to parameter changes must be explored prior to parameter selection for updating. If the model sensitivities to different types of parameters are to be compared simultaneously, the use of relative sensitivities is advised. The relative sensitivity matrix, $\mathbf{S}_r$, is obtained as follows:

$$\mathbf{S}_r = \left[\frac{\partial r_i}{\partial p_j} \right] [P_j] \tag{11.7}$$

where the column vector $\mathbf{r} = \{r_i\}$ represents all the selected responses, the column vector $\mathbf{p} = \{p_j\}$ represents all the selected parameters, the matrix element $\partial r_i / \partial p_j$ is the differential sensitivity coefficient, and $[P_j]$ is a diagonal square matrix holding the actual parameter values. If the differential coefficient $\partial r_i / \partial p_j$ is a positive value, then it means that an increase in the value of the parameter p_j will also result in an increase in the value of the response r_i and if the ratio has a negative value, and increase in the value of parameter p_j will result in a decrease in the value of the response r_i.

The relative sensitivity can also be normalized with respect to the response value as follows:

$$\mathbf{S}_n = \mathbf{S}_r [R_j]^{-1} \tag{11.8}$$

where $\mathbf{S}_n$ is the normalized relative sensitivity matrix, and $[R_j]$ is a diagonal square matrix holding the actual response values. The normalized sensitivity matrix $\mathbf{S}_n$ can be used to compare the effect of changing parameters on the dynamic response of the FE model. It is suggested that a large number of parameters be selected for the sensitivity analysis, but a reduced number of parameters be selected for the automated model updating, Dascotte [4]. The results of a sensitivity analysis can be used to reduce the computational effort required to update a finite element model of a structure.

As an illustration of how OMA results can be used to update a finite element model of a building, let us consider the case of the Heritage Court Tower (HCT) building located in downtown Vancouver, British Columbia in Canada. The automatic updating of the HCT's FE model was carried out utilizing the commercially available model updating program FEMtools [5].

The information presented in the design drawings of the building was used to formulate the geometry and material properties of the model. Since the experimental results indicated that the motions at the ground floor level of the building were negligible compared with the motions at the upper floors, it was decided to model only the superstructure of the building and assume a fixed base condition at the ground level. The main structural elements (concrete core shear walls, gravity load columns, header beams, and load transfer beams at the second floor) were all included in the model. Beams and columns were modeled as 3D beam-column elements and shear walls were modeled as 4-node plate elements. Flat slab floors were modeled mostly as 4-node plate elements. The exterior cladding of the building was also modeled as simplified 4-node thin plates placed near the perimeter of the structure. All setbacks and structural section changes throughout the height of the building were taken into account. The concrete material properties were determined form the design specifications included in the drawings. In total, the model consisted of 348 beam-column elements and 818 plate elements, 1456 nodes, 7 different material properties, and 184 different element geometries. This resulted in an FE model with 8736 degrees of freedom. Two views of the FE model of the building are presented in Figure 11.1: a 3D view of the complete model, and a wire-frame representation emphasizing the core shear walls distribution.

A sensitivity analysis showed that there are 13 different parameters that the program could use for updating the models. The correlation of responses and computation of MAC values between the experimental and analytical models was done at 40 points (four points per floor, at 10 different levels).

The resulting modal frequencies after 13 iterations of updating are presented in Table 11.1. The table includes the experimental frequencies (EMA values), as well as the FEM frequencies before and after updating. The last column of the table shows the MAC values of the updated model. From this table, it can be seen that some of the frequencies of the updated model are for all practical purposes the same as the experimental ones. The largest difference is about 12% for the third mode, but this is still acceptable for practical purposes in this case. The MAC before the updating are very low and only two are over 70%, but the MAC values after updating are also acceptable, and all are over 70%. In fact, only one value is below 80%. Considering the complexity of the actual building and the assumptions made in the development of the FE of the building, the results of the model updating are very useful from the point of view that the confidence on the updated model is much higher than before, and that the updated model can be used for further studies of the building, such as estimating its response to earthquake excitations or severe wind loading.

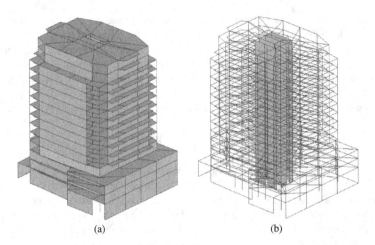

(a)   (b)

Figure 11.1 FE model of HCT building (courtesy of Ventura et al. [6]). (a) Complete Model. (b) Details of Model

Table 11.1 Comparison of first six natural frequencies of the HCT building before and after model updating

Mode No.	OMA Frequency (Hz)	FEM before		FEM updated	
		Frequency (Hz)	MAC (%)	Frequency (Hz)	MAC (%)
1	1.23	1.33	79	1.20	83
2	1.27	1.74	60	1.40	82
3	1.44	2.07	57	1.63	85
4	3.87	4.08	79	3.88	84
5	4.25	4.38	55	4.25	73
6	5.35	5.66	64	5.62	81

The resulting mode shapes of the updated model are shown in Figure 11.2 and a comparison of modes from the reduced FE model and the experimental test is shown in Figure 11.3. A 3D plot of the MAC matrix before and after the model updating is presented in Figure 11.4. The MAC matrix comparison clearly shows how the automatic updating process accomplished a good matching of experimental and analytical modes and how the modes of the initial model changed to match the experimental modes. Since not all the floors of the building were measured, the spatial representation of the higher modes of vibration might not be very accurate and the reliability of the experimental model may not be as high for the higher modes as for the lower modes of vibration. This is why only six modes of vibration were selected for the updating study. However, once a good correlation between experimental and analytical results was obtained, four more experimental modes were added to the analysis and a further refinement to the model was accomplished (see Ventura et al. [6]).

11.2.4 Structural Health Monitoring

SHM is an expression being around for more than two decades now. The development for SHM was mainly motivated by the need to inspect for damage in structures at defined intervals based on

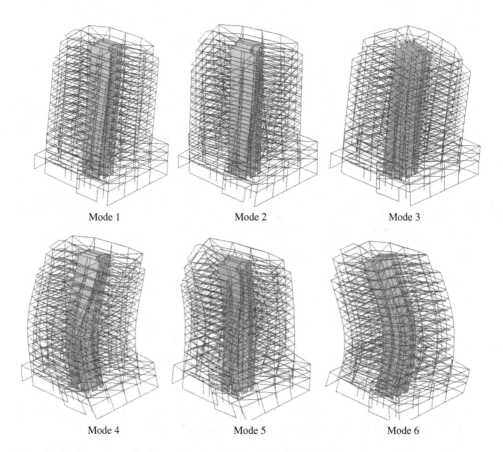

Figure 11.2   First six mode shapes of updated FE model of HCT building (Source: Courtesy of Ventura et al. [6])

nondestructive testing (NDT) techniques (see for instance Boller [7], Worden and Friswell [8], and Farrar et al. [9]). The advances in sensing hardware, better understanding of structural materials, and advanced signal processing software have helped NDT techniques to become an integral part of materials and structures. The result is that today many conventional inspection processes have been automated. Based on the extensive literature that has been developed on SHM over the last 20 years it can be argued that this field has matured to the point where several fundamental axioms, or general principles, have emerged. As stated by Worden et al. [10] these axioms are as follows:

- *Axiom I:* All materials have inherent flaws or defects.
- *Axiom II:* The assessment of damage requires a comparison between two system states.
- *Axiom III:* Identifying the existence and location of damage can be done in an unsupervised learning mode, but identifying the type of damage present and the damage severity can generally only be done in a supervised learning mode.
- *Axiom IVa:* Sensors cannot measure damage. Feature extraction through signal processing and statistical classification is necessary to convert sensor data into damage information.
- *Axiom IVb:* Without intelligent feature extraction, the more sensitive a measurement is to damage, the more sensitive it is to changing operational and environmental conditions.

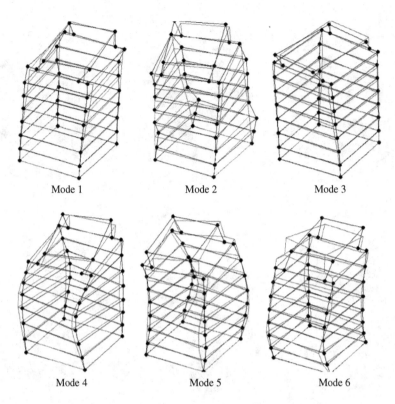

Figure 11.3   Comparison of reduced FE mode shapes of updated model and experimental determined mode shapes of HCT building (Source: Courtesy of Ventura et al. [6])

- *Axiom V:* The length and time scales associated with damage initiation and evolution dictate the required properties of the SHM sensing system.
- *Axiom VI:* There is a trade-off between the sensitivity to damage of an algorithm and its noise rejection capability.
- *Axiom VII:* The size of damage that can be detected from changes in system dynamics is inversely proportional to the frequency range of excitation.

In recent years, a number of researchers and practicing engineers, including the authors, have explored the possible uses of OMA for SHM by making use of the axioms described before. A full treatment of the use of OMA for SHM is outside the scope of this book, but the general aspects of how OMA can be used for SHM can be briefly discussed here. Regarding SHM, the reader is referred to the *Encyclopedia of Structural Health Monitoring* [7] for a detailed treatment of the topic. In a monitoring program normally the following OMA steps are involved.

1. Initial testing and analysis (virgin state)
2. Monitoring (over time)
3. Damage detection (if needed)

Step 1 might include several data sets and also additional testing to estimate scaled modes so that load identification can be carried out during monitoring. For practical reasons, step two is limited to a

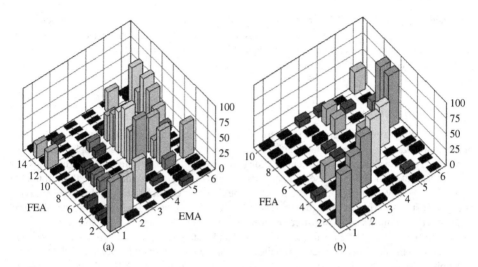

Figure 11.4 Comparison of 3D plots of MAC matrices for first six mode shapes of HCT building.[4] (a) MAC matrix before updating. (b) MAC matrix after updating (Source: Courtesy of Ventura et al. [6])

single data set and thus, normally spatial information is limited to a relatively small number of points at selected locations on structure in such a way that the needed information, or at least a good part of it, can be obtained. However, the limited information can be estimated at more points if mode shapes are known either from step one or from a finite element analysis, see for instance Section 12.3. During monitoring, usually OMA identification must be carried out automatically, thus ways to deal with automated identification becomes a key issue.

When one is monitoring a structure, the loading on the structure might change significantly. The environment will also change (temperature, humidity, etc.). Since the modal parameters might depend on these factors, it is important to gather data in such a way that this influence can be estimated and removed from the interpretation of results. Finally, it is important to evaluate possible nonlinearities in the system.

During step 2, changes in modal parameters might be used to indicate structural change (damage) directly, but OMA combined with signal integration (to obtain displacements) might also be used to find stress histories. From the stress histories, one could estimate accumulated fatigue damage.

If a possible damage has been indicated, step 3 might be implemented. In such case, several data sets might be used to obtain detailed spatial information about the damaged state, preferable at least the same spatial information as in step 1.

11.3 Case Studies

In this section, we present three case studies that illustrate a number of important aspects of OMA testing.[5] The first case shows how OMA can be used to determine the dynamic properties of a building under

---

[4] At the time the original paper [6] was published, the term OMA was not commonly used, so the figures are denoting the experimental results that were obtained by OMA as EMA.

[5] The case studies presented in this section were selected from past publications and are from "truly" OMA field tests. The information was kindly provided by the authors and the publishers of the corresponding references. In order to preserve the originality of the work published, text and figures were reproduced from the original publication with just some editing. This was done with permission from the authors and publishers.

construction; the second case illustrates the use of OMA to determine the dynamic properties of a large complex cable stay bridge; and the third case shows the results of studies done on a cargo ship. All these cases complement the case of the HCT building that is being used in many of the examples in this book. The HCT case is presented in appendix B.

11.3.1 Tall Building

A series of ambient vibration studies was performed on a 32-story high-rise building, called City Tower, in downtown Vancouver, British Columbia, Canada. This residential building, completed in 1994, is a reinforced concrete structure with a centrally located shear wall core surrounded by ten columns around the perimeter of the building. These vertical elements are at each story connected by 200-mm-thick floor slabs. The objective of this testing was to determine dynamic characteristics of this high-rise building at different stages during its construction. The influence of nonstructural elements on modal characteristics of the building was also investigated. A final step in this investigation was to model the building mathematically and calibrate the model's geometric and material properties to correspond to the results obtained from the field measurements. Details of this study are presented by Ventura and Schuster [11].

Ambient vibrations measurements were taken following the completion of 10, 15, 20, 25, and 32 stories during a 5-month period from January to June 1993. A final measurement was conducted in October 1993 when all of the major architectural components were in place.

11.3.1.1 Description of the Building

The building is a 32-story residential tower with three levels of underground parking (Figures 11.5 and 11.6). The residential floors are about $600\,m^2$ from ground to the 10th floor, stepping back to about $500\,m^2$ from the 10th to 25th floors. The floor plates are further reduced on the top penthouse floors. The parking levels are about $1100\,m^2$ split into halves that are staggered by half floors vertically. The tower floors are 190 mm normally reinforced concrete flat plates. Lateral forces are resisted by a centrally

Figure 11.5   South view of City Tower building

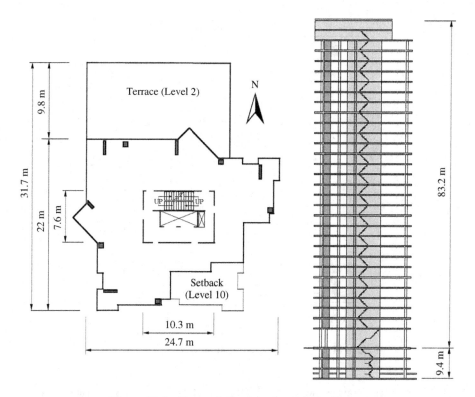

Figure 11.6   Plan and elevation views of City Tower

located reinforced concrete core formed by the stair shaft, elevator shaft. and the corridor leading to the residential suites. The core walls are 450 mm thick at the bottom reducing to 350 mm thick at the top floors.

11.3.1.2   Ambient Vibration Tests

The vibration mode shapes of interest were translational modes in the North–South (NS) and East–West (EW) directions, and torsional modes. To capture the lateral motion associated with these modes, two pairs of unidirectional accelerometers were positioned on selected floors as shown in Figure 11.7 (the arrows indicate location and orientation of the measurement, and the direction of the arrow is the positive polarity of the measurement).

The top floor of the building at the time of each test was selected for the location of the *reference* sensors before the measurements could begin; the cable used to connect the sensors to the data acquisition equipment had to be laid out. Following this, the data acquisition station was set up while the reference sensors were installed and connected. While a preliminary measurement was taken to verify that the equipment was working correctly, the remaining four roving sensors were installed and connected. Following each measurement, the roving sensors were systematically located from floor to floor until the test was completed. Since this entire study was comparative in nature, it was decided that the sensors would be placed in the same location for each of the six tests.

In the first test (10 stories completed), each floor was instrumented. In subsequent tests, only every second floor was instrumented. Rocking motion of the foundation was also of interest. In order to measure

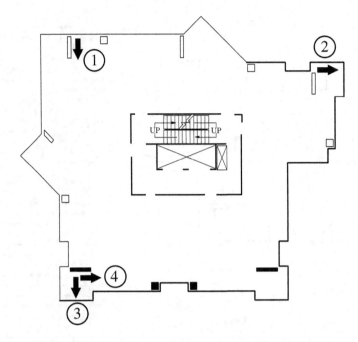

Figure 11.7   Typical sensor locations (see arrows marked 1–4)

this motion, four accelerometers were located at the base of the building, mounted at each corner of the core. These sensors were oriented in the vertical direction.

The OMA test was performed using the hybrid bridge evaluation system (HBES) developed at the University of British Columbia, Felber [2]. Acceleration measurements were obtained using eight Kinemetrics force-balanced accelerometers (model FBA-11). The sampling rate was 40 samples per second. Each record had a length of 32,768 data point (i.e., about 12 min of duration).

Figure 11.8 shows the average of the singular values of the spectral density matrices resulting from the application of the frequency domain decomposition (FDD) technique. This figure shows the averaged singular value plot for all the data sets of City Tower building with 32 stories using the computer program ARTeMIS [12]. Note that there are two closely spaced modes below 1 Hz and another 10 well-separated modes clearly visible below 10 Hz.

11.3.1.3 Typical Mode Shapes

Mode shapes obtained from the fifth test (32 stories completed) using the FDD technique are presented in Figure 11.9. Several of these modes are a combination of translational and rotational motion. For example, the first torsional mode also has a translational component in the N–S direction. The higher translational modes (third and fourth) also show significant rotational motion.

11.3.1.4 Frequency Trends

The first part of this investigation was to monitor the changes in the dynamic behavior of the structure as it was being constructed. Tabulated frequencies for the traditional and FDD techniques are shown in Table 11.2. In this table, CT10 means test results when 10 stories had been completed, CT15 are the

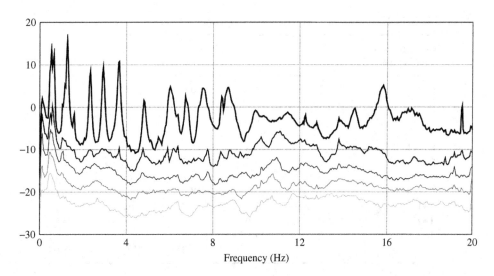

Figure 11.8   Average of the singular values of the spectral density matrices for all test setups of the City Tower building with all 32 stories completed using four projection channels

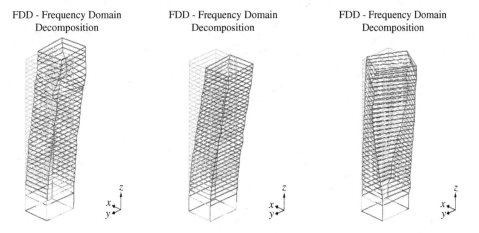

Figure 11.9   First three mode shapes of City Tower building with 32 stories completed obtained from the FDD technique

results when 15 stories had been completed, and so on. NS identifies a mode with main component along the North–South direction (x-axis), EW is for a mode with a main component along the East–West direction (y-axis), and T is for a mode whose main component is mainly torsion about the vertical axis of the building (z-axis). As to be expected, the modal frequencies decreased as the building increased in both height and mass. Modal frequency ratios between the higher modes and the corresponding fundamental mode were found to be consistent from test to test.

Trends for the modal frequencies were estimated by plotting the frequencies in Table 11.2 against the story height of the building. This plot is shown in Figure 11.10. As can be seen, the change in frequency

Table 11.2 Frequencies (Hz) for City Tower from five stages in its construction

Corresponding mode shape	CT10	CT15	CT20	CT25	CT32
NS•1	2.54	1.56	0.96	0.70	0.55
NS•2	**	6.27	4.08	3.06	2.33
NS•3		10.86*	8.05*	6.28	4.82
NS•4					7.46
EW•1	3.13	1.86	1.19	.85	0.65
EW•2		8.18	5.51	3.87	2.92
EW•3			11.13*	8.64*	6.76
EW•4			18.91		10.01*
T•1	3.95	2.56	1.90	1.50	1.29
T•2	10.66	7.68	5.18	4.37	3.66
T•3		11.62	9.51	7.52	6.03
T•4			12.25	13.11	8.62

*Questionable value.
**Not possible to identify with confidence.

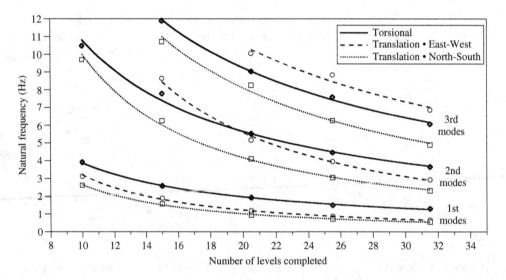

Figure 11.10 Frequency trends for the tower during its construction phase

for both translational and torsional modes follows a similar trend. Furthermore, translational modal frequencies appear to be more sensitive to building height than torsional modal frequencies. Worth noting is the crossover between the second EW translational frequencies and the second torsional frequencies. The intersection around 19 stories coincides with the discrepancies observed in Table 11.2 for the CT20 case. It is clear that modal interference affects the modal identification procedure.

The good quality of the data permitted a reliable estimation of modal characteristics of the building during different stages of construction. From the analysis of the data, it was observed that the asymmetrical mass distribution of the story floors lead to significant rotational motion and modal coupling.

Frequencies associated with translational modes decreased at a faster rate than those of torsional modes as the building height increased. It is interesting to note that the main frequencies along each axis (first mode in each direction) are less sensitive to changes in building height when compared to the higher modal frequencies, which are more sensitive to the changes in building height.

11.3.2 Long Span Bridge

The Vasco da Gama Bridge is the second Tagus river crossing in Lisbon, with a total length of 17.3 km, including three interchanges, a 5 km long section on land and a continuous 12.3 km long bridge. This bridge includes a cable-stayed component over the main navigational channel with a main span of 420 m and three lateral spans on each side (62 m + 70.6 m + 72 m), resulting in a total length of 829.2 m (Figure 11.11). The bridge deck is 31 m wide and is formed by two lateral prestressed concrete girders, 2.6 m high, connected by a cast in situ slab 0.25 m thick and by transversal steel I-girders every 4.42 m. The bridge is continuous along the total length and is fully suspended at 52.5 m above the river by two vertical planes of 48 stays connected to each tower. The two H-shaped towers are 147 m high above a massive zone at their base used as protection against ship collision. The stay cables consist of bundles of parallel self-protected strands covered by an HDPE sheath with a double helical rib in the cable cover to prevent wind-induced vibrations. Damper devices placed inside the steel guide pipe of the cables at the deck anchorages are also used to minimize cable vibrations. Additionally, set of hysteretic steel dampers connecting the pylons and the deck have been used to limit displacements and dissipate energy during seismic activity. The OMA testing reported in the following is due to Cuhna, et al. [13].

11.3.2.1   Ambient Vibration Tests

The ambient vibration tests were conducted using six triaxial 16-bit strong motion recorders. Two recorders served as references, permanently located at section 10, on the North one third span on both sides of the bridge. The main geometry of the bridge and the sections where sensors have been placed during the OMA testing are shown in Figure 11.12. Other two recorders were placed at section 15, serving also as references to confirm results. The other two recorders were roved throughout the bridge deck and the towers at a total of 29 measurement locations. Since the expected frequency range

Figure 11.11   Overall view of the Vasco da Gama cable-stayed bridge (Source: Courtesy of Cuhna, et al. [13])

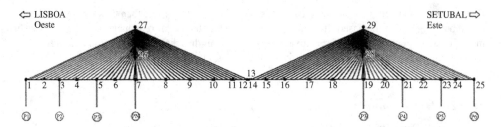

Figure 11.12 Measurement locations on the Vasco da Gama cable-stayed bridge (Source: Courtesy of Cuhna, et al. [13])

of interest was very low (0–1 Hz), the measurement time for each setup was chosen to be 16 min. The sampling rate was 50 samples per second. The recorders were operating independently, but were programmed and synchronized by a portable PC. The excitation source was wind, of which the speed varied between 1 and 22 m/s during the complete OMA test campaign. This resulted in large differences of bridge acceleration magnitudes and, inevitably, in quality differences of the acquired data.

The data was decimated 20 times and high-pass filtered to remove any offset and drift. The decimation (antialiasing) filter was an eighth-order Chebyshev type 1 low-pass filter cutting off at 1 Hz and the high-pass filter was a second-order Butterworth filter cutting off at 0.01 Hz. After decimation, the number of samples in each record was 2402 data points with a sampling interval of 400 ms, corresponding to a sampling frequency of 2.5 Hz and a Nyquist frequency of 1.25 Hz. Subsequently, the data was processed in order to estimate spectral densities with 256 frequency lines and a frequency line spacing of 4.883 MHz. This was achieved using an overlapping of 66.67% and applying a Hanning window. The identification of modal parameters of the bridge was developed, in this work, using the FDD and the Stochastic Subspace Identification (SSI) methods, implemented in the software package ARTeMIS [12].

11.3.2.2 Results

The results from the FDD method can be seen in Figure 11.13. Plots of nonzero singular values related with the setup involving measurements at sections 4 and 10 are shown in this figure. It can be seen that 12 modes are reasonably well identified. Table 11.3 summarizes the identified natural frequencies, as well as the standard deviations of the corresponding estimates. No damping was estimated because of the fact that the frequency resolution adopted was relatively low, so it was not possible to achieve a significant number of averages in order to prevent leakage bias on the damping estimates when using the enhanced FDD technique for damping analysis.

In most of the data sets, proper models were identified by the SSI method with a model order of 60–100, that is, models containing 30–50 modes. This means that the correct estimation of the 12 modes previously identified by the FDD technique involved the consideration of at least 5–8 times more noise modes. The search for the best models was based on the construction of stabilization diagrams. Figure 11.14 shows a typical stabilization plots associated to the data sets from sections 6 and 10. It can be seen that all the 12 modes are reasonably well represented; though some extra poles (stabilizing like physical poles) are present around 0.60 and 0.90 Hz. Table 11.3 summarizes the identified natural frequencies and modal damping factors, as well as the standard deviations of the corresponding estimates. The relative uncertainties on the natural frequencies are of the order of 1%, which may be related to the fact that frequencies have probably shifted from data set to data set due to changes in temperature and loading

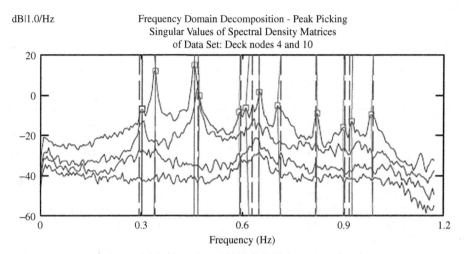

ARTeMB Extractor, Of9-4ess-9649-64e1, ARTX-0330E-270803PRO, Enterprise License

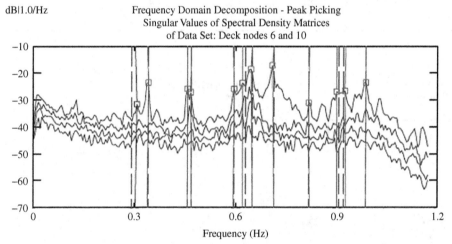

ARTeMB Extractor, Of9-4ess-9649-64e1, ARTX-0330E-270803PRO, Enterprise License

Figure 11.13 FDD plot for data sets 4 and 6 – sections 4 and 10, and 6 and 10 (Source: Courtesy of Cuhna et al. [13])

conditions. However, the uncertainty associated with choosing the right pole among a large number of noise poles may have also have contributed to the uncertainty.

The modal damping ratios are typically of the order 0.3–1%. However, since the data is rather limited, and since the SSI is only asymptotically unbiased, it can be expected that these estimates are biased, and thus lower damping values should be expected if longer time series were obtained. Also it should be noted that the damping estimate for mode 9 seems unrealistically high.

As shown in Table 11.3, the application of the FDD and SSI methods led to similar estimates of natural frequencies of the bridge. The FDD and SSI estimates of mode shapes were compared using the MAC

Table 11.3   Identified natural frequencies and modal damping ratios

Mode	FDD estimates		SSI estimates			
	Frequency (Hz)	Rel. standard deviation (%)	Frequency (Hz)	Rel. standard deviation (Hz)	Damping ratio (%)	Rel. standard deviation (%)
1	0.303	1.65	0.303	1.56	1.25	44
2	0.338	0.69	0.339	0.27	0.33	73
3	0.458	0.36	0.458	0.14	0.26	64
4	0.470	0.40	0.469	0.30	0.29	58
5	0.593	0.68	0.596	0.78	0.80	81
6	0.620	1.29	0.627	0.98	0.84	67
7	0.649	0.46	0.650	0.32	0.60	58
8	0.712	0.74	0.714	0.76	0.89	46
9	0.818	0.30	0.818	0.26	4.52	450
10	0.899	0.55	0.899	0.75	0.74	57
11	0.925	0.81	0.919	0.54	0.72	53
12	0.987	0.45	0.988	0.47	1.11	233

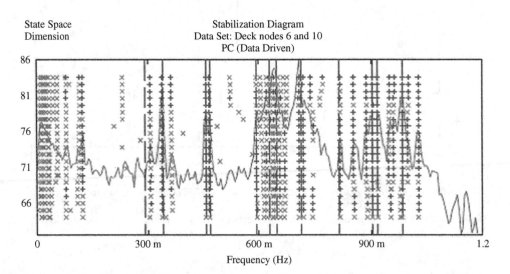

Figure 11.14   Stabilization diagram for data set 6 – measurement sections 6 and 10 (Source: Courtesy of Cuhna, et al. [13])

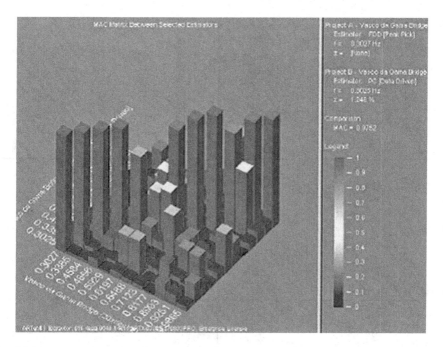

Figure 11.15 MAC matrix plot for mode shapes estimated by FDD and SSI methods (Source: Courtesy of Cuhna et al. [13])

matrix and the results are shown in Figure 11.15. This figure shows that modes 1–4, 7–9, and 11–12 were estimated with a MAC value higher than 0.9, whereas modes 5, 6, and 10 (especially mode 6) had a lower MAC value. A comparison of selected mode shapes obtained from each method is presented in Figure 11.16.

11.3.3   Container Ship

The prediction of the dynamic behavior of ship structures in normal operating conditions is important to avoid resonances and high vibration levels. For model correlation purposes and forced vibration calculations experimentally determined natural frequencies, corresponding mode shapes and damping values are necessary. Experimental modal analysis can be carried out at a shipyard. But it is known that on ships under operation the boundary conditions change and this leads to different dynamic parameters. The dynamic behavior is also influenced by added hydrodynamic masses due to the actual draft of the ship and the surrounding deep water conditions.

The case study presented here is due to Rosenow and Scholttman [14] and the purpose of the original study was to evaluate the applicability of classical and operational modal analysis for parameter estimation of ship structures. The first step of this undertaking was to apply both testing techniques at a ship yard. Results were compared with regard to the identified mode shapes and damping values. The study showed that differences in resonant frequencies between prediction and the respective experimental estimations are caused by changed ballast and draft conditions during the experimental investigations. Additional information and details about the tests can be found in reference [14].

Experimental investigations under ship yard conditions as well as during test trails were carried out on the container vessel shown in Figure 11.17. For the tests described here, a data acquisition system

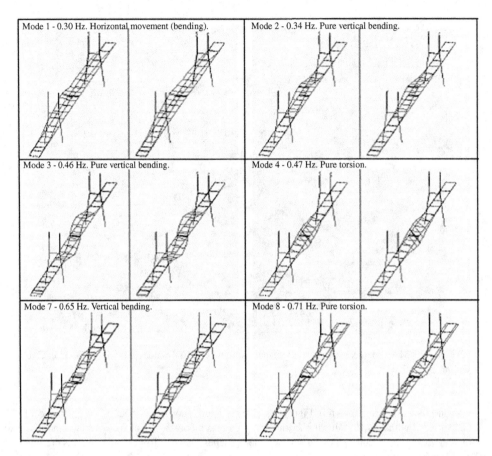

Mode 1 - 0.30 Hz. Horizontal movement (bending).

Mode 2 - 0.34 Hz. Pure vertical bending.

Mode 3 - 0.46 Hz. Pure vertical bending.

Mode 4 - 0.47 Hz. Pure torsion.

Mode 7 - 0.65 Hz. Vertical bending.

Mode 8 - 0.71 Hz. Pure torsion.

Figure 11.16 Most relevant mode shapes identified by the FDD (left) and the SSI (right) methods (Source: Courtesy of Cuhna et al. [13])

Dyn-X from Bruel&Kjaer with 2 * 24-bit AD-converter (dynamic range 160 dB) in combination with seismic accelerometers (piezoelectric, sensitivity: 10 V/g) from PCB was employed.

In a first step, this modal testing technique was applied at a ship yard for evaluation purposes. OMA analysis at the ship yard was conducted during weekends, during labor free periods. Therefore, only ambient vibrations resulting from wind and wave loads were available. Small wave loads were acting during all experimental investigations. Since wave height decreases by decreasing wave length, the white noise excitation level cannot be assumed to be constant over frequency. And this will be a case in which lower modes should be more excited due to the nature of the excitation.

The measurement setups are shown in Figure 11.18. Overall six data sets were acquired using 14 sensors including four reference sensors for linking purposes. The vibration response was recorded over a period of 60 min, sampled at 128 samples per second. The software *Artemis Extractor* was used for analysis. Both the EFDD and the SSI methods incorporated in this program were used to analyze the data. Using the EFDD, in a first step singular value decomposition (SVD) of the power spectral density (PSD) matrices is carried out. Sixty minute, long duration measurements were necessary in order to reduce random and leakage errors in the PSD estimation, especially for damping identification of the first modes (starting from 0.9 Hz).

Figure 11.17   Investigated container vessel (Source: Courtesy of Rosenow and Scholttman [14])

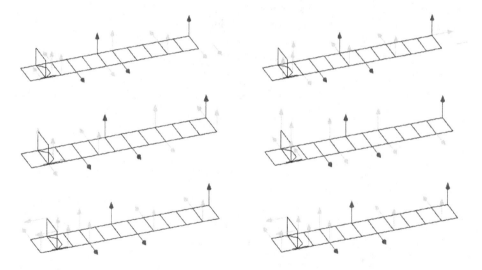

Figure 11.18   Measurement model at ship yard: acquisition of six data sets using four references (Source: Courtesy of Rosenow and Scholttman [14])

The singular values of PSD matrices from the FDD analysis of all data sets for the lower frequency range are shown in Figure 11.19. Because of the spectral characteristic of ambient wave loading, especially low frequency modes of the ship structure are well excited.

The results obtained from the SSI method are shown exemplary in Figure 11.20.

Since the time series of all measurement points were recorded during six measurement runs (six data sets) with an overall duration of the measurement (including repositioning of sensors, etc.) of 10 hours, changes in the ambient excitation occurred. These changes are noticeable by inspecting the signals of

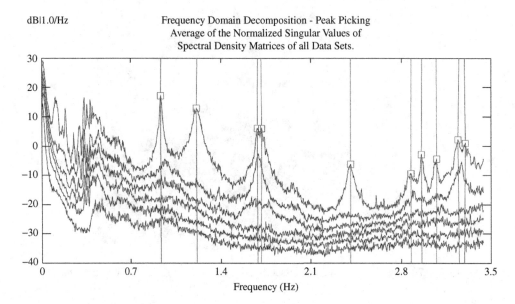

Figure 11.19 Singular values of spectral density matrices of all data sets up to 3.5 Hz from ship yard measurement (Source: Courtesy of Rosenow and Scholttman [14])

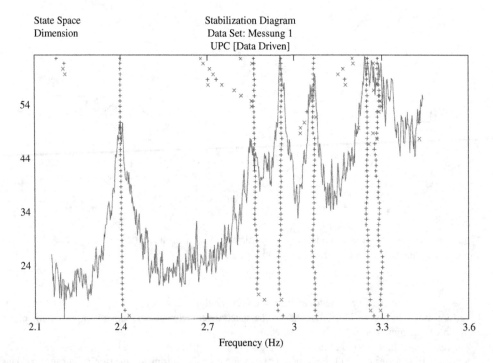

Figure 11.20 Stabilization diagram for the SSI technique (Source: Courtesy of Rosenow and Scholttman [14])

Table 11.4 Natural frequencies f (Hz) and damping ratios $\varsigma(\%)$ identified by classical and operational modal analysis at the ship yard

| Mode number | FEM | Class modal analysis | | Operational modal analyses (OMA) | | | |
| | | | | FDD | | SSI | |
	f (Hz)	f (Hz)	$\varsigma(\%)$	f (Hz)	$\varsigma(\%)$	f (Hz)	$\varsigma(\%)$
1	0.92	1.07	1.28	1.22	1.6	1.22	1.44
2	1.16	0.89	–	0.93	0.63	0.93	0.62
3	1.53	1.55	1.32	1.74	0.67	1.72	0.97
4	2.28	1.57	1.03	1.69	0.62	1.69	0.59
5	2.96	2.79	0.83	3.29	0.82	3.28	1.19
6	3.30	2.26	0.61	2.41	0.95	2.41	0.83

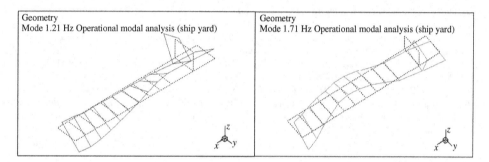

Geometry
Mode 1.21 Hz Operational modal analysis (ship yard)

Geometry
Mode 1.71 Hz Operational modal analysis (ship yard)

Figure 11.21 Mode shapes 1 and 3 identified using operational modal analysis at the ship yard (Source: Courtesy of Rosenow and Scholttman [14])

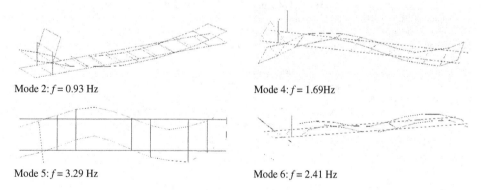

Mode 2: $f = 0.93$ Hz

Mode 4: $f = 1.69$Hz

Mode 5: $f = 3.29$ Hz

Mode 6: $f = 2.41$ Hz

Figure 11.22 Mode shapes 2, 4, 5, and 6 identified using operational modal analysis at the ship yard (Source: Courtesy of Rosenow and Scholttman [14])

the reference sensors of different data sets. Combined with the effects of changed sensor positions, the estimated modal parameters differ between single data sets for several modes and, therefore, a proper pole selection becomes more difficult.

The study described by Rosenow and Scholttman [14] includes results from test using classical modal analysis and from a detailed FE model of the ship. Only the OMA results are discussed here; more specifically the results using the EFDD and SSI estimation techniques are presented (Table 11.4). The mode shapes of mode 1 and mode 3 are shown in Figure 11.21. Further experimental mode shapes can be seen in Figure 11.22. Because of a high spatial resolution of mode shapes obtained by ship yard investigations 15 modes in the frequency range up to 13 Hz could be identified and correlated.

As a result of the analysis, Rosenow and Scholttman concluded that applying OMA under ship yard conditions good results can be achieved, but when using SSI methods resonant regions with harmonics have to be analyzed separately. Because of the high modal density, a bandpass filtering of the time series was found necessary.

References

[1] Allemang, R.: *The modal assurance (MAC) criterion –twenty years of use and abuse*. J. Sound Vibr., V. 37, p. 14–23, 2003.
[2] Felber, A.J.: *Development of the hybrid bridge evaluation system*. Ph.D. Dissertation, Department of Civil Engineering, University of British Columbia, Vancouver, BC, Canada, 1993.
[3] Peeters, B.: *System identification and damage detection in civil engineering*. Ph.D. Thesis, Department of Civil Engineering, Katholieke Universiteit Leuven, Belgium, 2000.
[4] Dascotte, E.: *Tuning of large-scale finite element models*. In Proceedings of the International Modal Analysis Conference-IX, Florence, Italy, V. 2, p. 1025–1028, 1991.
[5] Dynamic Design Solutions. *FEMtools Users' Guide*. Version 2.0, Leuven, Belgium (www.femtools.com).
[6] Ventura, C.E., Brincker, R., Dascote, E. and Andersen, P.: *FEM updating of the heritage court building structure*. In Proceedings of the International Modal Analysis Conference-XIX: A Conference on Structural Dynamics, Kissimmee, FL, V. 1, p. 324–330, 2001.
[7] Boller, C.: *Structural health monitoring – an introduction and definitions*. Encyclopedia of Structural Health Monitoring, 2009.
[8] Keith, W. and Friswell, M.I.: *Modal–vibration-based damage identification*. Encyclopedia of Structural Health Monitoring, 2009.
[9] Farrar, C. and Keith, W.: An introduction to structural health monitoring. Springer Vienna 2010-01-01, doi: 10.1007/978-3-7091-0399-9_1.
[10] Keith, W., Farrar, C.R., Manson, G. and Park, G.: The Fundamental Axioms of Structural Health Monitoring. *Philos. Trans. Roy. Soc. A*, V. 463, No. 2082, p. 1639–1664, (2007). doi: 10.1098/rspa.2007.1834.
[11] Ventura, C.E. and Schuster, N.: *Structural dynamic properties of a reinforced concrete high-rise building during construction*. CJCE, V. 23, No. 4, p. 950–972, 1996.
[12] Artemis Extractor Software, Structural Vibration Solutions, Inc., ©1999–2003 Structural Vibration Solutions, Inc., Denmark.
[13] Cunha, A., Caetano, E., Brincker, R. and Andersen, P.: *Identification from the natural response of Vasco Da Gama bridge*. In Proceedings of IMAC-22: A Conference on Structural Dynamics, Hyatt Regency Dearborn, Dearborn, Michigan, USA. Society for Experimental Mechanics, p. 202–209, January 26-29, 2004.
[14] Rosenow, S.-E. and Schlottmann, G.: *Parameter identification of ship structures using classical and operational modal analysis*. In Proceedings of International Conference on Noise and Vibration Engineering (ISMA), p. 3141–3156, 2006.

12

Advanced Subjects

"The most remarkable thing about my mother is that for thirty years she served the family nothing but leftovers."

– Calvin Trillin

This final chapter is the wrap-up of all the rest. Here, we will discuss the most important subjects that have not been dealt with in the preceding chapters.

All the subjects mentioned in the past are, therefore, in a way more important than the ones mentioned here, because by definition, they are more central to the OMA theory and applications. However, some were left out because they are less important, but some were left out because they may not be central to typical OMA applications. However they might be considered to be so in the future.

The subjects considered in the following are

- Closely spaced modes: where we will consider the importance of the techniques to accurately estimate the individual modes and ways to validate identified closely spaced modes.
- Uncertainty estimation: which is of importance whenever we need to get an estimate of the uncertainty on the modal parameters obtained from OMA.
- Mode shape expansion: which is important whenever we need to know the mode shape vectors at points where sensors have not been installed.
- Modal indicators and automated identification: which are of importance for structural health monitoring.
- Modal filtering: where each modal components is isolated by using the mode shape vectors as modal filters.
- Mode shape scaling: which is important for building frequency response functions and for some applications like damage detection.
- Force identification: which is possible if we have an estimate of the frequency response function of the system.
- Estimation of stress and strain: which is possible either by force estimation or by mode shape expansion.

12.1 Closely Spaced Modes

In Section 5.3.3, we have seen that when modes are closely spaced, the mode shapes become sensitive to small changes of the structural system. When modes are very closely spaced, and the frequency difference

Introduction to Operational Modal Analysis, First Edition. Rune Brincker and Carlos Ventura.
© 2015 John Wiley & Sons, Ltd. Published 2015 by John Wiley & Sons, Ltd.
Companion Website: www.wiley.com/go/brincker

between two[1] closely spaced modes approaches zero, the sensitivity of the corresponding mode shapes goes to infinity, see Eq. (5.181), which expresses the rotation of the eigenvectors in the their subspace.

Further, we have seen in Section 5.3.3 that when the shapes of two closely spaced modes are changing, they are only rotating in the subspace defined by the two mode shape vectors. In other words, small changes of the structural system might introduce large changes of the individual mode shapes, but no significant changes of the subspace defined by the two mode shape vectors.

12.1.1 Implications for the Identification

Taking the case of two closely spaced modes as our focus, we will discuss what this means in terms of identification.

Taking the aforementioned properties of the closely spaced modes into account means that when we are considering a case with closely spaced modes, only the subspace is of any physical significance. The reason for this is that the individual mode shape vectors with their – so to speak "over sensitivity" to very small changes of the system – cannot be considered important for normal applications.

In this case, we are free to choose any basis for the subspace, because one set of mode shapes is as good as another one, as long as they are in the subspace.

This means that using identification techniques that mix adjacent mode shape by linear combinations of the vectors can actually be used effectively for OMA, because the mixing does not affect the results as long as the mixing is done only for mode shape vectors that are closely spaced.

An unwanted mixing of the mode shapes might be introduced when modal participation vectors are estimated instead of the mode shapes. This situation may arise when the correlation functions are not being interpreted as free decays. Considering the analytical expression for the correlation function (CF) matrix in Eq. (6.90)

$$
\mathbf{R}_y(\tau) = \begin{cases} 2\pi \sum_{n=1}^{N} \left(\mathbf{b}_n \boldsymbol{\gamma}_n^T e^{-\lambda_n \tau} + \mathbf{b}_n^* \boldsymbol{\gamma}_n^H e^{-\lambda_n^* \tau} \right); & \tau \le 0 \\ 2\pi \sum_{n=1}^{N} \left(\boldsymbol{\gamma}_n \mathbf{b}_n^T e^{-\lambda_n \tau} + \boldsymbol{\gamma}_n^* \mathbf{b}_n^H e^{-\lambda_n^* \tau} \right); & \tau \ge 0 \end{cases} \tag{12.1}
$$

We see that for negative times, the columns of the CF matrix are proportional to the mode shape vectors $\mathbf{b}_n$, but for positive times, the columns are proportional to the modal participation vectors $\boldsymbol{\gamma}_n$. If we then use an identification technique that extracts the mode shapes from the columns of the positive time part of the CF matrix, then we will in fact not estimate the right mode shapes, but we will estimate the modal participation vectors and, thus, will obtain a biased estimate for the mode shape vectors.

However, as the modal participation vectors are only influenced significantly by the mode shape vector that are close in terms of frequency as shown in Section 6.2.6, then the bias in the results is mainly a "rotation in the subspace" as explained earlier and, therefore, this bias is not so serious for practical applications.

12.1.2 Implications for Modal Validation

It is obvious that when the individual mode shapes are rotating more or less arbitrarily in the subspace defined by a set of closely spaced modes, then it does not make much sense to calculate the MAC between one of the mode shapes and the corresponding mode shape from a model.

---

[1] In the general case, there might be more than two closely spaced modes, but for simplicity, we are considering just two.

The problem of mode shape validation can, in this case, be solved by evaluating the correlation of an experimentally obtained mode shape $\mathbf{a}$ with the corresponding group of closely spaced modes in a model as proposed by D'Ambrogio and Fregolent [1]. For simplicity, let us assume that we want to correlate this experimentally obtained mode shape with a set of two closely spaced mode shapes $\mathbf{b}_1$, $\mathbf{b}_2$ in the model.

Using the traditional definition of the MAC as given in Chapter 11, but now taking the vector from the model as an unknown linear combination of the two considered mode shapes and maximizing the result, we define the higher order MAC for the case of a subspace with two vectors as

$$S2MAC = \max_{\alpha,\beta} \frac{\left(\mathbf{a}^H\left(\alpha\mathbf{b}_1 + \beta\mathbf{b}_2\right)\right)^2}{\mathbf{a}^H\mathbf{a}\left(\alpha\mathbf{b}_1 + \beta\mathbf{b}_2\right)^H\left(\alpha\mathbf{b}_1 + \beta\mathbf{b}_2\right)} \tag{12.2}$$

As indicated by the max function, S2MAC is found for the combination of α and β that maximizes the MAC value between $\mathbf{a}$ and the fixed length linear combination $\alpha\mathbf{b}_1 + \beta\mathbf{b}_2$. If we assume that $\mathbf{a}, \mathbf{b}_1, \mathbf{b}_2$ and the linear combination $\alpha\mathbf{b}_1 + \beta\mathbf{b}_2$ are unit vectors, then Eq. (12.2) reduces to

$$S2MAC = \max_{\alpha,\beta} \left(\mathbf{a}^H\left(\alpha\mathbf{b}_1 + \beta\mathbf{b}_2\right)\right)^2 \tag{12.3}$$

It is shown in D'Ambrogio and Fregolent [1] that the analytical solution of Eq. (12.3) for real-valued mode shape vectors is given by

$$S2MAC = \frac{\left(\mathbf{a}^T\mathbf{b}_1\right)^2 - 2\left(\mathbf{a}^T\mathbf{b}_1\right)\left(\mathbf{b}_1^T\mathbf{b}_2\right)\left(\mathbf{a}^T\mathbf{b}_2\right) + \left(\mathbf{a}^T\mathbf{b}_2\right)^2}{1 - \left(\mathbf{b}_1^T\mathbf{b}_2\right)^2} \tag{12.4}$$

When correlating two closely spaced experimental modes with the corresponding closely spaced modes from a model, the correlation can also be evaluated by estimating the angles between the subspaces. Just like the generalized angle between two vectors $\mathbf{a}$ and $\mathbf{b}$ can be found from $\cos^2(\theta) = MAC(\mathbf{a}, \mathbf{b})$, see Eq. (3.21), the angles between the subspaces having a similar geometrical interpretation can be defined and found, for instance, using the Matlab function subspace.m.[2]

12.2   Uncertainty Estimation

Estimation of the uncertainty of the estimated modal parameters is always possible one way or the other and should always be considered. The simplest and best way to perform this estimation is to repeat the test and corresponding identification; the second best is to try to identify the uncertainty from the least squares problem in case such problem can be defined.

12.2.1   Repeated Identification

As discussed easlier in this book OMA testing includes important user choices that might affect the uncertainty of the resulting modal parameters.

Therefore, the best way to have an idea about the uncertainty on modal parameters is simply to repeat the test and then calculate the empirical uncertainty from the different estimates. Nothing is better than this, and this approach should be used whenever possible.

---

[2] This function only estimates the largest so-called principal angle, but there are as many angles as the number of mode shapes in the subspace.

If stabilization diagrams are used in the identification process, a set of stable modal quantities can be used to estimate the uncertainty on the considered modal quantity; see the discussion about stabilization diagrams in Chapter 11 and in Section 12.4.2 about the generalized problem of stabilization.

In cases of several data sets, the uncertainty on natural frequency and damping can be calculated using the pole estimates from the data sets. However, in many cases, all we have is one single response, or we may have several data sets, but we also need uncertainty estimates for the mode shapes. This cannot be estimated from a single test with different data sets, because from such test, we only get one single mode shape estimate for each mode. In such case – if least squares regression is used to identify the modal parameters – the covariance matrix can be estimated and can be used for subsequent uncertainty estimation.

12.2.2   Covariance Matrix Estimation

Whenever least squares regression is used to obtain modal parameters, an estimate of the uncertainty can be obtained from the fitting errors. To illustrate the principle, let us consider the problem of finding the modal participation vectors in time domain as given by Eq. (9.5)

$$\mathbf{H} = 2\pi \mathbf{\Gamma} \mathbf{M} \tag{12.5}$$

where $\mathbf{M}$ is a matrix containing the modal parameters to be fitted to the response in the matrix $\mathbf{H}$, and $\mathbf{\Gamma}$ is a matrix containing the modal participation vectors for the different modes. Taking the transpose of this equation, we have

$$\mathbf{H}^T = 2\pi \mathbf{M}^T \mathbf{\Gamma}^T \tag{12.6}$$

Now, we can rename the column vectors in $\mathbf{H}^T$ as $\mathbf{H}^T = \begin{bmatrix} \mathbf{y}_1, \mathbf{y}_2, \ ... \end{bmatrix} = \begin{bmatrix} \mathbf{y}_n \end{bmatrix}$ and in $\mathbf{\Gamma}^T$ as $\mathbf{\Gamma}^T = \begin{bmatrix} \mathbf{a}_1, \mathbf{a}_2, \ ... \end{bmatrix} = \begin{bmatrix} \mathbf{a}_n \end{bmatrix}$. The response $\mathbf{y}_1$ is the transposed row vectors of the correlation matrix stacked on top of each other for different time lags to form one column vector. The parameter vector $\mathbf{a}_1$ describes the participation of the different modes at DOF number one. We then get n overdetermined sets of equations

$$\mathbf{y}_n = 2\pi \mathbf{M}^T \mathbf{a}_n \tag{12.7}$$

Each of these overdetermined set of equations is a least squares regression problem of the form given by Eq. (3.90) where the design matrix for this case is given by $\mathbf{X} = 2\pi \mathbf{M}^T$. As we can conclude from the arguments for Eq. (3.91), we must add a noise vector to adequately describe the difference between the model (as given by the term $2\pi \mathbf{M}^T \mathbf{a}_n$) and the data (as given by the vector $\mathbf{y}_n$), and the least squares problem takes the form

$$\mathbf{y}_n = 2\pi \mathbf{M}^T \mathbf{a}_n + \mathbf{\varepsilon}_n \tag{12.8}$$

where the noise vector $\mathbf{\varepsilon}_n$ is equal to the aforementioned difference between the data vector and the model vector. Assuming that all the components in $\mathbf{\varepsilon}_n$ have identical statistical properties that can be described by zero mean variables with a certain variance σ_n^2, then according to Eq. (3.112), the covariance matrix of the elements of the parameter vector is given by

$$\begin{aligned} \mathbf{C}_n &= \sigma_n^2 \left(\mathbf{X}^T \mathbf{X} \right)^{-1} \\ &= 4\pi^2 \sigma_n^2 \left(\mathbf{M} \mathbf{M}^T \right)^{-1} \end{aligned} \tag{12.9}$$

Or, using only the diagonal elements of the covariance matrix $\mathbf{C}_n$, we obtain a vector that contains the uncertainty (the variance) on each of the components in the estimate $\hat{\mathbf{a}}_n$

$$\mathbf{c}_n = 4\pi^2 \sigma_n^2 \mathrm{diag}\left(\left(\mathbf{M} \mathbf{M}^T \right)^{-1} \right) \tag{12.10}$$

where $\mathrm{diag}((\mathbf{M}\mathbf{M}^T)^{-1})$ means the column vector formed by the diagonal elements of the matrix $(\mathbf{M}\mathbf{M}^T)^{-1}$. The vector $\mathbf{c}_n$ describes the fitting of the different modes at DOF number n, and if we compute the average of all elements in the vector, we obtain a single measure of the fitting at this DOF.

A similar technique can be used for finding the uncertainties for mode shapes and poles using, for instance, the poly reference time domain technique (see Section 9.2).

12.3   Mode Shape Expansion

We cannot place our sensors at all the DOFs of a structure because it is normally not possible, or practical, to do so on large complex structures. Further, we may only have a limited number of sensors available, and for these two reasons, in many cases, we may not have information about all the points where we would like to know the mode shapes. This problem can be solved by subsequent mode shape expansion.

As a first approximation, one can always use the mode shapes from a reasonable finite element (FE) model in a linear combination to perform an expansion. However, the following question arises: which modes should be included in such a linear combination? The ideas on how to do this are outlined in the following sections.

12.3.1   FE Mode Shape Subspaces

It follows from the structural modification theory, see Section 5.3.1, that two systems A and B are related through a linear combination of the corresponding mode shape matrices $\mathbf{A}$ and $\mathbf{B}$ as given by Eq. (5.158)

$$\mathbf{A} = \mathbf{B}\mathbf{T} \tag{12.11}$$

where the matrix $\mathbf{T}$ defines the linear combinations of the mode shapes $\mathbf{b}_n$ in $\mathbf{B}$ that is used to form the mode shapes $\mathbf{a}_n$ in $\mathbf{A}$. The exact linear relation given by Eq. (12.11) is only valid if a complete set of modes are included in both mode shape matrices $\mathbf{A}$ and $\mathbf{B}$.

In Eq. (12.11), we can think about system A as the experimental system (perturbed system) and system B as the structural system defined by an FE model (unperturbed system). The number of DOF's in the two mode shape matrices $\mathbf{A}$ and $\mathbf{B}$ must be the same, and therefore, either the large number of DOF's in FE mode shape matrix $\mathbf{B}$ must be reduced to the DOF's active in the experiment, or the DOF's in the experimental mode shape matrix $\mathbf{A}$ are expanded to all DOF's in the FE model.

It follows from the sensitivity equations outlined in Section 5.3.2 that even though not all mode shapes are included in the mode shape matrices, a similar relationship like Eq. (12.11) exists approximately, (see Eq. (5.167)). The condition is that for all mode shapes in the mode shape matrix $\mathbf{A}$, a reasonable number of modes must be included in the mode shape matrix $\mathbf{B}$.

An estimate $\hat{\mathbf{T}}$ of the linear transformation matrix $\mathbf{T}$ can be found by left multiplying both sides of the equation by the pseudo inverse of the mode shape matrix $\mathbf{B}$, that this

$$\hat{\mathbf{T}} = \mathbf{B}^+\mathbf{A} \tag{12.12}$$

The matrix $\mathbf{B}$ has the same number of rows as the matrix $\mathbf{A}$, because the matrix $\mathbf{B}$ has been reduced to the DOF's used in the experiment as explained earlier. The number of modes in $\mathbf{B}$ might not be the same as the number of modes in $\mathbf{A}$. This is because a larger number of modes in the FE model matrix $\mathbf{B}$ are often needed in order to perform a good modeling of all the experimental modes; more about this in Section 12.3.3. In order to have an overdetermined system of equations that leads to a reasonable solution using Eq. (12.12), the number of DOF's in the reduced mode shape matrix $\mathbf{B}$ must be larger than the number of mode shapes in $\mathbf{B}$. It should be noted that it is often better to find the estimate $\hat{\mathbf{T}}$ for the transformation matrix mode-by-mode as explained in Section 12.3.3.

When a transformation matrix $\hat{\mathbf{T}}$ has been obtained, a linear relationship of the form (12.11) is then established by inserting the estimated transformation matrix into Eq. (12.11) to obtain the following estimate $\hat{\mathbf{A}}$ of the mode shape matrix $\mathbf{A}$

$$\hat{\mathbf{A}} = \mathbf{B}\hat{\mathbf{T}} \tag{12.13}$$

Once the linear relationship (12.13) has been established, the linear transformation is valid for any set of DOF's in system B, as the estimate $\hat{\mathbf{A}}$ of the mode shape matrix $\mathbf{A}$ always includes the same DOF's as the mode shapes in $\mathbf{B}$. The mode shapes in system A can, therefore, be expanded to any set of unknown[3] DOF's just by including these DOF's in the mode shape matrix $\mathbf{B}$ from the FE model.

12.3.2   FE Mode Shape Subspaces Using SEREP

An equation similar to Eq. (12.13) also follows directly and even more explicitly from the model reduction principle SEREP introduced in Section 5.3.4. In SEREP, originally we are only considering one system, but in a full form including all the DOF's and in a reduced form with a reduced number of DOF's. Following SEREP and taking system A as the reduced system and system B as the system with a full set of DOF's, then it follows from Eq. (5.196) that according to SEREP, any mode shape $\mathbf{a}$ in A can be expressed by the corresponding mode shape $\mathbf{b}$ in B by

$$\mathbf{a} = \mathbf{T}\mathbf{b}_a \tag{12.14}$$

where $\mathbf{b}_a$ is the mode shape vector $\mathbf{b}$ in the reduced (active) coordinate set and the transformation matrix $\mathbf{T}$ is now given by[4] Eq. (5.192)

$$\mathbf{T} = \mathbf{B}\mathbf{B}_a^+ \tag{12.15}$$

where $\mathbf{B}$ is the mode shape matrix in the full coordinate set, $\mathbf{B}_a$ is the same mode shape matrix in the reduced (active) set of coordinates, and finally, $\mathbf{B}_a^+$ is the pseudo inverse of the latter mode shape matrix. It might not seem like Eq. (12.14) expresses the mode shape $\mathbf{a}$ as a linear combination of modes from system B, but it does with a proper choice of the involved variables. This is seen by inserting Eq. (12.15) into Eq. (12.14) to give

$$\mathbf{a} = \mathbf{B}\mathbf{B}_a^+\mathbf{b}_a \tag{12.16}$$

The last vector in the expression

$$\hat{\mathbf{t}} = \mathbf{B}_a^+\mathbf{b}_a \tag{12.17}$$

is an estimate of a linear transformation vector assuming the linear relationship

$$\mathbf{a} = \mathbf{B}_a\mathbf{t} \tag{12.18}$$

Solving this equation, we obtain the estimate

$$\hat{\mathbf{t}} = \mathbf{B}_a^+\mathbf{a} \tag{12.19}$$

This means that in SEREP, we must use the approximation $\mathbf{a} = \mathbf{b}_a$ in Eq. (12.17) in order to use SEREP for mode shape expansion.

The accuracy of the estimate given by Eq. (12.13) is different for the different mode shapes in $\mathbf{A}$ and is also strongly dependent upon the mode shapes included in the mode shape matrix $\mathbf{B}$. The problem of which mode shapes should be included in system B and how to estimate the transformation matrix in order to obtain a good estimate for the linear transformation matrix $\mathbf{T}$ is further discussed in the following paragraph.

---

[3] Unknown in the experiment.

[4] Note that the transformation matrix in Eq. (12.14) is not the same as the transformation matrix in Eq. (12.11).

12.3.3   Optimizing the Number of FE Modes (LC Principle)

For simplicity, let us consider just a single mode shape vector **a** from the experimental system to be estimated by a linear combination given by the vector **t** of mode shapes from an FE model included in the mode shape matrix **B** as given by Eq. (12.18)

$$\mathbf{a} = \mathbf{Bt} \tag{12.20}$$

where it is understood from Eq. (12.18) that we solve this equation by taking the pseudo inverse of **B** to find the estimate for the linear transformation

$$\hat{\mathbf{t}} = \mathbf{B}^+ \mathbf{a} \tag{12.21}$$

and the estimate of the experimental mode shape is

$$\hat{\mathbf{a}} = \mathbf{B}\hat{\mathbf{t}} \tag{12.22}$$

where the degrees of freedom in **B** must be the same as in **a**. However, once we have arrived at Eq. (12.22), the number of DOF's in **B** can be chosen freely by determining the number of DOF's in the estimate $\hat{\mathbf{a}}$.

Let us discuss if an optimal set of modes exist for the modes to be included in **B** such that the errors in the estimate $\hat{\mathbf{a}}$ are minimum.

We can include all the measured DOF's in the experiment in the vector **a**, or we can include some of them. This also controls the number of DOF's in the mode shape matrix **B**; thus, it is natural to talk about the number of DOF's included when solving Eq. (12.21). As in Eq. (12.21) we will try to fit **a** with the modes from **B**, and we will denote this set of DOF's as the fitting set.

First, let us consider the case when the fitting set includes all DOF's from the experiments. If only a single mode shape is included in **B**, then the question is only which mode should be included in order to obtain the best estimate. This is simple, because the obvious and true answer is that it must be the mode shape from system B that correlates best with **a**; in other words, the one with the largest MAC value. This mode shape from system B is denoted as "the primary mode shape." In this case, the estimate in Eq. (12.22) is simply equal to the primary mode shape from the FE model.

The next mode shape to include in **B** is also easy to define using the LC principle, see Section 5.3.2. According to this principle, the next mode shape to include is the mode shape associated with the mode closest to the primary mode in terms of frequency proximity. The third mode shape to include is the mode shape associated with the mode second-closest to the primary mode in terms of frequency proximity – and so on.

We can also follow a different scheme that might seem easier to understand. The second mode shape to include is the mode shape that results in the largest increase in the MAC value between the estimated mode shape $\hat{\mathbf{a}}$ and the measured mode shape **a**. This approach gives nearly the same set of modes as that obtained following the first mentioned scheme. This follows from the LC principle outlined in Section 5.3.2.

We can continue to include mode shapes from system B into the mode shape matrix **B**, and each time we include a new mode shape vector, we increase the MAC value between the estimate $\hat{\mathbf{a}}$ and the measured mode shape **a**. Finally, when we have included so many mode shapes from system B that the number of mode shapes in the mode shape matrix **B** is equal to the number of DOF's in the experiment, the MAC value is unity. The reason is that in this case, the solution to Eq. (12.21) is exact because then the mode shape matrix **B** is square and the pseudo inverse becomes equal to the inverse of **B**. This is illustrated in Figure 12.1a.

This approach obviously may not make any sense, because including all possible modes in the mode shape matrix **B** will give zero errors on the measured coordinates, but this is not the case for the unmeasured DOF's. In fitting theory, this is known as the problem of overfitting. Overfitting occurs when the errors in the fitting set are reduced, while the errors on the points not included in the fitting set may increase unacceptably.

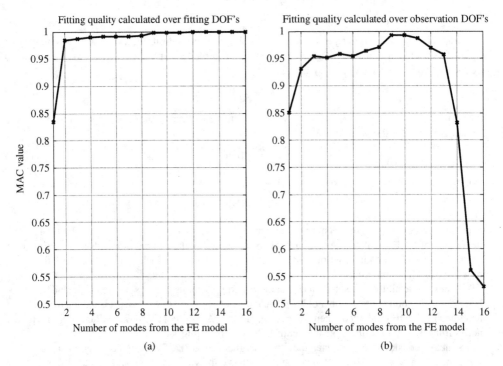

Figure 12.1 Measures of the best fit for a plate structure. (a) MAC value calculated on the fitting set of DOF's. (b) MAC value calculated on the observation set of DOF's. As it appears, the measure in the left plot is an increasing function whereas the measure in the right plot has a clear optimum

The problem can be solved by not including all DOF's in the fitting set, the remaining DOF's are denoted as the observation set of DOF's and then estimating the MAC value on the observation set. The observation set can contain all DOF's in the experiment by roving the observation set over all DOF's in the experiment. If we follow this procedure when estimating the linear transformation in Eq. (12.21), then the estimated MAC value is optimum for a certain number of modes in the mode shape matrix **B**. This is illustrated in the right plot of Figure 12.1.

The two fitting cases are shown in Figure 12.1 for a specific plate structure. Figure 12.1a shows the overfitting case where all DOF's from the experiment are included in the fitting set, and Figure 12.1b shows the case where some DOF's are used for fitting and some are used for observation of the fitting quality. In this example, the structure is a $580 \times 320 \times 3$ mm steel plate supported in free-free conditions and modeled with plate elements using 81 nodes placed in a 9×9 grid equally distributed over the plate. From the 81 DOF's in the FE model, 16 DOF's were randomly chosen as the fitting set and another 16 DOF's were randomly chosen as the observation set. The unperturbed set of modes (system B) was taken as the perfect FE model, and the perturbed system (system A) was defined by making a random change of the mass matrix and adding 5% Gaussian noise to the eigenvectors. Figure 12.1 shows the results for mode no. 4 in the experimental set of modes, and we see from Figure 12.1b that 9 or 10 modes from the FE should be included in order to obtain a "best fit" with the estimate given by Eq. (12.22).

It is clear from the considerations in Section 5.3.2 on sensitivity analysis and the LC principle that only the mode shapes from system B close to the primary mode should be included in the mode shape matrix **B** when performing the estimate given by Eq. (12.22). More information about the LC principle and how to obtain the mode shape cluster of local modes around the principal mode shape can be found in Brincker et al. [2].

12.4   Modal Indicators and Automated Identification

In this section, we shall briefly discuss the issue of oversized models and noise modes, the ways to deal with the noise modes by modal indicators, and then, finally, briefly discuss the need for automated identification, mainly in relation to structural health monitoring.

12.4.1   Oversized Models and Noise Modes

Whenever parametric modal analysis is performed, the question of a proper oversizing of the models and how to deal with the so introduced noise modes must be addressed. We shall not go deep into this problem, but only mention a few things that might help understanding this issue.

Oversizing of a modal model means that the model contains more modes than we are actually looking for. It is not unusual that the modal model contains 5–10 times more modes than the experiment data.

There are two main reasons for the need of model oversizing. One reason is that often our models are biased, and the other one is that often our models do not include any noise modeling.

Let us start with the last problem; modeling of the noise. Noise is always present in the data, and a way to deal with the modeling of the noise is simply to include more modes in the model so that these extra "noise modes" help the model fit better to the data. The more noise we have in the data, the more noise modes are needed in order to model the noise. On the other hand, if the modal model also includes a noise modeling, the oversizing needed to deal with the noise might be significantly reduced.

The first mentioned problem, the bias, is a problem that is always present to a certain degree. Of course, under the assumption of a linear and invariant system, we can formulate bias free models like, for instance, the poly reference or the Ibrahim time domain technique; see Chapter 9, for more information on these techniques. However, because the real structural behavior is never totally linear and invariant, our models are always biased, and in order to minimize the bias on the modal estimates, it is well known that oversizing is needed.

The conclusion is that in order to obtain a good fit and to obtain unbiased or nearly unbiased modal estimates, we normally need to deal with oversized models and this introduces the noise modes. We need efficient ways to deal with the noise modes in order to clearly identify where the physical modes are, and what are just noise modes that we can discarded, before we start using the obtained modal data for further analysis.

In any case, the user must gain experience with a certain identification scheme in order to develop guidelines for what is an appropriate oversized model.

12.4.2   Generalized Stabilization and Modal Indicators

It is not the intention with this section to give a full overview of modal indicators, but rather to give some examples on these quantities and how they can be used, for instance, in order to build stabilization diagrams.

In all modal identification – in traditional modal analysis and in OMA – it is customory to build stabilization diagrams in order to illustrate, and decide, if a mode is physical or not. The most common stabilization diagram is a plot where the horizontal axis is frequency and the vertical axis is model order. For an introduction to the traditional stabilization diagram, the reader is referred to Chapter 11 on applications.

The traditional idea of stabilization diagrams can be generalized if we say that the horizontal axis is a modal quantity, such as

- Natural frequency
- Damping

- Mode shape (projected onto a fixed vector)
- Modal participation
- Mode shape scaling

which will be denoted as $x(n,m)$. The vertical axis is a parameter describing a variation in the modal identification process such as

- Model order
- Response data filtering
- Number of data point used in the correlation or spectral density function

which we will call the "identification condition" and will be denoted by $y(n,m)$. The parameter n defines the values on the vertical axis (in traditional stabilization diagrams, the model order) and m defines the considered mode. The idea in such a diagram is to plot modal data in the $x(n,m)$, $y(n,m)$ diagram and to look for the changes of the modal parameter $x(n,m)$ (in traditional stabilization diagrams, the natural frequency of the considered mode). Figure 12.2a illustrates the idea of the generalized stabilization diagram.

A stabilization criterion is then used to discriminate between unstable and stable modal parameters; the stable ones are considered physical and the unstable are considered as noise modes. A modal parameter is often considered stable at a certain level n if the change from $y(n-1,m)$ to $y(n,m)$ is smaller than a certain value Δx, that is,

$$|y(n-1,m) - y(n,m)| < \Delta x \qquad (12.23)$$

In a stabilization diagram, the user is looking for modes that are situated along vertical lines because if a physical mode is present, the stabilization criteria given by Eq. (12.23) will select only the modes along vertical lines as the stable modes.

Two modal quantities, say $x_1(n,m)$ and $x_2(n,m)$, can also be depicted in the same diagram, see Figure 12.2b. In this case, the plot is called a clustering diagram, because in this case, the user is looking for clusters, that is, small areas in the x_1, x_2 plane where a high density of modal data is present.

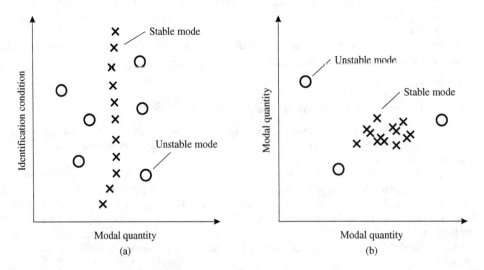

Figure 12.2 (a) Generalized stabilization diagram depicting an identification condition on the vertical axis together with a modal quantity on the horizontal axis. (b) A clustering plot where two modal quantities are depicted in the same plot for different identification conditions

In both diagrams, the stable modes can be indicated graphically helping the user find out which modes are to be chosen as the physical ones.

When a certain mode has been chosen to be a physical mode, and a large set of stable values has been determined by Eq. (12.23), then the mean value can be estimated and used for further analysis as the best representative for the considered modal quantity. Further, the standard deviation of the stable values can be estimated as a measure of the uncertainty on the considered modal quantity.

If all – or at least several of the aforementioned parameters – are considered at the same time, a better decision can be made concerning the validity of modal estimates.

It is not a tradition in modal analysis to look at the modal participation factor (MPF). However, as we have seen in Chapter 9 on time domain identification and in Chapter 10 on frequency domain identification, the MPF can always be estimated. Choosing between many identified modes in a certain frequency band, it is clear that we should choose one of the modes with the highest MPF, because a physical mode will nearly always have a high MPF.

Several of the techniques mentioned in Chapters 9 and 10 have their own definition of modal indicators. Application of these modal indicators is not limited to the specific technique where they were initially defined, but can be used on modal parameters from any technique.

One of the first suggested modal indicators was the modal confidence factor (MCF) for the Ibrahim time domain (ITD) technique, see Section 9.4. He suggested that two modal estimates $\mathbf{a}' = \{a_n'\}$; λ' and $\mathbf{a}'' = \{a_n''\}$; λ'' where $\mathbf{a}$ is the mode shape and λ is the continuous time pole, should be compared by calculating the MCF as the ratio, Ibrahim [3],

$$\mathrm{MCF} = \left(a_n' \exp\left(\lambda'\Delta t\right)\right) / \left(a_n'' \exp\left(\lambda''\Delta t\right)\right) \tag{12.24}$$

where Δt is some time interval. This could also be generalized to the inner product

$$\mathrm{MCF} = \left(\mathbf{a}' \exp\left(\lambda'\Delta t\right)\right)^H \mathbf{a}'' \exp\left(-\lambda''\Delta t\right) \tag{12.25}$$

where it is assumed that the two mode shapes vectors have unit length so that the MCF becomes a kind of MAC value including the poles. The MCF is a good example of combining several modal parameters into one unified modal indicator.

As another example of modal indicators we can consider the indicators defined for the ERA, see Section 9.5. The ERA defines two quantities for physical mode indication, Pappa et al. [4]. One is denoted as modal amplitude coherence that is closely related to the MCF mentioned earlier, and another indicator to describe the degree of "monophase" behavior of the identified mode shapes, that is, the degree to which all locations on the structure for a given mode is in-phase or out-of-phase with each other.

The aforementioned indicators assume that a mode has been estimated. Thus, they are all posteriori indicators that we can use for investigating the validity of a certain mode AFTER it has been estimated.

Sometimes, we would like to have an indicator if a mode might be present BEFORE we perform the actual identification of the mode. It would be useful to have some a priori indicators. An example of an a priori indicator is the modal coherence indicator suggested by Brincker et al. [5]

$$d\left(f_0\right) = \mathbf{u}_1(f)^H \mathbf{u}_1\left(f_0\right) \tag{12.26}$$

where $\mathbf{u}_1$ is the first singular vector of the spectral density matrix, f_0 is the frequency where we ask the question if a mode is expected around this frequency, and f is the adjacent frequency. If the response at the considered frequency is governed by a modal response, then the two singular vectors will be nearly identical and the inner product in Eq. (12.26) will be close to unity. If the response is governed by noise, then the two vectors are random, and the inner product will be close to zero.

When modal coherence is estimated, we can include several neighboring points; the more neighboring points we include and the more measurement channels we have, the stronger is the indicator. If we have a strong indication of modal presence, then we can search the region over which the modal coherence is high and define the corresponding modal domain, see the illustration in Figure 12.3.

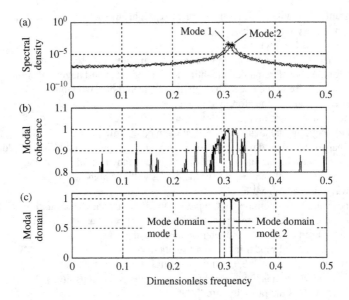

Figure 12.3 Example of modal coherence and modal domain. (a) SVD plot of the spectral density of the system with two closely spaced modes. (b) Modal coherence function d. (c) Modal domain of the two modes (Source: From Brincker et al. [5])

12.4.3 Automated OMA

Automated OMA means OMA that is performed without any user interaction. It is often used where the same test is repeated many times or where many data sets are available for the same OMA test. An important application of automated OMA is structural health monitoring (SHM) where the acquired response data have to be analyzed automatically so that changes of the modal data can be easily determined.

Automated identification has been an important subject in modal identification since the beginning of modal analysis, and we shall not point to any specific implementation or reference on this issue, but only mention a few things that are important to consider.

A successful automation of OMA is always based on a careful selection of the aforementioned modal indicators, preferably several well-selected indicators, which for the problem under consideration gives a clear indication about which modes are to be selected as the physical modes.

As the success depends on many things like

- The structure under consideration
- The test layout, especially the number and the placement of sensors
- The signal processing of the data
- The identification method used
- The modal indicators and their implementation in the physical mode selection process

it is essential that a solution is sought where all important links in the identification process are taken into consideration and that reasonable solutions are found for all the links, because in the process of automated identification, the weakest link determines the quality of the final result.

As mentioned earlier, the automated identification is very valuable in SHM applications. In these cases, it is essential that experience is obtained about how to perform the automated identification under the different environmental conditions that are going to be present during monitoring, and that this experience is used to adjust the automated procedures accordingly.

For instance, for a given environmental condition, a data base could be present where the expected modal parameters for the considered environmental condition can be looked up including confidence intervals so that this a priori information can be used in the automated identification procedure.

12.5 Modal Filtering

Modal filtering is used to obtain the modal coordinates from a measured random response. The ideas of modal filtering can be generalized to complex modes, but for simplicity, we will assume that the considered random response can be described by normal modes.

12.5.1 Modal Filtering in Time Domain

In this case, we are considering the random time domain response $\mathbf{y}(t)$, which for normal modes are given by

$$\mathbf{y}(t) = \mathbf{A}\mathbf{q}(t) \tag{12.27}$$

where the matrix $\mathbf{A}$ contains the mode shapes as column vectors, and $\mathbf{q}(t)$ is a column vector containing the corresponding modal coordinates. If the mode shape matrix is known, and if the number of DOF's of the mode shapes is larger than the number of mode shapes in the mode shape matrix $\mathbf{A}$, that is, if the matrix $\mathbf{A}$ has more rows than columns, then, by multiplying Eq. (12.27) with the pseudo inverse of $\mathbf{A}$ from the left, we can obtain a estimate $\hat{\mathbf{q}}(t)$ of the modal coordinates as

$$\hat{\mathbf{q}}(t) = \mathbf{A}^+\mathbf{y}(t) \tag{12.28}$$

It is important to recognize that this estimate is only meaningful if all the mode shapes that are responsible for the response are estimated well and included in the mode shape matrix $\mathbf{A}$. The accuracy of the estimated modal coordinates can be validated by plotting the modal coordinates in the frequency domain, because each modal coordinate – in case the estimate is good – must contain only a single modal peak; the peak that corresponds to the considered modal coordinate.

It is also worth noticing that the more overdetermined the problem is, that is, the more DOF's included in the mode shapes and the fewer modes included in the problem, the more noise is removed from the estimated modal coordinates. This means that in the time domain, band-pass filtering should be considered in order to perform the modal filtering in smaller frequency bands to remove as much noise as possible.

Having removed noise in a successful modal filtering, the synthesized response that is calculated from the modal coordinates as

$$\hat{\mathbf{y}}(t) = \mathbf{A}\hat{\mathbf{q}}(t) \tag{12.29}$$

is also cleaned for some of the noise that is always present in a measured response.

In case band-pass filtering is being used in the time domain to reduce the number of modes in the modal filtering, it might be advisable to perform the modal filtering in the frequency domain instead, simply because here the band-pass filtering can be omitted.

It should be noted that using an expanded set of mode shapes as described in Section 12.3 allows the user to estimate the operational response in any DOF defined in the FE model.

12.5.2   Modal Filtering in Frequency Domain

Taking the Fourier transform of Eq. (12.27), we get the similar equation in the frequency domain

$$\widetilde{\mathbf{y}}(\omega) = \mathbf{A}\widetilde{\mathbf{q}}(\omega) \tag{12.30}$$

where $\mathbf{y}(t) \leftrightarrow \widetilde{\mathbf{y}}(\omega)$ and $\mathbf{q}(t) \leftrightarrow \widetilde{\mathbf{q}}(\omega)$ are Fourier transform pairs, and exactly the same things can be said as for the time domain.

The main difference is that band-pass filtering is replaced by using different mode shape matrices $\mathbf{A}$ for different frequency bands and that the results are more easily validated because the modal coordinates estimated in parallel to Eq. (12.28) as

$$\widehat{\widetilde{\mathbf{q}}}(\omega) = \mathbf{A}^+\widetilde{\mathbf{y}}(\omega) \tag{12.31}$$

can be directly plotted in the frequency domain and checked for single-peak behavior.

The synthesized response in the frequency domain can be calculated from the modal coordinates as

$$\widehat{\widetilde{\mathbf{y}}}(\omega) = \mathbf{A}\widehat{\widetilde{\mathbf{q}}}(\omega) \tag{12.32}$$

A successful modal filtering is also characterized by a flat (no peaks present) spectrum of the difference $\widetilde{\mathbf{y}}(\omega) - \widehat{\widetilde{\mathbf{y}}}(\omega)$.

It should be noted that using an expanded set of mode shapes like described in Section 12.3 in Eq. (12.32) allows the user to estimate the operational response in any DOF defined in the FE model.

12.5.3   Generalized Operating Deflection Shapes (ODS)

When a signal is decomposed as given by Eqs. (12.29/12.32), we get some more possibilities for obtaining information about the structure from which the response has been measured.

It is normal in modal analysis to illustrate the measured response by an animation plot showing the movement of the measured DOF's. In the time domain, this is performed by showing 3D picture frames of the deformed structure during the testing as a movie, and in the frequency domain, similar picture frames are shown by creating a movement based on one of the columns of the spectral density matrix at a certain frequency. This is called operating deflection shapes and is used a lot for troubleshooting vibration problems.

Using mode shape expansion as mentioned in Section 12.3, we can expand the measured response using mode shapes from an FE model. The synthesized signals given by Eqs. (12.29/12.32) can be expanded to all DOF's in the FE model, and thus, we can show the 3D movement of all points in the considered structure.

Further, when all movements are known, this allows us to estimate the stresses and the strains in all points of the structure. See more about this in Section 12.8.

12.6   Mode Shape Scaling

The mode shape scaling factor (equivalent to knowing the modal mass) is also a modal parameter. This is often forgotten in modal analysis. The modal identification problem is not solved completely until this quantity is known, because it is needed in order to form the transfer and frequency response functions, see Sections 5.2.3 and 5.2.4 in Chapter 5.

In dealing with this subject, it is common to assume that the modal mass is a real-valued quantity, and here we will limit the treatment to this case.

In OMA, it is difficult to estimate the modal mass because we do not have a relation between the response and the force like in traditional modal analysis. Therefore, the mass must be estimated using some kind of dedicated testing or analysis. Several ways of estimating the modal mass are known in the literature, and the most important are mentioned next.

In this section, we will say (which is commonly used in most of the references mentioned below) that if the modal mass m is determined by the unit scaled mode shape $\mathbf{a}$ as

$$m = \mathbf{a}^T \mathbf{M} \mathbf{a} \tag{12.33}$$

where $\mathbf{M}$ is the mass matrix for the actual number of DOF's[5] used in the considered mode shape. The corresponding scaled mode shape $\boldsymbol{\alpha}$ is found as

$$\boldsymbol{\alpha} = \alpha \ \mathbf{a} \tag{12.34}$$

As the mass scaled mode shape produces a modal mass equal to unity when using Eq. (12.33), we have that the scaling factor α is related to the modal mass by the relation

$$\alpha = \frac{1}{\sqrt{m}} \tag{12.35}$$

12.6.1   Mass Change Method

In this subject, we shall not derive any of the equations, but just show them and provide the proper references to their origin. Concerning an overview of the different equations, the reader is referred to Lopez-Aenlle et al. [6].

In the mass change method, the idea is to perform two OMA tests: one test in its initial or undisturbed state and another test where additional mass is placed on the structure. The changes of the mode shapes and the natural frequencies are then used to calculate the scaling factors for the mode shape.

The first equation for the determination of the scaling factor based on mass change was given by Parloo et al. [7], who derived the following equation based on the sensitivity equations (see Section 5.3.2)

$$\alpha^2 = \frac{2\left(\omega_0 - \omega_I\right)}{\omega_0 \mathbf{a}^T \Delta \mathbf{M} \mathbf{a}} \tag{12.36}$$

where ω_0, ω_I are the natural frequencies for the considered mode before and after adding the mass, respectively, and $\Delta \mathbf{M}$ is the mass change matrix.

In the sensitivity equations, everything is linearized, which is not really a good idea mainly because the natural frequencies appear in the equations of motion by their squared value. Brincker and Andersen [8] proposed the following equation derived directly from the equation of motion

$$\alpha^2 = \frac{\omega_0^2 - \omega_I^2}{\omega_I^2 \mathbf{a}^T \Delta \mathbf{M} \mathbf{a}} \tag{12.37}$$

It should be noted that Eq. (12.36) follows directly from Eq. (12.37) using the approximation

$$\frac{\omega_0^2 - \omega_I^2}{\omega_I^2} \cong \frac{2\left(\omega_0 - \omega_I\right)}{\omega_0} \tag{12.38}$$

These equations assume that the influence of mass change on the mode shape is so small that it can be neglected. However, in order to obtain a good estimate of the scaling factor, mass changes must be large

---

[5] It is worth noting that a model with a large number of DOF's can be exactly reduced to a smaller number of DOF's if the number of modes included in the model is equal to the number of DOF's in the reduced system, see the section on SEREP in Chapter 5.

enough to achieve changes of the natural frequencies that are significantly larger than the estimation uncertainty. In order to be able to deal with this problem, the following equation was proposed by Lopez-Aenlle et al. [6]

$$\alpha^2 = \frac{\omega_0^2 - \omega_I^2}{\omega_0^2 \mathbf{a}_0^T \Delta \mathbf{M} \mathbf{a}_I}$$

(12.39)

where $\mathbf{a}_0$, $\mathbf{a}_I$ are the mode shapes before and after application of the mass change, respectively. The idea of using the mode shapes before and after the application of the mass change was introduced earlier by Bernal [9], who used a projection technique applied on the equation of motion to derive a similar formula that can be considered as an exact solution, Lopez-Aenlle et al. [10]

$$\alpha^2 = \frac{\left(\omega_0^2 - \omega_I^2\right) t}{\omega_I^2 \mathbf{a}_0^T \Delta \mathbf{M} \mathbf{a}_I}$$

(12.40)

where the quantity t is a projection term that is normally very close to unity[6] and can be found in the following way. Let all the unity scaled mode shapes before and after the mass change be gathered in the mode shape matrices $\mathbf{A}_0$, $\mathbf{A}_I$, respectively. We then know from the sensitively equations that with a good approximation, a linear relation exists between the two sets of mode shapes such that

$$\mathbf{A}_I = \mathbf{A}_0 \mathbf{T}$$

(12.41)

where the linear relation is described by the matrix $\mathbf{T}$. The diagonal elements in $\mathbf{T}$ that corresponds to the considered mode is then equal to the projection quantity t. Or more precisely, if the considered mode is mode n in the mode shape matrices, and the elements in the transformation matrix is given by $\mathbf{T} = [t_{rc}]$, then the projection term is $t = t_{nn}$.

12.6.2 Mass-Stiffness Change Method

Just like a mass change can be used to find the scaling factor, a stiffness change can be used in same manner. In Kathibi et al. [11], it is shown that Eq. (12.37) can be generalized to

$$\alpha^2 = \frac{\omega_0^2 - \omega_I^2}{\mathbf{a}^T \left(\omega_I^2 \Delta \mathbf{M} + \Delta \mathbf{K}\right) \mathbf{a}}$$

(12.42)

and in Lopez-Aenlle et al. [10], it is shown that the Bernal projection equation (12.40) can be similarly generalized to

$$\alpha^2 = \frac{\left(\omega_0^2 - \omega_I^2\right) t}{\mathbf{a}_0^T \left(\omega_I^2 \Delta \mathbf{M} + \Delta \mathbf{K}\right) \mathbf{a}_I}$$

(12.43)

All the aforementioned methods only provide one equation for each mode. However, in Lopez-Aenlle et al. [10], it is shown that the scaling factor for mode j can be found using information from both mode j and mode i as given by the following expression that can also be shown to be an exact solution

$$\alpha_j^2 = \frac{\left(\omega_{0j}^2 - \omega_{Ii}^2\right) t_{ji}}{\mathbf{a}_{0j}^T \left(\omega_{Ii}^2 \Delta \mathbf{M} + \Delta \mathbf{K}\right) \mathbf{a}_{Ii}}$$

(12.44)

In this equation, the quantities ω_{0j}, $\mathbf{a}_{0j}$ are the unperturbed eigenfrequency and mode shape of mode j, respectively; ω_{Ii}, $\mathbf{a}_{Ii}$ are the perturbed eigenfrequency and mode shape of mode i, respectively; and t_{ji} is

---

[6] Close to unity except in cases of closely spaced modes.

the matrix elements in the projection matrix $\mathbf{T} = \left[t_{ji} \right]$. For each scaling factor j, this gives us as many equations as we have elements in the jth row of $\mathbf{T}$ that are different from zero.

It should be noted that using a mass change strategy, from an experimental point of view, the idea is to provide as much change measured by the inner product $\mathbf{a}_{0j}^T \left(\omega_{ji}^2 \Delta\mathbf{M} + \Delta\mathbf{K} \right) \mathbf{a}_{ji}$ as possible for a given cost. We see that for modes with a low frequency, it takes more mass change to provide a given change of the inner product due to term $\omega_{ji}^2 \Delta\mathbf{M}$ where the mass matrix is weighted with the natural frequency squared. We see that the stiffness term does not have this weighting; thus, for low frequencies, it might be desirable to use a stiffness change instead of, or combined with, a mass change.

12.6.3   Using the FEM Mass Matrix

For large structures and, in general, for many civil engineering structures, both mass changes and stiffness changes might be difficult to apply in practice. However, in many cases, an FE model might be available, and in such case, it should be possible to use the mass matrix $\mathbf{M}$ from the FE model as a basis for mode shape scaling.

Two main approaches exist: one approach is to reduce the full mass matrix $\mathbf{M}$ to the active DOF's in the experiment by using SEREP to obtain the reduced mass matrix $\mathbf{M}_a$ and then calculate the scaling factor directly from Eqs. (12.33/12.35)

$$\alpha^2 = \frac{1}{\mathbf{a}^T \mathbf{M}_a \mathbf{a}} \tag{12.45}$$

The other approach is to expand the experimental mode shapes to all DOF's in the FE model by using expansion as indicated in Eq. (12.22).

$$\alpha^2 = \frac{1}{\hat{\mathbf{a}}^T \mathbf{M} \hat{\mathbf{a}}} \tag{12.46}$$

These two approaches are investigated further in Lopez-Aenlle and Brincker [12], and it is shown that often it is better to use the FE model mass matrix instead of the mass change approaches mentioned earlier. In the paper by in Lopez-Aenlle and Brincker [13], a simple formula for the modal mass is proposed using expansion according to Eq. (12.22)

$$m = \hat{\mathbf{a}}^T \mathbf{M} \hat{\mathbf{a}} = \hat{\mathbf{t}}^T \mathbf{B}^T \mathbf{M} \mathbf{B} \hat{\mathbf{t}} = \hat{\mathbf{t}}^T \hat{\mathbf{t}} \tag{12.47}$$

The result follows directly from Eq. (12.22) assuming that the FE mode shapes in the matrix $\mathbf{B}$ are mass scaled so that inner product $\mathbf{B}^T \mathbf{M} \mathbf{B}$ produces the identity matrix. It should be noted that in this approach, the mass matrix of the FE model is actually not needed – only the mass scaled mode shapes.

12.7   Force Estimation

As in the frequency domain, the response $\tilde{\mathbf{y}}(\omega)$ is simply the frequency response function (FRF) matrix $\tilde{\mathbf{H}}(\omega)$ times the external force vector $\tilde{\mathbf{x}}(\omega)$

$$\tilde{\mathbf{y}}(\omega) = \tilde{\mathbf{H}}(\omega) \tilde{\mathbf{x}}(\omega) \tag{12.48}$$

when the response $\tilde{\mathbf{y}}(\omega)$ is measured and all components to create the FRF matrix is known from OMA, then it would be strange if the unknown forces in the vector $\tilde{\mathbf{x}}(\omega)$ could not be determined.

The problem is known in the literature as the problem of inverse filtering. In the time domain version Eq. (12.48) turns into a matrix convolution, and this is known as the deconvolution problem. In OMA, the problem is often denoted as force estimation and load estimation, but in OMA, the literature on the subject is rather fragmented.

12.7.1   Inverting the FRF Matrix

The most direct way of solving the problem is to invert the FRF matrix in Eq. (12.48). When performing the inversion in the frequency domain the FRF matrix normally is rank deficient, and the issue of how to handle the deficient matrix inversion when taking the pseudo inverse becomes a central problem

$$\tilde{\mathbf{x}}(\omega) = \tilde{\mathbf{H}}^+(\omega)\tilde{\mathbf{y}}(\omega) \tag{12.49}$$

This issue has been investigated in, for instance, Lopez-Aenlle et al. [14] and Pedersen et al. [15].

12.7.2   Modal Filtering

The aforementioned problem of handling the deficiency of the FRF matrix becomes equivalent to choosing an appropriate set of mode shapes, frequency band by frequency band, in order to perform a good modal filtering as described in Section 12.5.2.

When the modal coordinates are estimated in the frequency domain, an estimate of each modal load can then be found by simple means for each modal coordinate $\tilde{q}_n(\omega)$

$$\tilde{q}_n(\omega) = H_n(\omega)\tilde{p}_n(\omega) \tag{12.50}$$

by assuming some variation of the modal force $\tilde{p}_n(\omega)$, where $H_n(\omega)$ is the modal FRF for the considered modal coordinate. Finally, the corresponding force distribution can be estimated from the relation between the external force and the modal force, see Eq. (5.78)

$$\tilde{\mathbf{p}}(\omega) = \mathbf{A}^T\tilde{\mathbf{x}}(\omega) \tag{12.51}$$

where $\mathbf{A}$ is the mode shape matrix and $\tilde{\mathbf{p}}(\omega)$ is a column vector holding the modal forces $\tilde{p}_n(\omega)$. The force distribution is then estimated as

$$\hat{\tilde{\mathbf{x}}}(\omega) = \left(\mathbf{A}^+\right)^T\tilde{\mathbf{p}}(\omega) \tag{12.52}$$

12.8   Estimation of Stress and Strain

Stress and strain can be estimated in OMA by two different approaches. One is to use force identification as described earlier, and when the external forces are known, the forces can be applied to a finite element model and stress and strain can be estimated. The second approach is to expand the mode shapes as explained in Section 12.3 and then use the expansion to obtain displacement of each element in the FE model and then use the shape function of the element to estimate strain and stress.

12.8.1   Stress and Strain from Force Estimation

Once the external forces $\hat{\mathbf{x}}(t)$ has been estimated in the DOF's measured in a test, the external load vector can then be applied as the right-hand side of the equation of motion related to the considered FE model

$$\mathbf{M}\ddot{\mathbf{y}}(t) + \mathbf{C}\dot{\mathbf{y}}(t) + \mathbf{K}\mathbf{y}(t) = \hat{\mathbf{x}}(t) \tag{12.53}$$

The FE model can be used to numerically determine stress and strain in any point of interest. It is here understood that the estimated external force $\hat{\mathbf{x}}(t)$ is to be expanded to all DOF's in the FE model by padding zeroes for all DOF's in the FE model that does not correspond to a DOF in the experiment. Alternatively, when finding the load distribution using Eq. (12.51), the modes shapes in the mode shape matrix can be expanded first to include all DOF's in the FE model as described in Section 12.3.

12.8.2   Stress and Strain from Mode Shape Expansion

When mode shapes are expanded to include all DOF's in the considered FE model, it also means that all DOF's in each element of the FE model are known. As for each element, there always exists a so-called shape function that relates the DOF's for each element with the state of stress and strain inside the element, and when the DOF's of the considered element have been estimated, the complete state of stress and strain inside the element is also known.

References

[1] D'Ambrogio, W. and Fregolent, A.: *Higher-order mac for the correlation of close and multiple modes. Mech. Syst. Signal Process.*, V. 17, No. 3, p. 599–610, 2003.

[2] Brincker, R., Skafte, A., López-Aenlle, M., Sestieri, A., D'Ambrogio, W. and Canteli, A.: *A local correspondence principle for mode shapes in structural dynamics. J. Mech. Syst. Signal Process.*, V. 45, No. 1, p. 91–104, 2014.

[3] Ibrahim, S.R.: *Modal confidence factor in vibration testing. J. Spacecraft*, V. 15, No. 5, p. 313–316, 1978.

[4] Pappa, R.S., Elliott, K.B. and Schenk, A.: *Consistent-mode indicator for the eigensystem realization algorithm. J. Guidance Contr. Dyn.*, V. 16, No. 5, p. 852–858, 1993.

[5] Brincker, R., Andersen, P. and Jacobsen, N.J.: *Automated frequency domain decomposition for operational modal analysis*. In Proceedings of IMAC-XXIV, 2007.

[6] Lopez-Aenlle, M., Brincker, R. and Canteli, A.F.: *Some methods to determine scaled mode shapes in natural input modal analysis*. In proceedings of the International Modal Analysis Conference (IMAC), January 31–February 3, Orlando, Florida, 2005.

[7] Parloo, E., Verboven, P., Guillaume, P. and VanOvermeire, M.: *Sensitivity-based operational mode shape normalization. Mech. Syst. Signal Process.*, V. 16, p. 757–767, 2002.

[8] Brincker, R. and Andersen, P.: *A way of getting scaled mode shapes in output only modal analysis*. In Proceedings of the International Modal Analysis Conference (IMAC) XXI, Orlando, USA, 2003.

[9] Bernal, D.: *Modal scaling from known mass perturbations. J. Eng. Mech.*, V. 130, No. 9, p. 1083–1088, 2004.

[10] Lopez-Aenlle, M., Brincker, R., Pelayo, F. and Canteli, A.F.: *On exact and approximated formulations for scaling-mode shapes in operational modal analysis by mass and stiffness change. J. Sound Vibr.*, V. 331, No. 3, p. 622–637, 2012.

[11] Khatibi, M.M., Ashory, M.R., Malekjafarian, A. and Brincker, R.: *Mass-stiffness change method for scaling of operational mode shapes. Mech. Syst. Signal Process.*, V. 26, p. 34–59, 2012.

[12] Aenlle, M.L. and Brincker, R.: *Modal scaling in operational modal analysis using a finite element model, Int. J. Mech. Sci.*, V. 76, p. 86–101, 2013.

[13] Aenlle, M.L. and Brincker, R.: *Modal scaling in OMA using the mass matrix of a finite element model*. In Proceedings of the XXXII Int. Modal Analysis Conference (IMAC), Orlando, February 3–6, 2014.

[14] Lopez-Aenlle, M., Brincker, R., Fernández, P.F. and Canteli, A.F.: *Load estimation from modal parameters*. In Proceedings of the 2nd International Operational Modal Analysis Conference, 2007.

[15] Pedersen, I.C.B., Mosegaard, S., Brincker, R. and López-Aenlle, M.: *Load estimation by frequency domain decomposition*. In Proceedings of the 2nd International Operational Modal Analysis Conference (IOMAC), p. 669–676, 2007.

Appendix A

Nomenclature and Key Equations

In this appendix, the most important notation for the book is summarized. This is done by presenting the important symbols used in each chapter of the book, proving a short description of each symbol, and indicating the equation number in which the symbol is used for the first time in the book.

Chapter 2. Random Variables and Signals

$x(t)$	Time series
X	Stochastic variable
$p(x)$	Probability density function, Eq. (2.1)
$E[X]$	Expectation, Eq. (2.3)
μ	Mean value, Eq. (2.3)
$E\left[\left(X - \mu_x\right)^2\right]$	Variance, Eq. (2.4)
σ	Standard deviation, Eq. (2.4)
$\operatorname{cor}\left[x(t), y(t)\right]$	Correlation, Eq. (2.20)
$R_x(\tau) = E[x(t)x(t + \tau)]$	Auto correlation, Eq. (2.27)
$R_x(\tau) = \frac{1}{T}\int_0^T x(t)x(t + \tau)\,dt$	Auto correlation, time averaging, Eq. (2.30)
$R_{xy}(\tau) = E\left[x(t)y(t + \tau)\right]$	Cross correlation, Eq. (2.31)
$\mathbf{R}(\tau) = E\left[\mathbf{y}(t)\,\mathbf{y}^T(t + \tau)\right]$	Correlation function matrix, Eq. (2.34)
$\mathbf{C} = E\left[\mathbf{y}(t)\,\mathbf{y}^T(t)\right] = \mathbf{R}(0)$	Correlation matrix, Eq. (2.37)

Chapter 3. Matrices and Regression

$\mathbf{A} = \left[a_{rc}\right]$	Matrix, Eq. (3.1)
$\mathbf{a} = \left\{a_r\right\}$	Column vector, Eq. (3.4)
$\mathbf{A}^T = \left[a_{cr}\right]$	Transpose matrix, Eq. (3.5)
$\mathbf{A}^H = \left[a_{cr}^*\right]$	Hermitian matrix, Eq. (3.6)
$\mathbf{a} \cdot \mathbf{b} = \mathbf{a}^T\mathbf{b}$	Inner product, Eq. (3.18)
$|\mathbf{a}|^2 = \mathbf{a}^H\mathbf{a}$	Length of vector, Eq. (3.20)
$\mathbf{a}^H\mathbf{b} = |\mathbf{a}|\,|\mathbf{b}|\cos(\theta)$	Generalized interior angle, Eq. (3.21)

Introduction to Operational Modal Analysis, First Edition. Rune Brincker and Carlos Ventura.
© 2015 John Wiley & Sons, Ltd. Published 2015 by John Wiley & Sons, Ltd.
Companion Website: www.wiley.com/go/brincker

$\mathbf{T} = \mathbf{a} \otimes \mathbf{b} = \mathbf{a}\mathbf{b}^T$ Outer product, Eq. (3.30)

$\mathbf{a}^H \mathbf{T} \mathbf{a} > 0$ Positive definite matrix, Eq. (3.35)

$\mathbf{T}\mathbf{v} = e\mathbf{v}$ Eigenvalue equation, Eq. (3.36)

$\mathbf{v}_m^T \mathbf{T} \mathbf{v}_n = 0$ Orthogonality, Eq. (3.40)

$\mathbf{T} = \left[\mathbf{v}_m\right]\left[e_m\right]\left[\mathbf{v}_m\right]^{-1}$ Eigenvalue decomposition, Eq. (3.47)

$f(\mathbf{T}) = \left[\mathbf{v}_m\right]\left[f(e_m)\right]\left[\mathbf{v}_m\right]^{-1}$ Function of a matrix, Eq. (3.54)

$\mathbf{A} = \mathbf{U}\mathbf{S}\mathbf{V}^H$ Singular value decomposition, Eq. (3.56)

$\mathbf{H} = \begin{bmatrix} \mathbf{y}(1) & \mathbf{y}(2) & \mathbf{y}(3) & \cdots \\ \mathbf{y}(2) & \mathbf{y}(3) & \mathbf{y}(4) & \cdots \end{bmatrix}$ Hankel matrix, Eq. (3.64)

$\det(\mathbf{A}) = \prod_{n=1}^{N} e_n$ Determinant of a matrix, Eq. (3.68)

$tr(\mathbf{A}) = \sum_{n=1}^{N} a_{nn}$ Trace of a matrix, Eq. (3.69)

$\|\mathbf{a}\|_p = \left(\sum_r |a_r|^p\right)^{1/p}$ p-norm, Eq. (3.72)

$\|\mathbf{A}\|_F = \sqrt{\sum_r \sum_c |a_{rc}|^2}$ Frobenius norm, Eq. (3.73)

$\mathbf{y} = \mathbf{X}\mathbf{a}$ Fitting equation, Eq. (3.92)

$\mathbf{X}^+ = \left(\mathbf{X}^T\mathbf{X}\right)^{-1}\mathbf{X}^T$ Pseudo inverse of a matrix, least squares, Eq. (3.99)

$\mathbf{X}^+ = \mathbf{U}\mathbf{S}^+\mathbf{V}^H$ Pseudo inverse of a matrix, SVD, Eq. (3.102)

$\mathbf{C}_a = \left(\mathbf{X}^T\mathbf{C}_\varepsilon^{-1}\mathbf{X}\right)^{-1}, \; \mathbf{C}_a = \sigma_\varepsilon^2\left(\mathbf{X}^T\mathbf{X}\right)^{-1}$ Parameter covariance matrix, Eqs. (3.112/3.114)

$\sigma_y^2(x) = \mathbf{f}^T(x)\,\mathbf{C}_a\mathbf{f}(x)$ Fitting covariance, Eq. (3.117)

Chapter 4. Transforms

$y(t)$ Signal (as a function of time, t)

$\omega_k = \frac{2\pi k}{T}$ Discrete circular frequency, Eq. (4.5)

T Period of vibration

$Y_k = A_k - iB_k$ Complex Fourier coefficients, Eq. (4.10)

$y(t) = \sum_{k=-\infty}^{\infty} Y_k e^{i\Delta\omega k t}$ Complex Fourier series expansion, Eq. (4.11)

$Y_k = \frac{1}{T}\int_{-T/2}^{T/2} y(t)\,e^{-i\Delta\omega k t}\,dt$ Fourier transform, periodic signal, Eq. (4.13)

$y(t) = \int_{-\infty}^{\infty} Y(\omega)\,e^{i\omega t}\,d\omega$ Fourier integral, Eq. (4.31)

$Y(\omega) = \frac{1}{2\pi}\int_{-\infty}^{\infty} y(t)\,e^{-i\omega t}\,dt$ Fourier transform, nonperiodic signal, Eq. (4.32)

$f_v = \frac{1}{2\Delta t} = \frac{f_s}{2}$ Nyquist frequency, Eq. (4.45)

$\Delta f = \frac{1}{T}$

Frequency resolution or frequency increment, Eq. (4.46)

$y_n = \sum\limits_{k=1}^{N} Y_k e^{i2\pi(k-1)(n-1)/N}$

Fourier series, discrete time, Eq. (4.53)

$Y_k = \frac{1}{N} \sum\limits_{n=1}^{N} y_n e^{-i2\pi(k-1)(n-1)/N}$

Discrete Fourier transform, Eq. (4.54)

$Y'(s) = \int\limits_{0}^{\infty} y(t) e^{-st} dt$

Laplace transform, Eq. (4.63)

$Y''(z) = \sum\limits_{n=0}^{\infty} y_n z^{-n}$

Z-transform, Eq. (4.96)

$y(n) = \sum\limits_{k=1}^{N} a_k y(n-k) + \sum\limits_{k=0}^{M} b_k x(n-k)$

ARMA model, Eq. (4.119)

Chapter 5. Classical Dynamics

$m\ddot{y}(t) + c\dot{y}(t) + ky(t) = x(t)$

Single DOF equation of motion (viscous damping), Eq. (5.5)

$\lambda = \frac{-c \pm i\sqrt{4mk-c^2}}{2m}$

Continuous time poles, Eq. (5.8)

$\omega_0 = \sqrt{\frac{k}{m}}; \quad \varsigma = \frac{c}{2\sqrt{mk}}$

Circular natural frequency and damping factor, Eq. (5.9)

$\omega_d = \omega_0 \sqrt{1-\varsigma^2}$

Damped circular natural frequency, Eq. (5.11)

$\lambda = -\varsigma\omega_0 + i\omega_d, \lambda^* = -\varsigma\omega_0 - i\omega_d$

Poles in terms of natural frequency and damping, Eq. (5.12)

$\delta = \omega \varsigma T_d = 2\pi \frac{\varsigma}{\sqrt{1-\varsigma^2}}$

Logarithmic decrement, Eq. (5.21)

$y(t) = \int\limits_{-\infty}^{\infty} h(t-\tau) x(\tau) d\tau$

General solution by convolution, Eq. (5.32)

$H(s) = \frac{1}{m(s-\lambda)(s-\lambda^*)}$

Transfer function, Eq. (5.37)

$h(t) = \frac{1}{m} \frac{e^{\lambda t} - e^{\lambda^* t}}{\lambda - \lambda^*}$

Impulse response, Eq. (5.38)

$H(\omega) = H'(i\omega) = \frac{1}{m} \frac{1}{(i\omega-\lambda)(i\omega-\lambda^*)}$

Frequency response function (FRF), Eq. (5.45)

$\mathbf{M}\ddot{\mathbf{y}}(t) + \mathbf{C}\dot{\mathbf{y}}(t) + \mathbf{K}\mathbf{y}(t) = \mathbf{x}(t)$

Multi-DOF equation of motion, Eq. (5.51)

$\mathbf{b}_n^T \mathbf{M} \mathbf{b}_m = 0, \mathbf{b}_n^T \mathbf{K} \mathbf{b}_m = 0$

Mode shape orthogonality, Eqs. (5.58–5.59)

$\mathbf{b}_n^T \mathbf{M} \mathbf{b}_n = m_n$

Modal mass, Eq. (5.61)

$\mathbf{b}_n^T \mathbf{K} \mathbf{b}_n = k_n$

Modal stiffness, Eq. (5.62)

$\mathbf{C} = \alpha\mathbf{M} + \beta\mathbf{K}$

Rayleigh's type proportional damping, Eq. (5.68)

$p_n(t) = \mathbf{b}_n^T \mathbf{x}(t)$

Modal load, Eq. (5.78)

$H_n(s) = \frac{1}{m_n(s-\lambda_n)(s-\lambda_n^*)}$

Modal transfer function, Eq. (5.80)

$\widetilde{\mathbf{y}}(s) = \widetilde{\mathbf{H}}(s)\widetilde{\mathbf{x}}(s)$

Input–output relation in the frequency domain, Eq. (5.86)

$\widetilde{\mathbf{H}}(s) = \begin{bmatrix}\mathbf{b}_n\end{bmatrix} \begin{bmatrix}H_n(s)\end{bmatrix} \begin{bmatrix}\mathbf{b}_n\end{bmatrix}^T$

Transfer function, Eq. (5.87)

$\mathbf{u}(t) = \left\{ \begin{array}{c} \dot{\mathbf{y}}(t) \\ \mathbf{y}(t) \end{array} \right\}$

State variable, Eq. (5.101)

$$\left.\begin{array}{r}\mathbf{A}\dot{\mathbf{u}}(t) + \mathbf{B}\mathbf{u}(t) = \mathbf{f}(t)\\ \mathbf{y}(t) = \mathbf{P}\mathbf{u}(t)\end{array}\right\}$$

State space equations, Eq. (5.102)

$$\boldsymbol{\varphi}_n = \left\{\begin{array}{c}\lambda\mathbf{b}_n\\ \mathbf{b}_n\end{array}\right\}$$

State space mode shapes, Eq. (5.107)

$$\boldsymbol{\varphi}_n^T\mathbf{A}\boldsymbol{\varphi}_m = 0, \; \boldsymbol{\varphi}_n^T\mathbf{B}\boldsymbol{\varphi}_m = 0$$

Generalized orthogonality, Eqs. (5.110–5.111)

$$a_n = \boldsymbol{\varphi}_n^T\mathbf{A}\boldsymbol{\varphi}_n$$

Generalized modal coefficient, Eq. (5.112)

$$b_n = \boldsymbol{\varphi}_n^T\mathbf{B}\boldsymbol{\varphi}_n$$

Generalized modal coefficient, Eq. (5.113)

$$\left.\begin{array}{r}\lambda_n\\ \lambda_n^*\end{array}\right\} = \sigma_n \pm i\omega_{dn}$$

Generalized eigenvalues, Eq. (5.115)

$$\omega_{0n} = \sqrt{\lambda_n\lambda_n^*}$$

Modulus of eigenvalue, Eq. (5.116)

$$\zeta_n = -\frac{\sigma_n}{\omega_{0n}}$$

Generalized damping ratio, Eq. (5.117)

$$\omega_{dn} = \omega_{0n}\sqrt{1 - \zeta_n^2}$$

Generalized circular damped natural frequency, Eq. (5.119)

$$\tilde{\mathbf{H}}(s) = \sum_{n=1}^{N}\left(\frac{\mathbf{b}_n\mathbf{b}_n^T}{a_n(s-\lambda_n)} + \frac{\mathbf{b}_n^*\mathbf{b}_n^{*T}}{a_n^*(s-\lambda_n^*)}\right)$$

General transfer function matrix, Eq. (5.132)

$$\tilde{\mathbf{H}}(i\omega) = \sum_{n=1}^{N}\left(\frac{\mathbf{b}_n\mathbf{b}_n^T}{a_n(i\omega-\lambda_n)} + \frac{\mathbf{b}_n^*\mathbf{b}_n^{*T}}{a_n^*(i\omega-\lambda_n^*)}\right)$$

General frequency response function (FRF) matrix, Eq. (5.133)

$$\mathbf{H}(t) = \sum_{n=1}^{N}\left(\frac{\mathbf{b}_n\mathbf{b}_n^T}{a_n}e^{\lambda_n t} + \frac{\mathbf{b}_n^*\mathbf{b}_n^{*T}}{a_n^*}e^{\lambda_n^* t}\right)$$

General impulse response function (IRF) matrix, Eq. (5.136)

$$\mathbf{A} = \mathbf{B}\mathbf{T}$$

Transformation, two complete sets of modes, Eq. (5.158)

$$\mathbf{A} \cong \mathbf{B}\mathbf{T}$$

Transformation, two incomplete sets of modes, Eq. (5.167)

$$\theta \cong \frac{\omega}{2m\Delta\omega}\left(\mathbf{b}_1^T\Delta\mathbf{M}\mathbf{b}_2 - \frac{1}{\omega^2}\mathbf{b}_1^T\Delta\mathbf{K}\mathbf{b}_2\right)$$

Rotation of closely spaced modes, Eq. (5.181)

$$\mathbf{B}^T\left(-\Delta\mathbf{M} + \frac{1}{\omega^2}\Delta\mathbf{K}\right)\mathbf{B} = \mathbf{T}\mathbf{D}\mathbf{T}^{-1}$$

Eigenvalue equation for closely spaced modes, Eq. (5.187)

$$\mathbf{T} = \begin{bmatrix}\mathbf{B}_a\mathbf{B}_a^+\\ \mathbf{B}_d\mathbf{B}_a^+\end{bmatrix} = \mathbf{B}\mathbf{B}_a^+$$

SEREP transformation, Eq. (5.197)

$$\mathbf{M}_a = \mathbf{T}^T\mathbf{M}\mathbf{T}; \quad \mathbf{K}_a = \mathbf{T}^T\mathbf{K}\mathbf{T}$$

SEREP reduction, Eq. (5.201)

$$\mathbf{y}(n) = \sum_{k=1}^{N}\mathbf{A}_k\mathbf{y}(n-k) + \sum_{k=0}^{M}\mathbf{B}_k\mathbf{x}(n-k)$$

ARMA model vector, Eq. (5.203)

$$\mathbf{u}_d(n) = \left\{\begin{array}{c}\mathbf{y}(n-N+1)\\ \vdots\\ \mathbf{y}(n-1)\\ \mathbf{y}(n)\end{array}\right\}$$

Discrete time state vector, Eq. (5.205)

$$\mathbf{A}_C = \begin{bmatrix}0 & \mathbf{I} & 0 & 0\\ \vdots & 0 & \ddots & \vdots\\ 0 & \vdots & & \mathbf{I}\\ \mathbf{A}_N & \mathbf{A}_{N-1} & \cdots & \mathbf{A}_1\end{bmatrix}$$

Companion matrix, Eq. (5.206)

$$\mathbf{A}_C\boldsymbol{\varphi}_d = \mu\boldsymbol{\varphi}_d$$

Discrete time eigenvalue equation, Eq. (5.210)

$$\left.\begin{array}{r}\dot{\mathbf{u}}(t) = \mathbf{A}\mathbf{u}(t) + \mathbf{B}\mathbf{x}(t)\\ \mathbf{y}(t) = \mathbf{P}\mathbf{u}(t)\end{array}\right\}$$

Unsymmetrical state space formulation, Eq. (5.211)

$$\mathbf{u}(t) = \exp(\mathbf{A}t)\mathbf{u}(0) + \int_0^t \exp(\mathbf{A}(t-\tau))\mathbf{B}\mathbf{x}(\tau)\,d\tau \qquad \text{General solution, Eq. (5.214)}$$

$$\mathbf{D} = \exp(\mathbf{A}\Delta t) \qquad\qquad\qquad\qquad \text{Discrete time system matrix, Eq. (5.216)}$$

$$\mathbf{y}(n) = \mathbf{P}\mathbf{D}^n\mathbf{u}_0 \qquad\qquad\qquad\qquad \text{Discrete time free decays, Eq. (5.217)}$$

$$\mathbf{A}_C = \begin{bmatrix} \mathbf{P} \\ \mathbf{P}\mathbf{D} \end{bmatrix} \mathbf{D} \begin{bmatrix} \mathbf{P} \\ \mathbf{P}\mathbf{D} \end{bmatrix}^{-1} \qquad\qquad \text{From state space to companion matrix, Eq. (5.222)}$$

Chapter 6. Random Vibrations

$$G_x(\omega) = \frac{1}{2\pi} \int_{-\infty}^{\infty} R_x(\tau) e^{-i\omega\tau}\,d\tau \qquad\qquad \text{Auto spectral density, Eq. (6.12)}$$

$$R_x(\tau) = \int_{-\infty}^{\infty} G_x(\omega) e^{i\omega\tau}\,d\omega \qquad\qquad \text{Correlation function obtained from spectral density,}$$
$$\text{Eq. (6.13)}$$

$$R_x(0) = \mathrm{E}\left[x^2\right] = \int_{-\infty}^{\infty} G_x(\omega)\,d\omega \qquad\qquad \text{Parseval's theorem, Eq. (6.14)}$$

$$\left.\begin{aligned} G_{xy}(\omega) &= \frac{1}{2\pi} \int_{-\infty}^{\infty} R_{xy}(\tau) e^{-i\omega\tau}\,d\tau \\[2mm] R_{xy}(\tau) &= \int_{-\infty}^{\infty} G_{xy}(\omega) e^{i\omega\tau}\,d\omega \end{aligned}\right\} \quad \text{Cross correlation spectral density relations, Eq. (6.16)}$$

$$R_{xy}(0) = \mathrm{E}\left[xy\right] = \int_{-\infty}^{\infty} G_{xy}(\omega)\,d\omega \qquad \text{General form of Parseval's theorem, Eq. (6.17)}$$

$$G_y(\omega) = G_x(\omega) H^*(i\omega) H(i\omega) \qquad\qquad \text{SISO fundamental theorem, Eq. (6.36)}$$

$$\begin{aligned} \mathbf{G}_y(\omega) &= \tilde{\mathbf{H}}^*(i\omega)\,\mathbf{G}_x(\omega)\,\tilde{\mathbf{H}}^T(i\omega) \\ &= \tilde{\mathbf{H}}(-i\omega)\,\mathbf{G}_x(\omega)\,\tilde{\mathbf{H}}(i\omega) \end{aligned} \qquad \text{MIMO fundamental theorem, Eq. (6.44)}$$

$$R_x(\tau) = \mathrm{E}[x(t)x(t+\tau)] = 2\pi G_{x0}\delta(\tau) \qquad \text{White noise input, Eq. (6.45)}$$

$$G_{x0} = \frac{\sigma_x^2}{2B} \qquad\qquad\qquad\qquad \text{White noise scaling, Eq. (6.47)}$$

$$R_x(\tau) = 2\pi \frac{\sigma_x^2}{2B}\delta(\tau) = \pi \frac{\sigma_x^2}{B}\delta(\tau) \qquad \text{White noise correlation function, Eq. (6.48)}$$

$$\left.\begin{aligned} \mathbf{R}_x(\tau) &= 2\pi \frac{\delta(\tau)}{2B}\sigma_x^2\mathbf{I} \\[2mm] \mathbf{G}_x(\omega) &= \frac{1}{2B}\sigma_x^2\mathbf{I} \end{aligned}\right\} \quad \text{Vector white noise solutions, Eq. (6.51)}$$

$$\boldsymbol{\gamma}_n = \mathbf{B}_n\mathbf{G}_x\frac{\mathbf{b}_n}{a_n} \qquad\qquad\qquad \text{Modal participation vector, Eq. (6.85)}$$

$$\mathbf{R}_y(\tau) = \begin{cases} 2\pi \sum_{n=1}^{N} \left(\mathbf{b}_n\boldsymbol{\gamma}_n^T e^{-\lambda_n\tau} + \mathbf{b}_n^*\boldsymbol{\gamma}_n^H e^{-\lambda_n^*\tau}\right); & \tau \leq 0 \\[4mm] 2\pi \sum_{n=1}^{N} \left(\boldsymbol{\gamma}_n\mathbf{b}_n^T e^{\lambda_n\tau} + \boldsymbol{\gamma}_n^*\mathbf{b}_n^H e^{\lambda_n^*\tau}\right); & \tau \geq 0 \end{cases} \quad \text{Correlation modal decomposition, Eq. (6.90)}$$

$$\mathbf{G}_y(\omega) = \sum_{n=1}^{N} \left(\frac{\mathbf{b}_n\boldsymbol{\gamma}_n^T}{-i\omega-\lambda_n} + \frac{\mathbf{b}_n^*\boldsymbol{\gamma}_n^H}{-i\omega-\lambda_n^*} + \frac{\boldsymbol{\gamma}_n\mathbf{b}_n^T}{i\omega-\lambda_n} + \frac{\boldsymbol{\gamma}_n^*\mathbf{b}_n^H}{i\omega-\lambda_n^*}\right) \quad \text{Spectral density modal decomposition, Eq. (6.91)}$$

$$\mathbf{R}_y(\tau) = \mathbf{B}\mathbf{R}_q(\tau)\mathbf{B}^T \qquad\qquad \text{Normal time-domain mode modal decomposition, Eq.}$$
$$\text{(6.119)}$$

$$\mathbf{G}_y(\omega) = \mathbf{B}\mathbf{G}_q(\omega)\mathbf{B}^T \qquad\qquad \text{Normal frequency-domain mode modal decomposition, Eq.}$$
$$\text{(6.125)}$$

Chapter 7. Measurement Technology

$D = 20\log_{10}\left(\frac{y_{\max}}{y_{\min}}\right)$ Dynamic range, Eq. (7.1)

$SN = 20\log_{10}\left(\frac{y_{\max}}{\sigma_n}\right)$ Signal-to-noise ratio, Eq. (7.2)

$T = \frac{200}{\zeta f}$ Minimum recording time as per ANSI S2.47, Eq. (7.3)

$V = 1 - \left|1 - \det\left(\hat{\mathbf{A}}^+\mathbf{A}\right)\right|$ Value of sensor placement, Eq. (7.7)

$f_s > 2.4 f_{\max}$ Recommended sampling frequency, Eq. (7.11)

$T_{tot} > \frac{20}{2\zeta f_{\min}} = \frac{10}{\zeta f_{\min}}$ Minimum recording time for OMA test, Eq. (7.16)

$U = Sa,\ U = Sv$ Sensor sensitivity, Eqs. (7.21-22)

$\delta U_0 = U_n\sqrt{2B}$ Noise in frequency band, Eq. (7.23)

$D = 20\log\left(\frac{U_m}{U_n\sqrt{2B}}\right)$ Dynamic range in frequency band, Eq. (7.24)

$\frac{1}{U}\frac{dU}{dT} = \frac{1}{S_0}\frac{dS_0}{dT}$ Scaling factor temperature sensitivity, Eq. (7.29)

$\frac{da_0}{dT} = -\frac{1}{S_0}\frac{dU}{dT}$ Bias temperature sensitivity, Eq. (7.30)

$2^{LSB} = \delta U_{0ADC}/\Delta U_{ADC}$ Least significant bit (LSB) in data acquisition, Eq. (7.52)

$N_{\text{eff}} = N_{\text{bit}} - LSB$ Number of effective bits (data acquisition), Eq. (7.53)

$\begin{aligned} D_{ADC} &= (N_{\text{eff}} - 1)20\log 2 \\ &\approx 6\left(N_{\text{eff}} - 1\right) dB \end{aligned}$ Data acquisition dynamic range, Eq. (7.54)

$\delta U_0 = \sqrt{4kTBR}$ Thermal noise voltage, Eq. (7.59)

$\delta a_{in} = \frac{\sqrt{4kTBc}}{M}$ Acceleration background noise, Eq. (7.60)

$G_n(f) = \left(1 - \gamma_{y_1 y_2}(f)\right) G_y(f)$ Noise floor estimation, Eq. (7.71)

Chapter 8. Signal Processing

$\left.\begin{aligned} \mu(\tau) &= \frac{1}{T}\int_{\tau-T/2}^{\tau+T/2} y(t)\,dt \\ \\ \sigma(\tau)^2 &= \frac{1}{T}\int_{\tau-T/2}^{\tau+T/2} (y(t) - \mu(t))^2 dt \end{aligned}\right\}$ Moving average and standard deviation, Eq. (8.1)

$\gamma = \frac{1}{\sigma^4 T}\int_0^T (y(t) - \mu)^4 dt$ Kurtosis for detection of harmonics, Eq. (8.5)

$H(\omega) = \begin{cases} 1/i\omega; & \text{for } \omega \geq \omega_1 \\ (a/i)\,W(\omega); & \text{for } \omega < \omega_1 \end{cases}$ Integration filter, Eq. (8.28)

$\hat{\mathbf{R}}(\tau) = \frac{1}{T-\tau}\int_0^{T-\tau} \mathbf{y}(t)\mathbf{y}^T(t+\tau)\,dt$ Unbiased correlation function estimation, Eq. (8.32)

$\hat{\mathbf{R}}(k) = \frac{1}{N-k}\mathbf{Y}_{(1:N-k)}\mathbf{Y}^T_{(k+1:N)}$ Unbiased correlation function estimation, Eq. (8.36)

$\mathbf{H}\mathbf{H}^T = \sum_n \mathbf{H}_n\mathbf{H}_n^T$ Segmented Hankel (Toeplitz) matrix estimation, Eq. (8.41)

$\hat{\mathbf{G}}(\omega) = \frac{1}{S}\sum_{s=1}^{S}\tilde{\mathbf{y}}_s(\omega)^*\tilde{\mathbf{y}}_s^T(\omega)$ SD matrix using Welch estimation, Eq. (8.44)

$\hat{R}_{0xy}(\tau) = \frac{T}{T-\tau}\hat{R}_{xy}(\tau)$ Unbiased Welch estimation, Eq. (8.53)

$\hat{D}_{xy}(\tau) = \frac{1}{S}\sum_{s=1}^{S}x\left(t_s+\tau\right)\Big|_{C_{y(t_s)}}$ Random decrement signature, Eq. (8.54)

$D_{xy}(\tau) = \frac{\hat{R}_{yx}(\tau)}{\sigma_y^2}y_0 - \frac{\hat{R}_{yx}(\tau)}{\sigma_y^2}\dot{y}_0$ Random decrement (RD) solution, Eq. (8.62)

$\mathbf{d}_r(\tau) = \mathrm{E}\left[\mathbf{y}(\tau+t)\Big|_{C_{y_r(t)}}\right]$ Vector random decrement, Eq. (8.67)

$\left.\begin{array}{l}\mathbf{D}_s(\tau) = \left(\mathbf{D}(\tau)+\mathbf{D}^T(-\tau)\right)/2 \\[2mm] \mathbf{D}_u(\tau) = \left(\mathbf{D}(\tau)-\mathbf{D}^T(-\tau)\right)/2\end{array}\right\}$ Symmetrical, unsymmetrical random decrement form, Eq. (8.71)

$\hat{\mathbf{G}}(\omega) = \frac{1}{R}\sum_{r=1}^{R}\tilde{\mathbf{d}}_r(\omega)^*\tilde{\mathbf{d}}_r^T(\omega)$ Random decrement spectral density, Eq. (8.79)

$\mathbf{G}_{yy}(\omega) = \sum_{m=1}^{N}\left(\frac{\gamma_m\mathbf{b}_m^T}{i\omega-\lambda_m} + \frac{\gamma_m^*\mathbf{b}_m^H}{i\omega-\lambda_m^*}\right)$ Half spectrum, Eq. (8.80)

Chapter 9. Time Domain Identification

$\mathbf{R}_y(k) = 2\pi\sum_{n=1}^{2N}\gamma_n\mathbf{a}_n^T e^{-\lambda_n k\Delta t}$ Modal participation vector fitting problem, Eq. (9.1)

$\hat{\boldsymbol{\Gamma}} = \frac{1}{2\pi}\mathbf{HM}^+$ Modal participation vector, Eq. (9.6)

$\pi_n = \frac{p_n^2}{\mathbf{p}^T\mathbf{p}}$ Modal participation factor, Eq. (9.9)

$\mathbf{y}(n) - \mathbf{A}_1\mathbf{y}(n-1) - \mathbf{A}_2\mathbf{y}(n-2) - \cdots - \mathbf{A}_{na}\mathbf{y}(n-na) = 0$ AR model free decay, Eq. (9.10)

$\left.\begin{array}{l}\mathbf{H}_1^T\mathbf{A}^T = \mathbf{H}_2^T \\[2mm] \Updownarrow \\[2mm] \hat{\mathbf{A}} = \left(\mathbf{H}_1^{T+}\mathbf{H}_2^T\right)^T = \mathbf{H}_2\mathbf{H}_1^+\end{array}\right\}$ AR matrix fitting solution, Eq. (9.15)

$\mathbf{y}(k) + \mathbf{A}_1\mathbf{y}(k-1) + \mathbf{A}_2\mathbf{y}(k-2) = \mathbf{e}(k) + \mathbf{B}_1\mathbf{e}(k-1)$ Vector ARMA (2,1) model, Eq. (9.26)

$\hat{\mathbf{A}}_1 = \mathbf{H}_2\mathbf{H}_1^T\left(\mathbf{H}_1\mathbf{H}_1^T\right)^{-1}$ Ibrahim time domain solution, Eq. (9.42)

$\hat{\mathbf{D}} = \mathbf{S}_n^{-1/2}\mathbf{U}_n^T\mathbf{H}(1)\mathbf{V}_n\mathbf{S}_n^{-1/2}$ ERA system matrix solution, Eq. (9.61)

$\mathbf{O} = \mathbf{T}_{21}\mathbf{T}_{11}^+\mathbf{H}_1$ SSI projection matrix estimation, Eq. (9.68)

Chapter 10. Frequency-Domain Identification

$\mathbf{G}_y(k) = \sum_{n=1}^{2N}\frac{\gamma_n\mathbf{a}_n^T}{ik\Delta\omega-\lambda_n}$ Modal participation vector fitting problem, Eq. (10.1)

$\hat{\boldsymbol{\Gamma}} = \mathbf{HM}^+$ Modal participation solution, Eq. (10.6)

$\mathbf{G}_y(f) = G_q(f)\mathbf{aa}^T$ Single mode spectral density, Eq. (10.11)

$\mathbf{G}_y(f) = \mathbf{A}\left[g_n^2(f)\right]\mathbf{A}^H$ FDD estimation, Eq. (10.16)

$\mathbf{v}_n^H\mathbf{G}_y(f)\mathbf{v}_n = g_n^2(f)$ FDD modal filtering, Eq. (10.32)

$\left(\mathbf{I} - \sum_{n=1}^{na} \mathbf{A}_n e^{-inw(k)} \right) \mathbf{Y}\left(f\left(k\right)\right) = \mathbf{Y}_0$ Constant matrix AR fitting problem, Eq. (10.39)

$\hat{\mathbf{A}} = \left(\mathbf{H}_1^{T+} \mathbf{H}_2^T \right)^T = \mathbf{H}_2 \mathbf{H}_1^+$ Constant matrix AR solution, Eq. (10.45)

Chapter 11. Applications

$\mathrm{MAC}\,(\mathbf{a}, \mathbf{b}) = \frac{|\mathbf{a}^H \mathbf{b}|^2}{(\mathbf{a}^H \mathbf{a})(\mathbf{b}^H \mathbf{b})}$ Modal assurance criteria (MAC) value, Eq. (11.1)

$\mathbf{a} = \begin{Bmatrix} \mathbf{a}_1' \\ \mathbf{a}_1'' \\ \hat{\alpha}_{12}\mathbf{a}_2'' \\ \vdots \\ \alpha_{1D}\mathbf{a}_D'' \end{Bmatrix}$ Mode shape merging, Eq. (11.6)

$\mathbf{S}_r = \left[\frac{\partial r_i}{\partial p_j} \right] \left[P_j \right]$ The relative sensitivity matrix, Eq. (11.7)

$\mathbf{S}_n = \mathbf{S}_r \left[R_j \right]^{-1}$ Normalized sensitivity matrix, Eq. (11.8)

Chapter 12. Advanced Topics

$S2MAC = \max_{\alpha,\beta} \left(\mathbf{a}^H \left(\alpha \mathbf{b}_1 + \beta \mathbf{b}_2 \right) \right)^2$ S2MAC for closely spaced modes, Eq. (12.3)

$\hat{\mathbf{A}} = \mathbf{B}\hat{\mathbf{T}}$ Mode shape expansion, Eq. (12.13)

$\mathbf{a} = \mathbf{T}\mathbf{b}_a$ Mode shape expansion using SEREP, Eq. (12.14)

$\hat{\mathbf{a}} = \mathbf{B}\hat{\mathbf{t}}$ Mode shape expansion using the LC principle, Eq. (12.22)

$d\left(f_0\right) = \mathbf{u}_1(f)^H \mathbf{u}_1\left(f_0\right)$ Modal coherence, Eq. (12.26)

$\hat{\mathbf{q}}\left(t\right) = \mathbf{A}^+ \mathbf{y}\left(t\right)$ Modal filtering in TD, Eq. (12.28)

$\hat{\tilde{\mathbf{q}}}\left(\omega\right) = \mathbf{A}^+ \tilde{\mathbf{y}}\left(\omega\right)$ Modal filtering in FD, Eq. (12.31)

$\alpha^2 = \frac{\left(\omega_0^2 - \omega_r^2 \right) t}{\omega_r^2 \mathbf{a}_0^T \Delta \mathbf{M} \mathbf{a}_l}$ Scaling factor, Eq. (12.39)

$\alpha_j^2 = \frac{\left(\omega_{0j}^2 - \omega_{li}^2 \right) t_{ji}}{\mathbf{a}_{0j}^T \left(\omega_{li}^2 \Delta \mathbf{M} + \Delta \mathbf{K} \right) \mathbf{a}_{li}}$ Scaling factor, mass, and stiffness change, Eq. (12.44)

$m = \hat{\mathbf{a}}^T \mathbf{M} \hat{\mathbf{a}} = \hat{\mathbf{t}}^T \mathbf{B}^T \mathbf{M} \mathbf{B} \hat{\mathbf{t}} = \hat{\mathbf{t}}^T \hat{\mathbf{t}}$ Modal mass, Eq. (12.47)

$\tilde{\mathbf{x}}\left(\omega\right) = \tilde{\mathbf{H}}^+\left(\omega\right) \tilde{\mathbf{y}}\left(\omega\right)$ Force estimation, FRF inversion, Eq. (12.49)

$\hat{\tilde{\mathbf{x}}}\left(\omega\right) = \left(\mathbf{A}^+ \right)^T \tilde{\mathbf{p}}\left(\omega\right)$ Force estimation, modal filtering, Eq. (12.52)

Appendix B

Operational Modal Testing of the Heritage Court Tower

B.1 Introduction

The results of the operational modal testing test conducted by researchers from the University of British Columbia at the Heritage Court Tower (HCT) Two building are presented in this appendix. This study was carried out in order to investigate the dynamic characteristics of the building by way of operational modal testing. The HCT is of particular interest as it is a shear core medium rise building, which has most of its lateral resisting elements concentrated at the center core of the building. These types of buildings may exhibit increased torsional response when subjected to strong earthquake motion depending on the lateral to torsional frequency ratio (see for instance, Paulay [1]).

The purpose of this study was to determine key dynamic characteristics of HCT for comparison with results from the dynamic analysis and quasi-static analysis of the building. The dynamic characteristics of interest were the natural frequencies and corresponding mode shapes of the building.

A complete operational modal testing test was conducted on April 28, 1998 with the aid of the Hybrid Bridge Evaluation System (HBES) developed at the University of British Columbia (Felber [2]).

B.2 Description of the Building

The HCT Two is a relatively regular 15 story reinforced concrete shear core building located at the corner of Hamilton and Robson in Vancouver, British Columbia, Canada. In-plan view, the building is essentially rectangular in shape with only small projections and setbacks. The elevator and stairs are concentrated at the center core of the building and form the main lateral resisting elements for wind and seismic lateral and torsional forces.

The building's ground floor is for commercial use and has a greater story height than the residential stories above. The ground floor elevation varies substantially due to changes in the exterior grade elevations and the requirements for underground parking access. The bottom three stories are slightly larger in floor plan from the remainder of the tower. There is an approximately 3-m irregular setback to the main body of the tower on all four sides at the fourth floor level. The tower floor plan is repeated from the fourth to the twelfth floor with only interior variations at the 13th level. Another setback occurs at the 14th floor level and the top two floors form two level penthouse suites. A small mechanical room projecting above

Introduction to Operational Modal Analysis, First Edition. Rune Brincker and Carlos Ventura.
© 2015 John Wiley & Sons, Ltd. Published 2015 by John Wiley & Sons, Ltd.
Companion Website: www.wiley.com/go/brincker

Figure B.1   North building face from Robson street

the stair/elevator core area forms a small partial 16th story. The entire tower structure sits on four levels of reinforced concrete underground parking. The parking structure extends approximately 14 m beyond the tower in the south direction but is essentially flush with the first floor walls on the remaining three sides.

The building tower is stocky in elevation having a height to width aspect ratio of approximately 1.7 in the east–west direction and 1.3 in the north–south direction. Figure B.1 shows the north elevation facing Robson Street, while Figures B.2 and B.3 show the east elevation and the south elevation, respectively, from Hamilton Street.

Because the building sits to one side of the underground parking structure, coupling of the torsional and lateral modes of vibration may be expected primarily in the east–west direction. The modes in the north–south direction should be essentially uncoupled from torsion since the structural resisting elements of the building are relatively symmetrical in this direction and the building is flush with the parking structure in this direction.

B.3   Operational Modal Testing

Unlike forced vibration testing, the force applied to a structure in operational modal testing is not controlled. The structure is assumed to be excited by wind, traffic, and human activity. The measurements, in our case accelerations, are taken for a long duration to ensure that all the modes of interest are sufficiently excited. Measurements are taken at predetermined locations, which will capture the desired degrees of freedom of the structure.

Figure B.2   East building face from Hamilton street

Figure B.3   South building face from Hamilton street

When using information obtained from operational modal testing measurements the designer should be aware that the dynamic properties of a structure at low levels of vibration may be different, and in some cases significantly different, from those of the same structure subjected to severe dynamic loading, such as strong ground motion. These differences are generally between 10% and 20% for the natural periods of vibration, but could be in excess of 30% for reinforced concrete structures due to localized cracking of structural elements. For steel structures, the differences may not be as great. The fundamental periods estimated using simple building design code formulas will likely be longer than those found by the operational modal testing.

B.3.1   Vibration Data Acquisition System

The components of the vibration measurement hardware utilized in this project were as follows:

1. *Sensors and Cables:* Sensors convert the physical excitation into electrical signals. The hardware measurement system used for this project had sensor connections capable of measuring up to eight different signals from eight different sensor locations or directions. These sensors (force-balanced accelerometers, Kinemetrics, Model FBA-11) were capable of measuring accelerations of up to ±0.5 g with a resolution of 0.2 μg. Cables were used to transmit the electronic signals from sensors to a signal conditioner.
2. *Signal Conditioner:* The signal conditioner unit was used to improve the quality of the signals by removing undesired frequency contents (filtering) and amplifying the signals. For this test, the filter for all cards (Kinemetrics AM-3) was set for a cutoff frequency of 50 Hz.
3. *Analog/Digital Converter:* The amplified and filtered analog signals were converted to digital data using an analog to digital converter (Keithly Model 575 with an AMM2 board) prior to storing on the data acquisition computer. The analog to digital converter was controlled by a data acquisition computer using a custom program called AVDA (Operational modal testing Data Acquisition) developed by Schuster [3]. The analog to digital converter was capable of sampling up to eight channels at sampling frequencies between 0.2 and 2000 Hz. For this test, a sampling rate of 100 samples per second was selected.
4. *Data Acquisition Computer:* Signals converted to digital form were stored on the hard disk of the data acquisition computer in binary form. The data could then be transferred to a separate data analysis computer where numerical analysis of measured data can be done independently of the data acquisition processes. In this way, preliminary onsite data analysis could be carried out concurrently with data acquisition.

A photo of the data acquisition and analysis computers is shown in Figure B.4. The large box on the left is the signal conditioner and the analog/digital converter. The small computer on the top left is the data acquisition computer. The computer on the right is the data analysis computer, which allows the user to perform on-site data analysis, animate the mode shapes, and compute spectral characteristics of the recorded motions during the test. Figure B.5 shows a typical accelerometer setup in the building.

The custom data acquisition program AVDA was used to record the measurements at the HCT building. The measurements were typically taken in the northwest and northeast corners of the building.

B.4   Vibration Measurements

Excitation was provided by wind and human activity as the building was in the final stages of construction. Eight accelerometers were used for the operational modal testing measurements, two of which were allocated as reference sensors located on the 14th floor. Six accelerometers were used as roving sensors. Five setups were conducted to accurately capture the natural frequencies and mode shapes. Measurements

Figure B.4   Data acquisition and analysis computers

Figure B.5   Photo of accelerometer setup

were taken in two locations on every second floor beginning from the roof of the uppermost penthouse down to the second floor. Finally, measurements were taken in three locations on the ground level. Figure B.6 shows a typical layout of the approximate locations and the directions of the operational modal testing measurements on each instrumented floor recorded during the test. Figure B.7 similarly shows the layout of the instruments for the ground floor. A complete description of the test setups is presented as follows.

The vibration measurements were conducted by Dr. Carlos E. Ventura, Professor of Civil Engineering at The University of British Columbia with the assistance of a number of graduate students. Graduate student Cheryl Dyck was responsible for preliminary modeling, supervising setups during acquisition, and final analysis. The tests were conducted on a Sunday, and the temperature was about 18 degrees Celsius with mainly sunny conditions and sustained winds.

Better results would have been achieved if measurements had been taken at all floor levels rather than every second floor. This was not possible due to time constraints. Another contributing factor was that

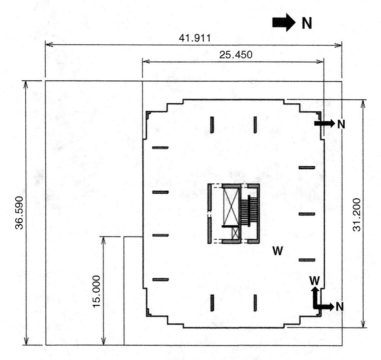

Figure B.6 Typical accelerometer locations and directions on every second floor (all dimensions are in meters)

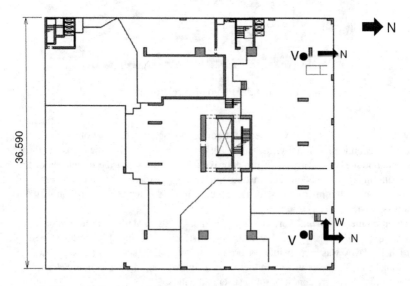

Figure B.7 Ground floor accelerometer locations and directions

the location of the reference sensors turned out to be near to the upper nodal points in both the second and third mode shapes. This affected somewhat the derived mode shapes because of the small modal contribution to the motions measured at the reference location.

Other machine activity was present during the testing and thus the data contains some spurious frequencies. These included workers walking in the vicinity of the sensors and drilling occurring sporadically during acquisition close to the sensors.

B.4.1   Test Setups

Table B.1 gives the location of each of the sensors during the test on April 28, 1998. The first column is the setup number. The second column lists the first five characters of the file name created. The remaining three characters identify the channel number. The eight columns (labeled Channel 1 through Channel 8) refer to the eight channels of the data acquisition computer. Each number-letter combination refers to sensor location and the direction of measurement, vertical (V), north (N), or west (W). See also Figures B.6 and B.7.

B.4.2   Test Results

Ventura and Horyna [4] presented the first set of results of these tests at the International Modal Analysis Conference (IMAC XVIII) in 2000. At that time, traditional peak-peak techniques using power spectral densities were used to identify the modes and frequencies of the building (see Felber [2] and Bendat and Piersol [5]). The authors identified nine modes of vibration in the frequency range between 1 and 10 Hz. The results were compared by the authors with those obtained from a detailed finite element analysis of the building. The test data was also made available to the international community in 1999 and several research teams were invited to use their own analysis techniques to identify the modal characteristics of the building and to participate in a round-robin project to discuss OMA frequency and time domain analyses techniques at the IMAC XVIII. A total of six research teams participated in the study. The participants obtained the modal characteristics of the building using their own techniques. A description of the approach of each team and their results is presented in the paper by Horyna and Ventura [6].

Table B.1   Operational modal testing Test Setup Locations and Directions for April 28, 1998 Test

Setup	File	Channel 1	Channel 2	Channel 3	Channel 4	Channel 5	Channel 6	Channel 7	Channel 8	Attenuation (dB)
1	HCT01	48N	40N	47N	47W	39W	39N	R39W	R39N	6*
2	HCT02	32N	36N	31N	31W	35W	35N	R39W	R39N	6
3	HCT03	24N	28N	23N	23W	27W	27N	R39W	R39N	6
4	HCT04	16N	20N	15N	15W	19W	19N	R39W	R39N	6
5	HCT05	11V	11N	10N	9V	10W	10V	R39W	R39N	6

*Setup 1 had an attenuation of 12 dB for Channel 8 only.
Notes:
Filter cut-off at 50 Hz Nyquist.
Sampling rate = 100 sps.
Number of points per dataset = 64K.
Number of allowable saturation points = 10 in each setup (signal was not stored if this number was exceeded).
Global gain = 5.

The first application of modal updating using OMA techniques in combination with advanced automatic modal updating techniques was performed with the data from this building in 2001 by Ventura et al. [7]. In this paper, the authors describe the procedure followed to perform the modal updating of the building finite element model. A starting model of the structure was developed from the information provided in the design documentation of the building. Different parameters of the model were then modified using an automated procedure to improve the correlation between measured and calculated modal parameters. Careful attention was placed to the selection of the parameters to be modified by the updating software in order to ensure that the necessary changes to the model were realistic and physically realizable and meaningful. The paper highlights the model updating process and provides an assessment of the usefulness of using an automatic model updating procedure combined with results from an OMA modal identification.

A recent analysis of the test data using the commercially available computer program ARTeMIS [8] is presented as follows.

B.5   Analysis of the HCT Cases

The data was first decimated to a Nyquist frequency of 10 Hz and the modal estimation was performed using the FDD peak picking technique, and the SSI (UPC) identification routine.

B.5.1   FDD Modal Estimation

Spectral density functions were estimated using 1024 frequency lines in the Nyquist band. Modes were estimated from the singular value decomposition SVD spectrum obtained as an average of the SVD

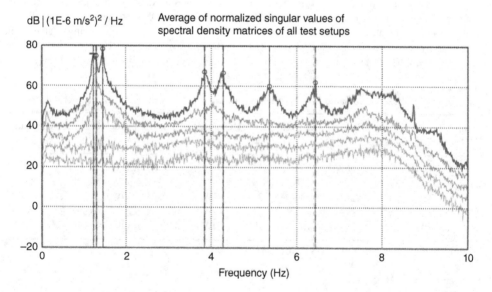

Figure B.8   FDD plot showing the SVD diagram found as an average of the SVD diagrams of the four data sets

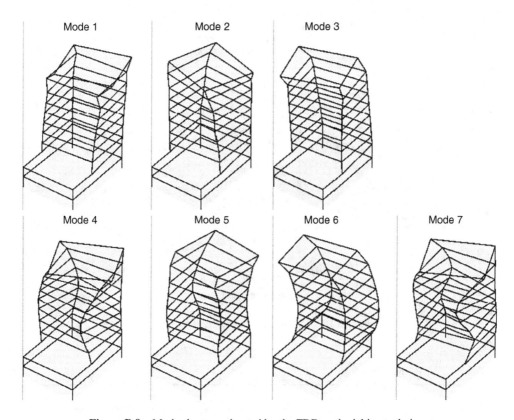

Figure B.9   Mode shapes estimated by the FDD peak picking technique

diagrams of the four data sets. The seven first modes were estimated as shown in Figure B.8. The resulting mode shapes in this frequency range are shown in Figure B.9.

B.5.2   SSI Modal Estimation

The SSI modal estimation was performed using a state space dimension of 100. The stabilization diagram for data set 1 is shown in Figure B.10. As it appears, the stabilization of the frequencies at the selected modes is good and a suitable model is easily found; in this case, a model with a state space dimension of about 20 was sufficient. The same is the case for the other data sets. The modes are then defined combining the selected models from the four data sets as shown in Figure B.11.

B.5.3   Modal Validation

Following the recommended practice in Chapter 11, the modal results from different techniques should be validated against each other. A simple way to perform such validation is to animate different estimates of the same mode shape and see how well they compare. However, this is usually done very effectively

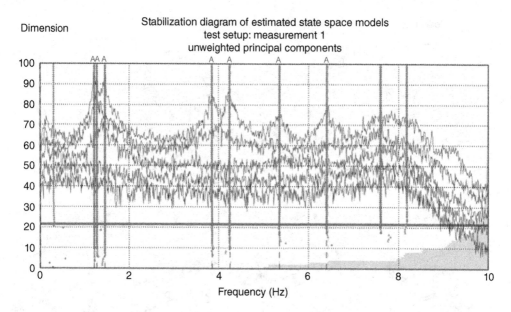

Figure B.10   SSI stabilization diagram for data set 1

Table B.2   MAC matrix table

		SSI (UPC) modes						
		1.228 Hz	1.287 Hz	1.453 Hz	3.856 Hz	4.260 Hz	5.354 Hz	6.401 Hz
FDD modes	1.230 Hz	**0.945**	0.044	0.018	0.092	0.036	0.002	0.086
	1.299 Hz	0.017	**0.963**	0.115	0.147	0.237	0.005	0.172
	1.455 Hz	0.02	0.091	**0.996**	0.051	0.026	0.175	0.069
	3.848 Hz	0.077	0.203	0.05	**0.989**	0.152	0.014	0.334
	4.277 Hz	0.057	0.207	0.021	0.093	**0.991**	0.007	0.068
	5.352 Hz	0.004	0.002	0.167	0.024	0.006	**0.993**	0.012
	6.436 Hz	0.098	0.173	0.094	0.304	0.051	0.047	**0.955**

using graphical animation tools, and thus the result cannot be shown here. Instead of this, here we will consider the MAC values between the two different mode shape estimates, and discuss the differences in natural frequencies.

The MAC values between the two sets of mode shapes are given in Table B.2. We see that the modes compare quite well with MAC values higher than 95% expect for the first mode that has MAC values that fall just below 95%. The low MAC value of the first three modes is probably due to the problems related to increased estimation uncertainty in case of closely spaced modes. Figure B.12 gives a graphical presentation of the same information as given in Table B.2.

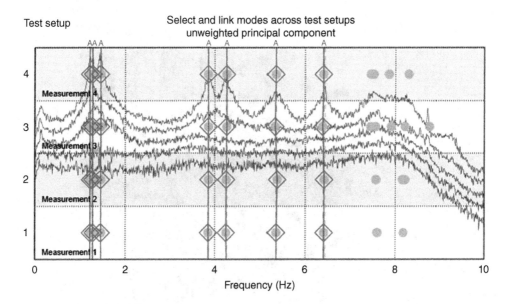

Figure B.11 Mode definition by combining the four selected models from the four data sets named measurement 1, 2, 3, and 4

Table B.3 Summary of modal results for the two techniques

FDD modal estimation		SSI (UPC) modal estimation				
Frequency (Hz)	Complexity (%)	Frequency (Hz)	Standard frequency (Hz)	Damping (%)	Standard damping (%)	Complexity (%)
1.230	16.4	1.228	0.003	2.03	0.433	3.9
1.299	12.7	1.287	0.002	2.21	1.299	1.0
1.455	0.6	1.453	0.003	1.40	0.414	0.4
3.848	2.0	3.856	0.005	1.25	0.116	0.3
4.277	1.9	4.260	0.008	1.51	0.289	0.4
5.352	2.9	5.354	0.016	1.81	0.440	1.3
6.436	17.1	6.401	0.012	1.52	0.211	2.0

In Table B.3, the modal results of natural frequencies, mode shape complexity, and damping ratios are summarized. We see that the natural frequency estimates agree quite well, and we see that the FDD in general has higher complexity of the mode shapes than the SSI. Since bias and random error on mode shape estimates tend to increase the mode shape complexity, this is an indication that in this case, most likely the SSI estimates are more accurate than the estimates from FDD technique. Since the SSI technique provides individual modal estimates for each data set, an average value with corresponding standard deviation for the natural frequency and the damping ratio can be calculated.

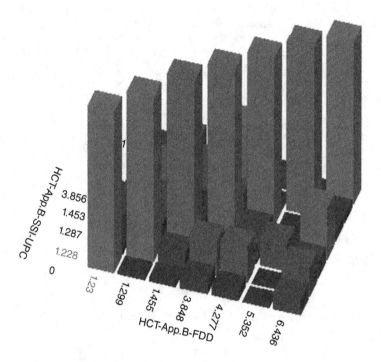

Figure B.12   MAC matrix 3D plot

References

[1] Paulay, T.: *Are existing seismic torsional provisions achieving the design aims?*, Earthquake Spectra, V. 13, No. 2, p. 259–279, 1997.
[2] Felber, A.J.: *Development of a hybrid evaluation system*, Ph.D. Thesis, University of British Columbia, Department of Civil Engineering, Vancouver, 1993.
[3] Schuster, N.D.: *Dynamic characteristics of a 30 story building during construction detected from operational modal testing measurements*. M.A.Sc. thesis, University of British Columbia, Canada, 1994.
[4] Ventura, C.E., and Horyna, T.: *Measured and calculated modal characteristics of the heritage court tower in Vancouver, B.C.*. In Proceedings of the XVIII International Modal Analysis Conference, San Antonio, Texas, p. 1070–1074, 2000.
[5] Bendat, J.S., Piersol, A.G.: *Engineering applications of correlation and spectral analysis*. New York, NY, John Wiley & Sons, 1993.
[6] Horyna, T. and Ventura, C.E.: *Summary of HCT building ambient vibration data analyses*. In Proceedings of the XVIII International Modal Analysis Conference, San Antonio, Texas, p. 1095–1098, 2000.
[7] Ventura, C.E., Brincker, R., Dascotte, E., and Andersen, P.: *FEM updating of the heritage court building structure*. In Proceedings of the XIX International Modal Analysis Conference, Orlando, Florida, p. 324–330, 2001.
[8] ARTeMIS Modal Software Version 5.3, Structural Vibration Solutions A/S, Aalborg Denmark, Copyright 1999–2013.

Appendix C

Dynamics in Short

In this appendix, all major properties and key equations in structural dynamics are derived based on the commonly accepted state space formulation approach.

In this presentation, we will follow the traditional state formulation given in Kozin and Natke [1], but we will change the notation slightly in order to be consistent with the general notation used in this book.

C.1 Basic Equations

The general N-degree-of-freedom (N-DOF) equation of motion of a discrete system is

$$\mathbf{M}\ddot{\mathbf{y}}(t) + \mathbf{C}\dot{\mathbf{y}}(t) + \mathbf{K}\mathbf{y}(t) = \mathbf{x}(t) \tag{C.1}$$

where the only assumptions that we make is that the mass, damping, and stiffness matrices $\mathbf{M}$, $\mathbf{C}$, and $\mathbf{K}$ are symmetric positive definite $N \times N$ matrices. Introducing the state space variable

$$\mathbf{z}(t) = \left\{ \begin{array}{c} \mathbf{y}(t) \\ \dot{\mathbf{y}}(t) \end{array} \right\} \tag{C.2}$$

in which the upper half of the state vector $\mathbf{z}(t)$ corresponds to the vector of displacement of the system and the lower half is the vector of velocities. Introducing the extended system matrices

$$\mathbf{A} = \begin{bmatrix} [0] & \mathbf{I} \\ -\mathbf{M}^{-1}\mathbf{K} & -\mathbf{M}^{-1}\mathbf{C} \end{bmatrix}; \quad \mathbf{B} = \begin{bmatrix} [0] \\ \mathbf{M}^{-1} \end{bmatrix}; \quad \mathbf{P} = \begin{bmatrix} \mathbf{I} & [0] \end{bmatrix} \tag{C.3}$$

where $\mathbf{A}$ is defined as the system matrix, $\mathbf{B}$ is defined as the load distribution matrix, and $\mathbf{P}$ is defined as the observation matrix leads to the following vector differential equation of first order

$$\dot{\mathbf{z}}(t) - \mathbf{A}\mathbf{z}(t) = \mathbf{B}\mathbf{x}(t)$$
$$\mathbf{y}(t) = \mathbf{P}\mathbf{z}(t) \tag{C.3}$$

The upper part is the equation of motion and the lower part is the observation equation. This state space formulation is commonly used in control theory, see for instance Kozin and Natke [1] or Kailath [2].

Introduction to Operational Modal Analysis, First Edition. Rune Brincker and Carlos Ventura.
© 2015 John Wiley & Sons, Ltd. Published 2015 by John Wiley & Sons, Ltd.
Companion Website: www.wiley.com/go/brincker

C.2 Basic Form of the Transfer and Impulse Response Functions

Taking the Laplace transform of the equation of motion of both sides of Eq. (C.3) leads to

$$s\widetilde{\mathbf{z}}(s) - \mathbf{A}\widetilde{\mathbf{z}}(s) = \mathbf{B}\widetilde{\mathbf{x}}(s) \Rightarrow$$
$$\widetilde{\mathbf{z}}(s) = (s\mathbf{I} - \mathbf{A})^{-1}\mathbf{B}\widetilde{\mathbf{x}}(s) \tag{C.5}$$

where $\mathbf{z}(t) \leftrightarrow \widetilde{\mathbf{z}}(s)$ is a Laplace transform pair. The transfer function matrix between the load vector $\widetilde{\mathbf{x}}(s)$ and the state space response $\widetilde{\mathbf{z}}(s)$ is given by

$$\widetilde{\mathbf{H}}_{zx}(s) = (s\mathbf{I} - \mathbf{A})^{-1}\mathbf{B} \tag{C.6}$$

The eigenvalue decomposition of the system matrix $\mathbf{A}$ is

$$\mathbf{A} = \mathbf{U}\left[\lambda_n\right]\mathbf{U}^{-1} \tag{C.7}$$

where $\mathbf{U}$ is the eigenvector matrix in which columns define the eigenvectors and $\left[\lambda_n\right]$ is a diagonal matrix of eigenvalues. This leads to the following expression for the matrix $s\mathbf{I} - \mathbf{A}$

$$s\mathbf{I} - \mathbf{A} = \mathbf{U}\left[s - \lambda_n\right]\mathbf{U}^{-1} \tag{C.8}$$

The expression for the transfer function Eq. (C.6) can then be formulated as

$$\widetilde{\mathbf{H}}_{zx}(s) = \mathbf{U}\left[s - \lambda_n\right]^{-1}\mathbf{U}^{-1}\mathbf{B} \tag{C.9}$$

Since for each of the eigenvalues we have the Laplace transform pair $e^{\lambda_n t} \leftrightarrow 1/\left(s - \lambda_n\right)$ we obtain the corresponding impulse response function matrix

$$\mathbf{H}_{zx}(t) = \mathbf{U}\left[e^{\lambda_n t}\right]\mathbf{U}^{-1}\mathbf{B} \tag{C.10}$$

Realizing that the exponential function of an arbitrary quadratic matrix $\mathbf{X}$ can be defined by its power series $e^{\mathbf{X}} = a_0 + a_1\mathbf{X} + a_2\mathbf{X}^2 + \cdots$ and using the eigenvalue decomposition of the matrix $\mathbf{X} = \mathbf{U}\left[d_n\right]\mathbf{U}^{-1}$ we can show that

$$e^{\mathbf{X}} = a_0 + a_1\mathbf{X} + a_2\mathbf{X}^2 + \cdots$$
$$= a_0 + a_1\mathbf{U}\left[d_n\right]\mathbf{U}^{-1} + a_2\mathbf{U}\left[d_n^2\right]\mathbf{U}^{-1} + \cdots \tag{C.11}$$
$$= \mathbf{U}\left[e^{d_n}\right]\mathbf{U}^{-1}$$

Therefore, the product of the first three matrices in Eq. (C.10) is simply equal to $e^{\mathbf{A}t}$ and thus Eq. (C.10) reduces to the following simple expression for the impulse response

$$\mathbf{H}_{zx}(t) = e^{\mathbf{A}t}\mathbf{B} \tag{C.12}$$

C.3 Free Decays

A general free response is obtained in state space by considering that if $\mathbf{z}(t) = \mathbf{u}e^{\lambda t}$ is a possible solution to the homogenous equation of motion of Eq. (C.3) then we have

$$\mathbf{A}\mathbf{u} = \lambda\mathbf{u} \tag{C.13}$$

and we see that $\mathbf{u}$ is an eigenvector and λ is the corresponding eigenvalue. The possible free decays defined by the eigenvalues and the eigenvectors of the system matrix $\mathbf{A}$ define the modes of the system.

A general free decay is a linear combination of all modes

$$\mathbf{z}(t) = c_1 \mathbf{u}_1 e^{\lambda_1 t} + c_2 \mathbf{u}_2 e^{\lambda_2 t} + \cdots = \mathbf{U}\left[e^{\lambda_n t}\right]\mathbf{c} = \mathbf{U}\left[e^{\lambda_n t}\right]\mathbf{U}^{-1}\mathbf{Uc}$$
$$= e^{\mathbf{A}t}\mathbf{Uc} \tag{C.14}$$

and we see that $\mathbf{U} = \left[\mathbf{u}_1, \mathbf{u}_2, \dots\right]$ is the state space mode shape (or eigenvector) matrix and $\mathbf{c} = \left\{c_n\right\}$ is the modal initial condition vector that defines a complete set of initial conditions for the transient problem in modal space. Evaluating the equation at time zero we see that $\mathbf{z}(0) = \mathbf{Uc}$, and the expression for a general free decay in state space can be written as

$$\mathbf{z}(t) = e^{\mathbf{A}t}\mathbf{z}(0) \tag{C.15}$$

where $\mathbf{z}(0)$ defines a complete set of initial conditions for the transient problem in state space. Defining the observable part of the mode shape matrix as $\mathbf{V} = \left[\mathbf{v}_1, \mathbf{v}_2, \dots\right]$ and inspecting Eq. (C.14) we see that the free decay can also be expressed as

$$\mathbf{z}(t) = \left\{\begin{matrix} \mathbf{y}(t) \\ \dot{\mathbf{y}}(t) \end{matrix}\right\} = \left[\begin{matrix} \mathbf{V} \\ \mathbf{V}\left[\lambda_n\right] \end{matrix}\right]\left[e^{\lambda_n t}\right]\mathbf{c} \tag{C.16}$$

From this equation, we see that the state space eigenvector matrix $\mathbf{U}$ is related to the observable part of the eigenvector matrix $\mathbf{V}$ representing the modal displacements of the system by

$$\mathbf{U} = \left[\begin{matrix} \mathbf{V} \\ \mathbf{V}\left[\lambda_n\right] \end{matrix}\right] \tag{C.17}$$

and we see that the eigenvector matrix for the velocities of the system is represented by the matrix $\mathbf{V}\left[\lambda_n\right]$.

From Eq. (C.7), we see that we have $2N$ eigenvectors and eigenvalues and the corresponding free decays define $2N$ modes for state space formulation given by Eq. (C.7). Since the original equation (1) is of second order, the $2N$ eigenvectors and eigenvalues appear in N complex conjugate pairs defining N physical modes of the form $y(t) = v e^{\lambda t} + v^* e^{\lambda^* t}$ securing that the modes of Eq. (C.1) is always real. We can also say that the initial equation (C.1) has N physical modes and the state space equation (C.7) has $2N$ modes.

C.4   Classical Form of the Transfer and Impulse Response Functions

The transfer function matrix $\widetilde{\mathbf{H}}_{yx}(s)$ between the input and the observable response is obtained by multiplying Eq. (C.9) with the observation matrix. That is

$$\widetilde{\mathbf{H}}_{yx}(s) = \mathbf{PU}\left[s - \lambda_n\right]^{-1}\mathbf{U}^{-1}\mathbf{B} \tag{C.18}$$

This matrix is symmetric. This follows from the assumption that the mass, damping, and stiffness matrices are symmetric and from the fact that the transfer function can also be expressed by the direct Laplace transformation of Eq. (C.1) as $\widetilde{\mathbf{H}}_{yx}(s) = \left(\mathbf{M}s^2 + \mathbf{C}s + \mathbf{K}\right)^{-1}$. Taking the transpose of Eq. (C.18) and using the fact that the middle matrix $\left[s - \lambda_n\right]^{-1}$ is diagonal and thus remains unchanged, we have that $\mathbf{U}^{-1}\mathbf{B} = (\mathbf{PU})^T$, but since $\mathbf{PU} = \mathbf{V}$ picks out the observable part $\mathbf{V}$ of the eigenvector we have

$$\mathbf{U}^{-1}\mathbf{B} = \mathbf{V}^T \tag{C.19}$$

and the transfer function reduces to

$$\widetilde{\mathbf{H}}_{yx}(s) = \mathbf{V}\left[s - \lambda_n\right]^{-1}\mathbf{V}^T \tag{C.20}$$

This expression can be formulated as a series of outer products of the observable eigenvectors

$$\tilde{\mathbf{H}}_{yx}(s) = \sum_{n=1}^{N} \left(\frac{\mathbf{v}_n \mathbf{v}_n^T}{s - \lambda_n} + \frac{\mathbf{v}_n^* \mathbf{v}_n^H}{s - \lambda_n^*} \right) \tag{C.21}$$

where we have used that the solution to the second order matrix equation (C.1) in terms of eigenvalues and eigenvectors is known to appear in complex conjugate pairs, one pair for each mode, and therefore the summation in Eq. (C.21) is reduced to the number of modes. Similarly from Eq. (C.10), we find the impulse response as

$$\mathbf{H}_{yx}(t) = \mathbf{PU}\left[e^{\lambda_n t}\right] \mathbf{U}^{-1} \mathbf{B}$$
$$= \mathbf{V}\left[e^{\lambda_n t}\right] \mathbf{V}^T \tag{C.22}$$
$$= \sum_{n=1}^{N} \left(\mathbf{v}_n \mathbf{v}_n^T e^{\lambda_n t} + \mathbf{v}_n^* \mathbf{v}_n^H e^{\lambda_n^* t} \right)$$

The classical forms of the transfer and impulse response functions are well known in the literature; see for instance Heylen et al. [3] and Ewins [4].

C.5   Complete Analytical Solution

From Eqs. (C.5–C.6) we have

$$\tilde{\mathbf{z}}(s) = \tilde{\mathbf{H}}_{zx}(s)\tilde{\mathbf{x}}(s) \tag{C.23}$$

Using the Laplace transform convolution theorem, we have the particular solution in the time domain

$$\mathbf{z}(t) = \mathbf{H}_{zx}(t) * \mathbf{x}(t)$$
$$= \int_{-\infty}^{\infty} \mathbf{H}_{zx}(t - \tau)\mathbf{x}(\tau)\,d\tau \tag{C.24}$$

or since the impulse response vanishes for negative times and the loading is assumed also to be zero for negative times we get

$$\mathbf{z}(t) = \int_0^t \mathbf{H}_{zx}(t - \tau)\mathbf{x}(\tau)\,d\tau \tag{C.25}$$

The complete analytical solution is then the sum of the complete solution to the homogenous equation given by Eq. (C.15), and the aforementioned equation for the particular solution is

$$\mathbf{z}(t) = e^{\mathbf{A}t}\mathbf{z}(0) + \int_0^t \mathbf{H}_{zx}(t - \tau)\mathbf{x}(\tau)d\tau \tag{C.26}$$

Until now we have not used the concepts of modal coordinates and modal loads, but these concepts introduce themselves in this solution. From Eq. (C.12) and the relation $e^{\mathbf{A}t} = \mathbf{U}\left[e^{\lambda_n t}\right]\mathbf{U}^{-1}$ we see that both the transient term and the particular solution are formulated as the eigenvector matrix $\mathbf{U}$ times a vector that defines the modal coordinates. The complete solution, formulated in modal coordinates, is then given by

$$\mathbf{z}(t) = \mathbf{U}\mathbf{q}_0(t) + \mathbf{U}\mathbf{q}(t)$$
$$\mathbf{q}_0(t) = \left[e^{\lambda_n t}\right]\mathbf{U}^{-1}\mathbf{z}(0) \tag{C.27}$$
$$\mathbf{q}(t) = \int_0^t \left[e^{\lambda_n(t-\tau)}\right]\mathbf{U}^{-1}\mathbf{B}\mathbf{x}(\tau)d\tau$$

where $\mathbf{q}_0(t)$ is the modal coordinate vector of the transient response, $\mathbf{q}(t)$ is the modal coordinate vector of the particular solution and where each modal DOF contribute with two modal coordinates that appear as each other complex conjugate. The vector $\mathbf{U}^{-1}\mathbf{z}(0)$ equal to $\mathbf{c}$ using that $\mathbf{z}(0) = \mathbf{Uc}$ is the modal initial condition vector of the transient modal response, see Eq. (C.14), and the vector $\mathbf{U}^{-1}\mathbf{Bx}(\tau)$ equal to $\mathbf{U}^{-1}\mathbf{Bx}(\tau) = \mathbf{V}^T\mathbf{x}(\tau)$ using Eq. (C.19) is known as the modal load.

C.6 Eigenvector Scaling

In the classical formulations, such as in Eqs. (C.20–C.22), scaling of the eigenvectors is an important issue.

However, by scaling the eigenvector using the diagonal matrix $\left[w_n\right]$ such as $\mathbf{U}\left[w_n\right]$ we see that the eigenvalue decomposition of the system matrix remains the same because the scaling matrix cancels out

$$\mathbf{A} = \mathbf{U}\left[w_n\right]\left[\lambda_n\right]\left[1/w_n\right]\mathbf{U}^{-1} = \mathbf{U}\left[\lambda_n\right]\mathbf{U}^{-1} \tag{C.28}$$

It can be concluded that all expressions that are based on a form that contains a similar kernel, (such as Eqs. (C.9, C.10, C.12, C.15, and C.18)) are invariant to any scaling. Therefore, scaling is not an issue in these cases.

Returning to the classical formulations of Eqs. (C.20–C.22) that are based on the outer products of the eigenvectors, the above is no longer true because using the scaling matrix on the eigenvectors of Eq. (C.20) we get

$$\begin{aligned}
\widetilde{\mathbf{H}}_{yx}(s) &= \mathbf{V}\left[w_n\right]\left[s - \lambda_n\right]^{-1}\left[w_n\right]\mathbf{V}^T \\
&= \mathbf{V}\left[w_n^2/(s - \lambda_n)\right]\mathbf{V}^T
\end{aligned} \tag{C.29}$$

The scaling factor w_n weights the corresponding modal contribution to the transfer function with the factor w_n^2. Therefore, in Eqs. (C.20–C.22) care must be taken to use the right scaling of the eigenvectors $\mathbf{v}_n$ in $\mathbf{V}$. The problem is that normally the scaling of the observable eigenvector from $\mathbf{PU} = \mathbf{V}_1$ and from Eq. (C.13) $\left(\mathbf{U}^{-1}\mathbf{B}\right)^T = \mathbf{V}_2$ are different, thus $\mathbf{V}_1 \neq \mathbf{V}_2$ due to the indeterminate scaling of the state space eigenvector matrix $\mathbf{U}$.

One way to deal with this problem is simply to accept that the two observable sets $\mathbf{V}_1$ and $\mathbf{V}_2$ of the eigenvectors are different and use Eqs. (C.20–C.22) in the following form similar to Eq. (C.20): $\widetilde{\mathbf{H}}_{yx}(s) = \mathbf{V}_1\left[s - \lambda_n\right]^{-1}\mathbf{V}_2^T$. The set $\mathbf{V}_1$ is used when multiplying from the left, and the set $\mathbf{V}_2$ is used when multiplying from the right. This will ensure the same indifference to the scaling as we have for the fundamental forms given by Eqs. (C.9, C.10, C.12, C.15, and C.18).

Another approach is to obtain the diagonal weight matrix $\left[w_n^2\right]$ with the diagonal elements taken from the matrix $\mathbf{V}_1^T\mathbf{V}_2$, scale one of the eigenvector sets to unity length to define the eigenvector set $\mathbf{V}$ and then use Eq. (C.29).

It is important to note that the procedure for the aforementioned scaling requires a full set of modes in order to obtain $\left(\mathbf{U}^{-1}\mathbf{B}\right)^T$, which is not possible when the eigenvectors have been estimated by testing. In this case, the weighting term can be $w_n^2 = 1/\left(2\lambda_n m_n + c_n\right)$ where $m_n = \mathbf{v}_n^T\mathbf{Mv}_n$ is the modal mass, and $c_n = \mathbf{v}_n^T\mathbf{Cv}_n$ is the modal damping as it is shown in Heylen et al. [3]. Unless we know the full set of modes or can find the scaling factors by experimental means, knowledge of the mass matrix and the damping matrix is required in general in order to scale the eigenvectors.

C.7 Closing Remarks

The frequency response function (FRF) matrix can be found from the aforementioned expressions for the corresponding transfer function matrices by restricting the complex Laplace variable s to the imaginary

352					Appendix C: Dynamics in Short

frequency variable $i\omega$. This is due to the similarity between the differentiation theorems for the two transforms.

If the reader is wondering about the orthogonality of the mode shapes, it is worth noting that since for the present formulation the system matrix $\mathbf{A}$ is not symmetric, no orthogonality exist in the classical sense, because the orthogonality is not a property related to the eigenvectors, but related to the symmetry of the system matrix. However, a set of orthogonal vectors related to $\mathbf{A}$ can be found as discussed in Section 5.2.5.

References

[1] Kozin, F. and Natke, H.G.: *System identification techniques. Struct. Saf.*, V. 3, p. 269–316, 1986.
[2] Kailath, T.: *Linear systems*. New Jersey, Prentice-Hall Inc., 1980.
[3] Heylen, W., Lammens, S. and Sas, P.: *Modal analysis theory and testing*. Katholieke Universiteit Leuven, Faculty of Engineering, 2007.
[4] Ewins, D.J.: *Modal testing: Theory, practice and application*. 2nd edition. Research Studies Press Ltd, 2000.

Index